Solar Origins of Space Weather and Space Climate

Irene González Hernández • Rudolf Komm •
Alexei Pevtsov • John Leibacher
Editors

Solar Origins of Space Weather and Space Climate

Previously published in *Solar Physics*, Volume 289,
Issue 2, 2014

Springer

Editors
Irene González Hernández
National Solar Observatory
Tucson, AZ, USA

Alexei Pevtsov
National Solar Observatory
Sunspot, NM, USA

Rudolf Komm
National Solar Observatory
Tucson, AZ, USA

John Leibacher
National Solar Observatory
Tucson, AZ, USA

ISBN 978-1-4939-4979-3 ISBN 978-1-4939-1182-0 (eBook)
DOI 10.1007/978-1-4939-1182-0
Springer New York Heidelberg Dordrecht London

Cover illustration: Potential-field source-surface model derived from magnetic-field observations by the Global Oscillation Network Group (GONG). Line-of-sight views of the tallest field lines (streamer belt, blue) and positive and negative coronal holes (green and red, respectively) are shown. The middle picture is from the point of view of Earth on 3 March 2014 and the left and right pictures are from the viewpoints of the NASA STEREO-B (behind, left) and -A (ahead, right) spacecraft. Image credit: G.J.D. Petrie/NISP/NSO/AURA/NSF.

Printed on acid-free paper

Springer is part of Springer Science+Business Media (www.springer.com)

Irene González Hernández, in memoriam

Irene González Hernández (image courtesy of Igor Suarez-Sola)

Dr. Irene González Hernández, born on 20 October 1969 in the Canary Islands, Spain, sadly passed away on 14 February 2014 in Tucson, Arizona, USA after a battle with cancer.

Most of Irene's professional career was dedicated to research in the field of helioseismology; initially analyzing the images taken by the *Taiwan Oscillation Network* (TON) instrument at the Observatorio del Teide and the *Michelson Doppler Imager* (MDI) onboard the *Solar and Heliospheric Observatory* (SOHO), and later joining the Global Oscillation Network Group (GONG). In 1998, she received her Ph.D. from the Universidad de La Laguna. In Summer 1998, Irene joined the National Solar Observatory (NSO) in Tucson, Arizona as a long-term visitor. She liked it so much that a few years later, in 2003, she moved there permanently and became an integral part of NSO/GONG. She remained an active member of the NSO until her very last days. Before moving to Tucson, she was granted an ESA postdoctoral fellowship from Queen Mary College, University of London for a year and then switched to the private sector at Morgan Stanley, UK as an IT analyst for three years.

Throughout her scientific career, Irene worked on a variety of research projects, but her passion was farside imaging, where seismic waves observed on the frontside of the Sun are used to detect the presence of active regions on the invisible-side of our star. Her clever way of calibrating the farside signatures of active regions in terms of active-region size and magnetic-field strength added a new dimension to farside imaging, which has now become an important tool for space-weather forecasting. She also worked on implementing farside imaging to improve the short-term prediction of solar-wind speed and UV irradiance, and studied large-scale flows below the solar surface.

In 2007, Irene received a Science Award of the Association of Universities for Research in Astronomy (AURA) in recognition of her contributions to helioseismology research. As part of a group of NSO scientists, she was also presented with a NASA Group Achievement Award in 2012 for contributions to the *Solar Dynamics Observatory* (SDO) Science Investigation.

On a personal level, Irene was a lovable, enthusiastic, and caring individual. She will be deeply missed by her friends and colleagues at NSO and in the solar community. She is survived by her parents and her husband, Francisco (Igor) Suarez-Sola.

Contents

DOI 10.1007/978-1-4939-1182-0_1
Reprinted from *Solar Physics* Journal, DOI 10.1007/s11207-013-0454-x

Solar Origins of Space Weather and Space Climate: Preface

I. González Hernández · R. Komm · A. Pevtsov · J.W. Leibacher

Published online: 21 November 2013

As the impact of space weather and climate on daily life is becoming more important, it is timely to discuss the latest research on the solar origin of these phenomena. Recent advances in helioseismology have demonstrated that keys for understanding many aspects of solar activity from flare and CME eruptions to cyclic variations may lie with the subphotospheric plasma dynamics. On the other hand, the advent of synoptic vector magnetic-field measurements opens up a new path to a better understanding of magnetic topology of space-weather source regions on the Sun, *e.g.* active regions, flares, chromospheric filaments, and CMEs. Furthermore, the space-weather research is rapidly maturing, and is now becoming capable of producing stable space-weather forecasts. Despite this recent progress, many questions remain to be answered, including the future directions for the research and applied components of space weather, the role of ground-based and space-borne instrumentation and networks, and the ways for transitioning the results of research into the operational forecast.

This topical issue is based on the presentations given at the 26th National Solar Observatory (NSO) Summer Workshop *Solar Origins of Space Weather and Space Climate* held at the National Solar Observatory/Sacramento Peak, New Mexico, USA from 30 April to 4 May 2012. The volume also includes contributions that were not presented at this meeting. More than 60 scientists from the US and abroad came to Sac Peak to exchange their ideas on topics of space weather, to share the results of recent studies, and to discuss the future developments. This unique forum brought together experts in different areas of solar and space physics to help in developing a full picture of the origin of solar phenomena that affect Earth's technological systems on a short and long-term basis. The workshop discussions served as a starting ground for future research, and sparked new collaborations in the field.

Solar Origins of Space Weather and Space Climate
Guest Editors: I. González Hernández, R. Komm, and A. Pevtsov

I. González Hernández (✉) · R. Komm · A. Pevtsov · J.W. Leibacher
National Solar Observatory, Tucson, AZ, USA
e-mail: irenegh@nso.edu

J.W. Leibacher
Institut d'Astrophysique Spatial, Orsay, France

The articles in this volume include theory, model, and observation research on the origin of the solar cycle, solar magnetic features, active-region evolution, eruptive events, as well as a discussion on how to incorporate the research into space-weather forecasting tools.

Acknowledgments

The 26th NSO workshop was sponsored by the SHINE Program of the US National Science Foundation Division of Atmospheric and Geospace Sciences, the US Air Force Office of Scientific Research, the Solar Physics Division of the American Astronomical Society, and the National Solar Observatory.

Articles

González Hernández, I., Komm, R., Pevtsov, A., Leibacher, J.W.: 2014, Solar Origins of Space Weather and Space Climate: Preface. *Solar Phys.* **289**, 437. doi:10.1007/s11207-013-0454-x.

Nelson, N.J., Brown, B.P., Brun, A.S., Miesch, M.S., Toomre, J.: 2014, Buoyant Magnetic Loops Generated by Global Convective Dynamo Action. *Solar Phys.* **289**, 441. doi:10.1007/s11207-012-0221-4.

Braun, D.: 2014, Helioseismic Holography of an Artificial Submerged Sound Speed Perturbation and Implications for the Detection of Pre-emergence Signatures of Active Regions. *Solar Phys.* **289**, 459. doi:10.1007/s11207-012-0185-4.

Komm, R., Gosain, S., Pevtsov, A.: 2014, Active Regions with Superpenumbral Whirls and Their Subsurface Kinetic Helicity. *Solar Phys.* **289**, 475. doi:10.1007/s11207-012-0218-z.

Gao, Y., Zhao, J., Zhang, H.: 2014, A Study of Connections Between Solar Flares and Subsurface Flow Fields of Active Regions. *Solar Phys.* **289**, 493. doi:10.1007/s11207-013-0274-z.

González Hernández, I., Díaz Alfaro, M., Jain, K., Tobiska, W.K., Braun, D.C., Hill, F., Pérez Hernández, F.: 2014, A Full-Sun Magnetic Index from Helioseismology Inferences. *Solar Phys.* **289**, 503. doi:10.1007/s11207-013-0339-z.

Fontenla, J.M., Landi, E., Snow, M., Woods, T.: 2014, Far- and Extreme-UV Solar Spectral Irradiance and Radiance from Simplified Atmospheric Physical Models. *Solar Phys.* **289**, 515. doi:10.1007/s11207-013-0431-4.

Lefevre, L., Clette, F.: 2014, Survey and Merging of Sunspot Catalogs. *Solar Phys.* **289**, 545. doi:10.1007/s11207-012-0184-5.

Murakőzy, J., Baranyi, T., Ludmány, A.: 2014, Sunspot Group Development in High Temporal Resolution. *Solar Phys.* **289**, 563. doi:10.1007/s11207-013-0416-3.

Gyenge, N., Baranyi, T., Ludmány, A.: 2014, Migration and Extension of Solar Active Longitudinal Zones. *Solar Phys.* **289**, 579. doi:10.1007/s11207-013-0424-3.

Pevtsov, A.A., Bertello, L., Tlatov, A.G., Kilcik, A., Nagovitsyn, Y.A., Cliver, E.W.: 2014, Cyclic and Long-Term Variation of Sunspot Magnetic Fields. *Solar Phys.* **289**, 593. doi:10.1007/s11207-012-0220-5.

Panasenco, O., Martin, S.F., Velli, M.: 2014, Apparent Solar Tornado-Like Prominences. *Solar Phys.* **289**, 603. doi:10.1007/s11207-013-0337-1.

Altrock, R.C.: 2013, Forecasting the Maxima of Solar Cycle 24 with Coronal Fe XIV Emission. *Solar Phys.* **289**, 623. doi:10.1007/s11207-012-0216-1.

Yeates, A.R.: 2014, Coronal Magnetic Field Evolution from 1996 to 2012: Continuous Non-potential Simulations. *Solar Phys.* **289**, 631. doi:10.1007/s11207-013-0301-0.

Choudhary, D.P., Lawrence, J.K., Norris, M., Cadavid, A.C.: 2014, Different Periodicities in the Sunspot Area and the Occurrence of Solar Flares, Coronal Mass Ejections in Solar Cycle 23 – 24. *Solar Phys.* **289**, 649. doi:10.1007/s11207-013-0392-7.

Kahler, S.W., Arge, C.N., Akiyama, S., Gopalswamy, N.: 2014, Do Solar Coronal Holes Affect the Properties of Solar Energetic Particle Events? *Solar Phys.* **289**, 657. doi:10.1007/s11207-013-0427-0.

Steenburgh, R.A., Biesecker, D.A., Millward, G.H.: 2014, From Predicting Solar Activity to Forecasting Space Weather: Practical Examples of Research-to-Operations and Operations-to-Research. *Solar Phys.* **289**, 675. doi:10.1007/s11207-013-0308-6.

DOI 10.1007/978-1-4939-1182-0_2
Reprinted from *Solar Physics* Journal, DOI 10.1007/s11207-012-0221-4

Buoyant Magnetic Loops Generated by Global Convective Dynamo Action

Nicholas J. Nelson · Benjamin P. Brown · A. Sacha Brun · Mark S. Miesch · Juri Toomre

Received: 10 October 2012 / Accepted: 25 December 2012 / Published online: 29 January 2013

Abstract Our global 3D simulations of convection and dynamo action in a Sun-like star reveal that persistent wreaths of strong magnetism can be built within the bulk of the convention zone. Here we examine the characteristics of buoyant magnetic structures that are self-consistently created by dynamo action and turbulent convective motions in a simulation with solar stratification but rotating at three times the current solar rate. These buoyant loops originate within sections of the magnetic wreaths in which turbulent flows amplify the fields to much higher values than is possible through laminar processes. These amplified portions can rise through the convective layer by a combination of magnetic buoyancy and advection by convective giant cells, forming buoyant loops. We measure statistical trends in the polarity, twist, and tilt of these loops. Loops are shown to preferentially arise in longitudinal patches somewhat reminiscent of active longitudes in the Sun, although broader in

Solar Origins of Space Weather and Space Climate
Guest Editors: I. González Hernández, R. Komm, and A. Pevtsov

N.J. Nelson · J. Toomre
JILA and Dept. Astrophysical & Planetary Sciences, University of Colorado, Boulder, CO 80309-0440, USA

N.J. Nelson
e-mail: nnelson@lcd.colorado.edu

B.P. Brown
Dept. Astronomy, University of Wisconsin, Madison, WI 53706-1582, USA

B.P. Brown
Center for Magnetic Self Organization in Laboratory and Astrophysical Plasmas, University of Wisconsin, 1150 University Avenue, Madison, WI 53706, USA

A. Sacha Brun
Laboratoire AIM Paris-Saclay, CEA/Irfu Université Paris-Diderot CNRS/INSU, 91191 Gif-sur-Yvette, France

M.S. Miesch (✉)
High Altitude Observatory, NCAR, Boulder, CO 80307-3000, USA
e-mail: miesch@hao.ucar.edu

extent. We show that the strength of the axisymmetric toroidal field is not a good predictor of the production rate for buoyant loops or the amount of magnetic flux in the loops that are produced.

Keywords Convection zone · Global dynamo · Magnetic buoyancy · Flux emergence

1. Flux Emergence and Convective Dynamos

Convective dynamo action in the interior of the Sun is the source of the magnetism that creates sunspots and drives space weather. This type of magnetism is not limited to the Sun, as magnetic activity is observed to be ubiquitous among Sun-like stars. To understand the origin of sunspots and starspots, one must explore the processes that generate magnetic structures and then transport them through the convection zone to the surface. Here we present the results of a global numerical simulation, called case S3, which self-consistently generates wreaths of strong magnetic field by dynamo action within the convective zone. Case S3 models the convection zone of a Sun-like star that nominally rotates at three times the current solar rate, or $3\,\Omega_\odot$. The wreaths reverse polarity in a cyclic fashion, yielding cycles of magnetic activity. Portions of these wreaths form buoyant magnetic structures, or loops, which rise through our convective envelope. Initial results on the behavior of a small number of these loops were reported by Nelson *et al.* (2011).

Here we discuss the properties of a much larger number of loops to achieve a statistical description of their properties. We find coherent magnetic structures with a variety of topologies, latitudinal tilts, twists, and total fluxes. Additionally, we observe only a weak correlation between the unsigned magnetic flux in a buoyant loop and the axisymmetric toroidal magnetic field at that latitude and time, indicating that the generation mechanism for these loops relies on local, coherent toroidal-field structures amplified by turbulent intermittency rather than large-scale instabilities of axisymmetric fields. We also find evidence for longitudinal intervals that preferentially produce buoyant loops, hinting at a possible origin for active longitudes for sunspots (Henney and Harvey, 2002), although our intervals are quite broad.

Our work builds upon a series of simulations that consider the dynamics within the deep convective envelopes of young suns that rotate faster than our current Sun. Strong differential rotation was found in hydrodynamic simulations involving a range of rotation rates up to $10\,\Omega_\odot$ (Brown *et al.*, 2008), including prominent longitudinal modulation in the strength of the convection at low latitudes. Turning to dynamo action achieved in an MHD simulation in such stars at $3\,\Omega_\odot$, Brown *et al.* (2010) reported that the convection can build global-scale magnetic fields that appear as wreaths of toroidal magnetic field of opposite polarity in each hemisphere. These striking magnetic structures persist for long intervals despite being embedded within a turbulent convective layer. At a faster rotation rate of $5\,\Omega_\odot$, self-consistently generated magnetic wreaths at low latitudes underwent reversals in global magnetic polarity and cycles of magnetic activity (Brown *et al.*, 2011). These cyclic reversals can also be achieved at lower rotation rates if the diffusion is decreased, as the reversals can only occur when resistive diffusion is not able to prevent reversals in the axisymmetric poloidal fields (Nelson *et al.*, 2013). As diffusion is decreased, the level of turbulent intermittency rises, leading to coherent magnetic structures that can become buoyant (Nelson *et al.*, 2011).

Although the simulation discussed here describes Sun-like stars that nominally rotate faster than the current Sun, the dynamo action realized here may not be only confined to rapidly rotating stars. The most important non-dimensional parameter for the generation of magnetic wreaths is the Rossby number (the ratio of convective vorticity to twice the frame rotation rate), which is small in both the Sun and our simulation here. While no simulation can achieve solar-like values of all relevant parameters, the ability to self-consistently capture a wide range of dynamics, including the buoyant transport of magnetic structures through the convective layer, provides us with a unique tool for exploring dynamo action in a solar-like context. Thus our work may be broadly applicable also to processes occurring in the solar interior.

1.1. Magnetism in Many Settings

Magnetic activity and cycles appears to be characteristic of many Sun-like stars. The best-studied example is clearly the Sun's 22-year magnetic-activity cycle. The interplay of turbulent convection, rotation, and stratification in the solar convection zone creates a cyclic dynamo that drives variations in the interior, on the surface, and throughout the Sun's extended atmosphere (Charbonneau, 2010). Yet the Sun is not alone in its magnetic variability. Solar-type stars generate magnetism almost without exception. Observations reveal a clear correlation between rotation and magnetic activity, as inferred from proxies such as X-ray and chromospheric emission (Saar and Brandenburg, 1999; Pizzolato *et al.*, 2003; Wright *et al.*, 2011). However, superimposed on this trend are considerable variations in the presence and the period of magnetic-activity cycles. There have been a number of attempts to monitor the magnetic-activity cycles of other stars using solar-calibrated proxies for magnetic activity (*e.g.* Baliunas *et al.*, 1995; Hempelmann, Schmitt, and Stępień, 1996; Oláh *et al.*, 2009). Improved observational techniques include spot-tracking from *Kepler* photometry (Meibom *et al.*, 2011; Llama *et al.*, 2012) and Zeeman-Doppler imaging (Petit *et al.*, 2008; Gaulme *et al.*, 2010; Morgenthaler *et al.*, 2012). These are beginning to provide assessments of the size, frequency, and magnetic flux of starspots and the topology and spatial variability of photospheric magnetic fields.

1.2. Theoretical Approaches to Solar and Stellar Dynamos

The solar dynamo is nonlinear, three-dimensional, and involves a wide range of scales in both space and time, but the basis for most theoretical explorations of the solar dynamo comes from mean-field theory (Parker, 1955; Moffatt, 1978; Krause and Raedler, 1980). In these models, the toroidal field is generated through the Ω-effect as differential rotation shears large-scale poloidal field into a band of toroidal field in each hemisphere. The poloidal field is created through a nonlinear interaction parameterized by the α-effect. A wide variety of mechanisms for the α-effect have been proposed, some of which rely on the rise of buoyant magnetic loops to form active regions. In the Babcock–Leighton model, for example, buoyant transport of toroidal magnetic flux provides the mechanism for the regeneration and reversal of the poloidal magnetic field (Babcock, 1961; Leighton, 1964). In mean-field models, magnetic buoyancy is parameterized, assuming that a constant fraction of magnetic flux escapes or that flux emergence is triggered when mean fields achieve a certain magnitude (see review by Charbonneau, 2010).

There have been two main numerical approaches to the study of dynamo action and the source of active regions. The first class of models tracks the rise of buoyant magnetic structures that have been inserted into stratified domains and then allowed to rise (*e.g.* Caligari, Moreno-Insertis, and Schüssler, 1995; Fan, 2008; Jouve and Brun, 2009; Weber, Fan, and Miesch, 2011), or alternatively use forced shear layers to create magnetic structures, which then rise buoyantly (*e.g.* Cline, Brummell, and Cattaneo, 2003;

Vasil and Brummell, 2009; Guerrero and Käpylä, 2011). The second class of models has focused on global-scale convective-dynamo processes that generate magnetic structures in the deep interior and may produce cycles of magnetic activity (*e.g.* Browning *et al.*, 2006; Brown *et al.*, 2010, 2011; Ghizaru, Charbonneau, and Smolarkiewicz, 2010; Racine *et al.*, 2011). Recently Miesch and Brown (2012) have explored 3D convective-dynamo action with a Babock–Leighton term to include flux transport by means of a parameterization of magnetic buoyancy. Our study here belongs to the second class, using convective-dynamo simulations to produce buoyant magnetic loops. The first account of such modeling was reported by Nelson *et al.* (2011, 2013).

2. Nature of the Simulation

We use the 3D anelastic spherical harmonic (ASH) code to model large-scale convective-dynamo action in the solar convective envelope. ASH solves the anelastic MHD equations in rotating spherical shells (Clune *et al.*, 1999; Brun, Miesch, and Toomre, 2004). ASH is limited to the deep interior due to the anelastic approximation, which limits us to low Mach-number flows. Additionally, we stayed away from the near-surface layers because we cannot resolve the small scales of granulation and super-granulation realized near the photosphere. Our simulation extends from 0.72 $R_\odot$ to 0.965 $R_\odot$, covering a density contrast of about 25 from top to bottom. The details of the numerical scheme used in case S3 are described by Brown *et al.* (2010), and the specific parameters are given by Nelson *et al.* (2013). Of special note, in case S3 the Rossby number is 0.581, which is in the same rotationally influenced regime as the giant-cell convection realized in the solar interior (Miesch, 2005). Thus the dynamics in case S3 may be broadly applicable to stars like the Sun, in which rotational influences on convective motions are significant.

To achieve very low levels of diffusion, we employed a dynamic Smagorinsky subgrid-scale (SGS) model that uses the self-similar behavior in the inertial range of the resolved turbulent cascade to extrapolate the diffusive effects of unresolved scales. In this model, the viscosity at each point in the domain is proportional to the magnitude of the strain-rate tensor, and the constant of proportionality is determined using the resolved flow and an assumption of self-similiar behavior. A detailed description of the dynamic Smagorinsky SGS model is provided in Appendix A of Nelson *et al.* (2013). Here we employ constant SGS Prandtl and magnetic Prandtl numbers of 0.25 and 0.5, respectively. In practice this permits a reduction in the average diffusion by about a factor of 50 compared to a simulation with identical resolution and a less complex SGS model, such as in Brown *et al.* (2011). This reduction in diffusion is critical not only in enhancing the turbulent intermittency of the magnetic field, but also in permitting the buoyant loops to rise through the convective layer without diffusive reconnection altering their magnetic topology.

Figure 1(a) shows a snapshot of the convective radial velocities [v_r] in case S3 at a single instant. The convection near the Equator is dominated by convective rolls aligned with the rotation axis, while the higher latitudes have more vortical motions. The rotational influence on the convective motions is key to achieving a pronounced differential rotation (Miesch, Brun, and Toomre, 2006). Case S3 maintains strong gradients in angular velocity Ω (Figure 1(b)), which are key to generating the large-scale magnetic wreaths through the Ω-effect. Figures 1(c) – (d) show snapshots of the wreaths, both on a spherical surface at mid-convection zone and in their axisymmetric component. The wreaths are dominated by non-axisymmetric fields and thus have a limited longitudinal extent, while clearly still retaining global coherence.

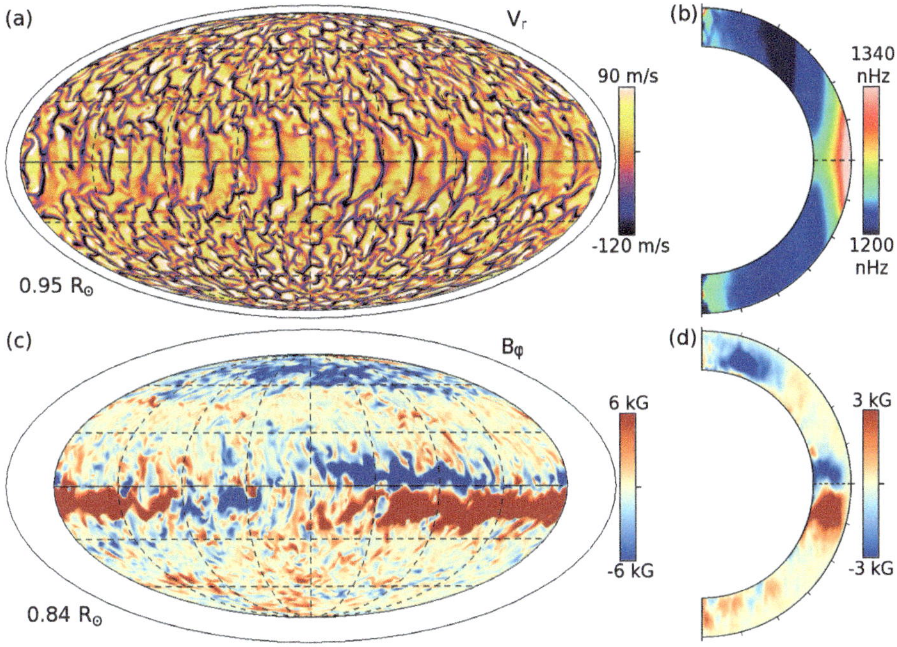

Figure 1 (a) Snapshot of radial velocities v_r at time $t_1 = 716$ days in case S3 on a spherical surface at 0.95 $R_\odot$ shown in Mollweide projection (Equator at center, lines of constant latitude parallel) in proportional size (outer ellipse represents photosphere). (b) Rotation rate [Ω] averaged in longitude and time. Strong differential rotation is achieved in radius and latitude over the simulated domain. (c) Companion snapshot of toroidal magnetic field [B_ϕ] at 0.84 $R_\odot$, with a strong coherent magnetic wreath in each hemisphere (blue negative, red positive, ranges labeled), with considerable small-scale fields also present. (d) Azimuthally averaged toroidal magnetic field [$\langle B_\phi \rangle$] at the same instant. Low-latitude wreaths are evident in both hemispheres.

Remarkably, the wreaths are generated and maintained in the bulk of the convective layer without a tachocline of shear. It had been reasonably postulated that coherent, large-scale fields in the convection zone would be shredded by the intense turbulence of the convective motions. However, the convective turbulence evidently does not destroy the wreaths. In fact, Nelson *et al.* (2013) showed that while the axisymmetric fields show some decrease in amplitude with increased turbulence, regions of extremely strong fields actually become more common due to increased turbulent intermittency. In regions of particularly strong magnetic fields, the convective motions are diminished by the Lorentz force, resulting in even less convective disruption of the wreaths.

The dynamic Smagorinsky procedure requires additional computational expense, limiting the temporal evolution of our simulations. Case S3 presented here was run for 3.4 million time steps, with an average of 40 seconds of simulated time per step. In total, case S3 covers about four years of simulated time, compared to the rotational period of 9.3 days and the convective over-turning time of about 50 days. Figure 2 shows the temporal evolution of the axisymmetric toroidal magnetic field [$\langle B_\phi \rangle$] in case S3 over about 1100 days. In this interval there are three reversals of global magnetic polarity. While the true polarity cycle involves two reversals, we term the interval between each reversal an activity cycle in the same way that the Sun's 11-year activity cycles are just about half of the true 22-year polarity cycle. These three activity cycles have durations of about 280 days, although the reversals are not generally synchronized between the two hemispheres. This nonuniform behavior hints

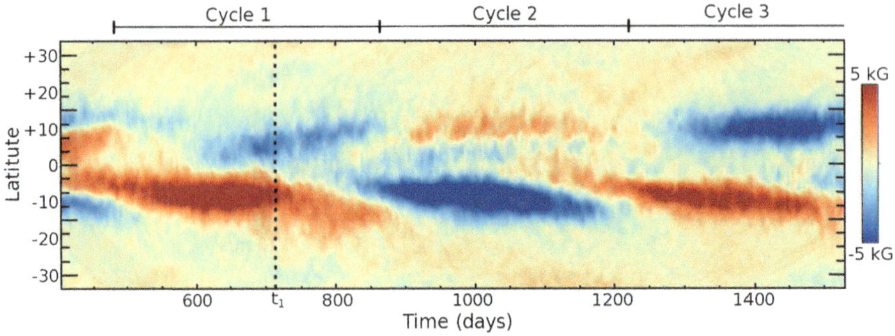

Figure 2 Evolution in time of the longitudinally averaged toroidal magnetic field [$\langle B_\phi \rangle$] at mid-convection zone shown in a time–latitude plot. Three magnetic reversals are realized, each with a period of about 280 days (reversals indicated by hash marks, cycles labeled 1 – 3 for convenience). Considerable asymmetry is seen between hemispheres in both the phase and amplitude of the reversals. The time [t_1] at which snapshots in Figures 1 and 3 are sampled is indicated by the dotted line at 716 days.

at the important role of asymmetries in the flows between the two hemispheres (DeRosa, Brun, and Hoeksema, 2012).

3. Identifying Magnetic Loops

In order to provide a consistent treatment, we define a magnetic loop as a coherent segment of magnetic field that extends from below 0.80 $R_\odot$ to above 0.90 $R_\odot$ and back down again (Nelson *et al.*, 2011). Additionally, we require that the buoyant loops have peak magnetic-field strengths greater than 5 kG above 0.90 $R_\odot$ at selected samples in time. To find magnetic loops fitting that description, we have developed a pattern-recognition algorithm that searches the 3D volume of our simulation. The most direct method of finding loops is to look for magnetic-field lines that pass through a region where $|B_\phi| > 20$ kG below 0.80 $R_\odot$, then pass through a region above 0.90 $R_\odot$ with $|B_\phi| > 5$ kG, and then again through a region where $|B_\phi| > 20$ kG below 0.80 $R_\odot$ over less than 50° in longitude. In practice, this can be done much more efficiently by recognizing that the loops start as primarily toroidal magnetic-field structures, but that as they rise into a region of faster rotation, the loops are tilted in longitude so that one side of the loop retains a strong component of B_ϕ while the other becomes almost totally radial. Thus we initially identified loop candidates by looking for this pattern of B_ϕ and B_r. The loop candidates were then verified using field-line tracings.

Case S3 uses 1024 grid points in longitude, 512 in latitude, and 192 in radius for eight evolution variables [velocity **v**, magnetic field **B**, entropy S, and pressure P], thus each snapshot in time requires over 3 GB of data. We are therefore limited in the number of time steps we can analyze. For the 278 days of Cycle 1 we ran our loop-finding procedure on snapshots of the simulation spaced roughly every four days. In doing so, we identified 131 buoyant loops. Additionally, we sampled Cycle 2 for 20 days and Cycle 3 for 40 days with the same four-day cadence and found 27 additional loops. We anticipate that we would be able to find many more loops if we carried out a more complete search through Cycles 2 and 3.

For a subset of the 158 loops found in case S3, we analyzed the dynamics of the rise of 22 loops in detail (11 from Cycle 1 and 11 more from Cycle 3). To do this, we used

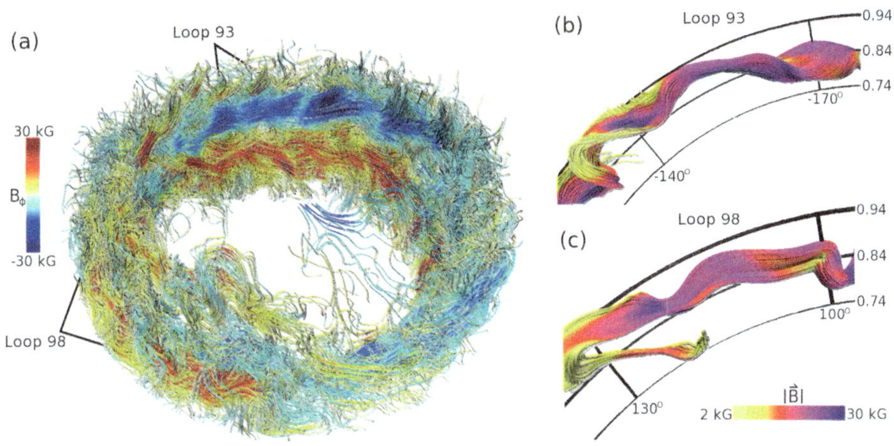

Figure 3 (a) Volume-renderings of magnetic-field lines at low latitudes colored by toroidal field [B_ϕ] (red positive, blue negative, amplitudes labeled). Strong magnetic wreaths exist in each hemisphere with considerable modulation in longitude. The location of two sample buoyant loops (labeled Loops 93 and 98) are indicated. In this view it is difficult to distinguish the loops from the surrounding magnetic fields. (b – c) Close-up views of Loops 93 and 98 at the same instant with only the field lines that comprise the buoyant loops rendered for visual clarity. Grid-lines in radius [in units of $R_\odot$] and longitude are provided. Color shows magnitude of magnetic-field strength (yellow weak, purple strong). Loop 93 is part of the negative polarity wreath in the northern hemisphere, while Loop 98 is part of the positive-polarity wreath in the southern hemisphere. Time shown corresponds to the snapshots in Figure 1 and t_1 in Figure 2.

data with a temporal resolution of about ten hours, which is sufficient to track loops backward in time from their peak radial position to their origins in the magnetic wreaths. We found that while the specific evolution of each of these 22 loops varies due to the chaotic nature of the turbulent convection, all 22 loops have significant acceleration due to magnetic buoyancy and are embedded in convective upflows that aid their rise. This agrees with the dynamics of the sample loop studied in detail by Nelson *et al.* (2011). While we cannot say with certainty that magnetic buoyancy was a significant factor in the rise of all 158 magnetic loops, we found that for all 22 of the loops studied at high time resolution the average ascent speed due to magnetic buoyancy alone is at least 28 % of the total average ascent speed. Thus we assume that magnetic buoyancy is at least an important factor in the rise of these loops.

Figure 3 displays the complex nature of the magnetic fields in case S3 with a volume-rendering of magnetic-field lines in the convection zone at low latitudes, forming two prominent magnetic wreaths of opposite polarity. We also indicate the location of two buoyant magnetic structures, labeled Loops 93 and 98. The simulation continuously exhibits magnetic fields throughout the convection zone, including strong, small-scale magnetic fields, coherent buoyant loops, and large-scale wreaths with global scale organization. Prior studies of magnetic buoyancy typically involved specified buoyant magnetic structures whose rise was studied in a largely unmagnetized domain. In contrast, our convection zone has on average 77 % of our simulated volume that contain magnetic fields in excess of 1.5 kG, and 21 % possess field amplitudes in excess of 5 kG. This makes identification of the buoyant loops difficult. Figures 3(b) – (c) show close-up renderings of only the field lines that comprise the buoyant Loops 93 and 98. We omitted rendering other field lines in these regions for visual clarity. Magnetic fields in the loops can be quite strong even near the top of our domain, with portions of Loop 93 exceeding 25 kG at 0.92 $R_\odot$.

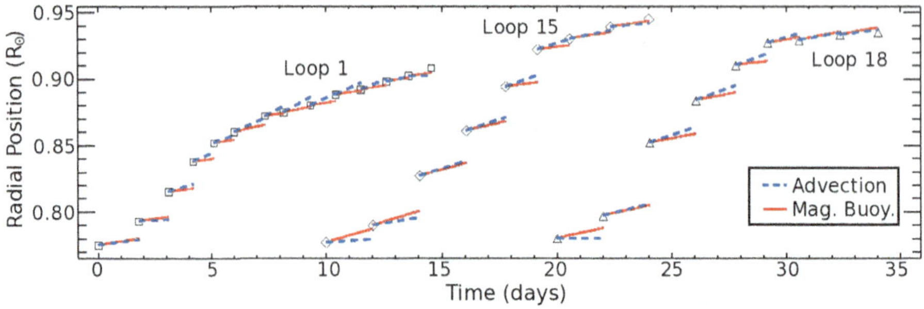

Figure 4 Location of three buoyant loops (labeled Loops 1, 15, and 18) as a function of time as they rise from the core of the toroidal wreaths in the lower convection zone through the simulated domain to their peak radial positions between 0.91 and 0.95 $R_\odot$. Times are given relative to the launch of the loops with offsets for clarity. Also plotted are the mean motions of the loops at each time interval due to magnetic buoyancy (red lines) and advection by the surrounding convective upflows (blue line). Additional motions due to forces such as thermal buoyancy, viscous drag, and magnetic tension are not plotted, and account for what may appear to be missing in this display.

4. Properties of Rising Loops

Unlike many previous models of buoyant magnetic transport in which convective turbulence is presumed to play a purely disruptive role, the buoyant loops in our models fall under the turbulence-enhanced magnetic-buoyancy paradigm discussed by Nelson *et al.* (2013). In this model, turbulent intermittency plays a key role in the formation of strong, coherent structures that are magnetically buoyant and can be advected by convective upflows. As was shown by Nelson *et al.* (2011), these loops rise through a combination of magnetic buoyancy and advection by giant-cell convection. Thus convection plays a key role both in the dynamo that generates the buoyant magnetic fields, and in the transport of the magnetic loops. Due to the cooperation between convective motions and magnetic buoyancy, the loops are able to rise from below 0.80 $R_\odot$ to above 0.90 $R_\odot$ in as little as 12 days, as suggested by Figure 4.

4.1. Dynamics and Timing of Loop Ascents

Buoyant loops are born from the much larger and less coherent magnetic wreaths shown in Figures 1 and 3. The wreaths in case S3 are not axisymmetric structures and are typically coherent over spans of between 90° and 270° in longitude. Wreaths exhibit a high degree of magnetic connectivity with the rest of the convection zone, with field lines threading in and out, suggesting rather leaky overall structures. Wreaths in case S3 generally have average field strengths of between 10 and 15 kG and are confined in the lower half of the convection zone by magnetic pumping. In the core of the wreaths convective motions can be limited by Lorentz forces to as little as a meter per second.

Portions of these wreaths can be amplified by intermittency in convective turbulence. Turbulence has been shown to generate strong, coherent structures in a variety of settings (Pope, 2000). In case S3 localized portions of the wreaths are regularly observed to attain field strengths of 40 kG and to be highly coherent over as much as 50° in longitude. These magnetic structures with strong fields are able to rise into regions where vigorous convective motions are present. Many structures are seen to emerge from the core of the wreaths only to be pummeled by a convective downflow, disrupted by a region of unusually strong turbulence, or limited by the development of a particularly unfavorable magnetic configuration.

Whether any given magnetic structure becomes a buoyant magnetic loop is therefore not due to the passing of some threshold, but largely a conspiracy of favorable events.

Figure 4 shows the radial location of the top of three different loops as they rise from roughly 0.77 $R_\odot$ to above 0.90 $R_\odot$. Also plotted are the contributions to the radially outward motion due to magnetic buoyancy and advection by convective upflows. The acceleration due to magnetic buoyancy is deduced by comparing the density in the region within the loop and the density of the surrounding convective plume. We do this to separate magnetic- and thermal-buoyancy effects. Each of these three loops starts in a region where convective motions are largely suppressed by Lorentz forces due to the very strong magnetic fields in the cores of the magnetic wreaths. As they begin to rise, the magnetic energy at the core of the wreath exceeds the kinetic energy of the flows locally by a factor of 10 to 100. As the loops rise, they enter regions of strong upflows and are advected upwards by the convective giant cells. Averaged over their entire ascent, magnetic buoyancy drives an average upward speed of about 50 $\mathrm{m\,s^{-1}}$ for these three loops, in addition to the surrounding upflows, which move at an average of about 80 $\mathrm{m\,s^{-1}}$. At their maximum radial extent, the loops are prevented from rising further by our impenetrable upper boundary condition.

Figure 5 shows three sample loops (labeled Loops 1, 2, and 3) as they rise over ten days. The loops remain coherently connected as they rise. Here again all three loops are aided by convective upflows while convective downflows pin the ends of the loops downward. The direction of motion is largely radial with a deflection of as much as 10° in latitude toward higher latitudes. This deflection is largely due to the roughly cylindrical differential–rotation contours realized in this simulation.

Loops expand as they rise through the stratified domain, but less than would be expected for a purely adiabatic rise. Without any diffusion or draining of material along the field lines, the cross-sectional area of the loops should be inversely proportional to the background pressure, leading to expansion by roughly a factor of 20. Instead, loops are seen to expand by a factor of five. This is consistent with previous studies of buoyant magnetic structures in which expansion of magnetic structures is seen to be inversely proportional to the square root of the change in pressure (Fan, 2001; Cheung *et al.*, 2010).

The expansion of the loops is slowed by draining flows of higher entropy fluid along magnetic-field lines, which serves to cool the material at the top of the loop. These divergent flows are too weak to be measured in individual loops due to the turbulent background, but when averaged over 158 loops, a mean divergent flow of 47 $\mathrm{cm\,s^{-1}}$ is obtained along the top of the loops. This compares well with estimates from a simple model (neglecting viscosity and thermal diffusion) that assumes that the draining flows are constant in time and uniform perpendicular to the axis of the loop.

Axial flows along loops are also seen as the loops rise through regions of faster rotation. When averaging over many loops, a net axial flow of 5.1 $\mathrm{m\,s^{-1}}$ is detectable in the retrograde direction, consistent with the fluid inside the loop tending to conserve its specific angular momentum as the loop moves radially outward. Loops often become distorted as this retrograde motion interacts with the surrounding prograde differential rotation as the loop rises across rotational contours (see Figure 1(b)).

The geometry of each loop that we examined is unique in its details, but Figure 6 shows three different perspectives on a single 3D volume-rendering of a typical loop. Loop 3, which is also shown in Figure 5, is located in the northern hemisphere, and its top is roughly centered at 76° N latitude and 12° W longitude. Its parent wreath-segment runs slightly Northwest to Southeast at this location and time, causing the western foot-point to be centered farther north than the eastern foot-point. The deflection away from the Equator is evident in Figures 6 (b) and (c) as the top of the loop is roughly 10° farther North than the foot-points.

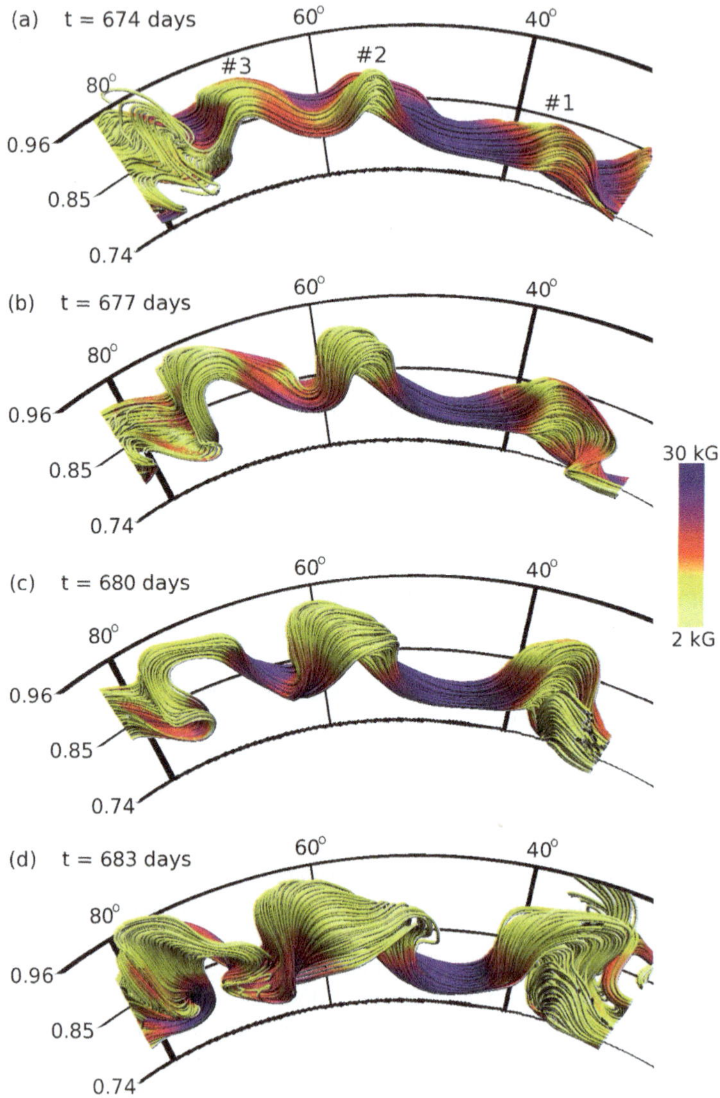

Figure 5 Sequence of volume-rendering of magnetic-field lines that comprise three buoyant loops (labeled Loops 1, 2, and 3) as they rise through the convective layer with three days between each frame (times indicated, progressing downward). Grid lines show radius [in units of $R_{\odot}$] and longitude. The expansion of each loop is evident here as they rise. Magnetic buoyancy and advection by convective upflows allow the loops to traverse the radial interval shown here in roughly 15 days. Loop 1 is also shown in Figures 4 and 7. Loop 3 is also shown in Figure 6.

The roughly five-fold expansion of the loop's cross-sectional area can be seen, particularly in Figure 6(c). This loop also shows an asymmetric top because of a downflow plume that impacts the eastern side of the top of the loop, causing the western side to extend farther in radius.

Loops start with a wide variety of field strengths and sizes and at a variety of initial radial positions. Most loops start between 0.75 and 0.78 $R_{\odot}$, although loops starting as low

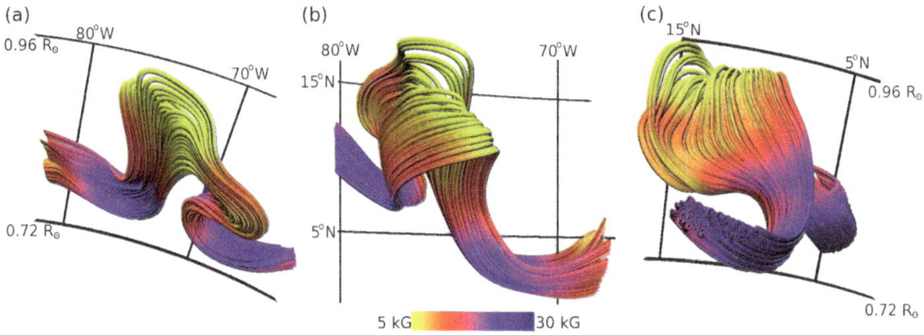

Figure 6 Three viewpoints of the same volume-rendering of magnetic-field lines in Loop 3 at $t = 683$ days (same as in Figure 5(d)). Color indicates amplitude of magnetic field (purple strong, yellow weak). Views are looking (a) South along the rotation axis with grid lines in radius (in units of the solar radius) and longitude, (b) radially inward with grid lines in longitude and latitude, (c) westward along the axis of the magnetic wreath.

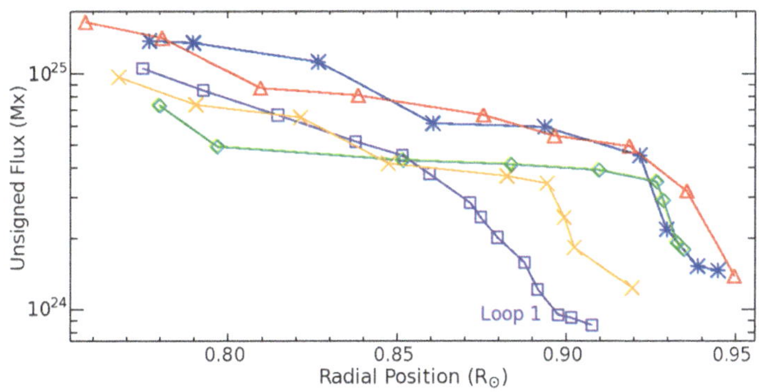

Figure 7 Unsigned magnetic flux in five sample loops as they rise through the convective layer, including Loop 1 (see Figures 4 and 5). Loops continuously lose magnetic flux through both diffusion and leakage of fluid. Their ascent is faster than the convective upflows in which they are embedded, leading to the loss of fluid and flux due to drag-like effects. Here and in the 22 loops for which detailed tracking is possible, the initial and final magnetic flux are not correlated.

as 0.73 $R_{\odot}$ are evident. When loops are traced backward in time to their starting location in order to identify the flux that will become buoyant and rise, we find that most progenitors of loops begin with about 10^{25} Mx of flux. The structures lose roughly 90 % of their flux as they rise to their peak radial positions between 0.90 and 0.96 $R_{\odot}$. Much of the flux is lost because convection in a stratified fluid requires a large percentage of the fluid to overturn prior to reaching the top of the domain. Figure 7 shows the magnetic flux as a function of the radial position of the top of the loop for five sample loops. Initial flux and initial radial location do not appear to be good predictors of either final radial location or final magnetic flux.

In the specific case of Loop 1, 92 % of the magnetic flux that it started with is lost over the course of its ascent while 69 % of the mass flux at 0.78 $R_{\odot}$ turns over below 0.91 $R_{\odot}$. The overturning mass flux carries away 61 % of the magnetic flux, because preferentially regions of lower field strength are lost. The next-largest contributor is resistive diffusion, which dissipates 19 % of the initial flux. The remaining 12 % of the flux is lost through

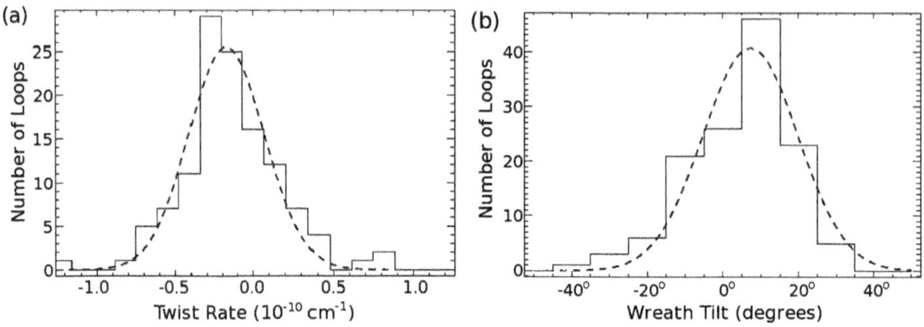

Figure 8 (a) Histogram of twist rate parameter [q_J] values for the 131 loops observed in Cycle 1 along with the best-fit Gaussian distribution of those values. The distribution shows a slight preference for negative twist rates, although the mean twist rate is $(-1.8 \pm 2.4) \times 10^{-11}$ cm^{-1}. (b) Histogram of latitudinal tilt [Δ_θ] values for the same 131 loops. Positive tilts indicate that the leading edge of the loop is closer to the Equator than the trailing edge, as with Joy's law. Tilts were calculated so that all values fall between $\pm 90°$ for this analysis. Positive tilts are preferred and the mean latitudinal tilt is $7.3 \pm 12.6°$ in latitude.

a combination of small-scale turbulent advection and shear. Eventually, diffusive reconnection realigns the fields so that the loops are no longer distinct from the surrounding MHD turbulence.

4.2. Statistical Distribution of Twist and Tilt

Previous MHD simulations of flux emergence have emphasized that magnetic structures must be twisted to remain coherent as they rise (see review by Fan, 2009). Twist in this context can be defined by a parameter q_A, which for a uniformly twisted flux tube is defined by

$$B_\parallel = a_\pm q_A \lambda |\nabla \times A_\parallel|, \tag{1}$$

where $B_\parallel$ and $A_\parallel$ are the magnetic field and magnetic vector potential along the axis of the flux tube, $a_\pm$ is 1 in the northern hemisphere and -1 in the southern hemisphere, and λ is the distance from the axis of the flux tube. For the tube to remain coherent as it rises, previous numerical simulations have suggested that twist must exceed some critical value Q_A (Moreno-Insertis and Emonet, 1996). Fan (2008) used 3D simulations of buoyant magnetic structures rising through a quiescent, stratified layer and found a critical level of twist $Q_A \approx -3 \times 10^{-10}$ cm^{-1}.

For our simulation, the loops are clearly not uniformly twisted flux tubes, so we calculated another measure of twist following the procedure used in observational studies (*e.g.* Pevtsov, Canfield, and Metcalf, 1995; Pevtsov, Maleev, and Longcope, 2003; Tiwari, Venkatakrishnan, and Sankarasubramanian, 2009). Sunspots often show large variations in the level and even sign of twist, so a weighted average of the twist parameter is employed, which we call q_J. We computed the twist parameter as

$$q_J = a_\pm \left[\frac{J_\phi}{B_\phi} \right], \tag{2}$$

where brackets denote an average over radius and latitude for a longitudinal cut taken through the loop, and $a_\pm$ is 1 in the northern hemisphere and -1 in the southern hemisphere. We restricted our averages to contiguous regions with the correct polarity and to those where fields are stronger than 2.5 kG. Figure 8(a) shows a histogram of values for the twist parameter $[q_J]$ for the 131 loops identified in Cycle 1, as well as the best-fit Gaussian to that distribution, which peaks at $\bar{q_J} = -1.8 \times 10^{-11}$ cm^{-1}. For comparison, Tiwari, Venkatakrishnan, and Sankarasubramanian (2009) reported an average twist parameter of $\bar{q_J} = -6.12 \times 10^{-11}$ cm^{-1} for a sample of 43 sunspots.

It is difficult to make a direct comparison between the two measures of twist mentioned here. In practice, our loops are poorly represented by uniformly twisted tubes. It is possible to compute the value of q_A at each point in the loop and create an average value, but we find that those averages are highly sensitive to the weighting of the points and the region over which the average is taken. Alternatively, we computed the value of q_J for the formulation employed in Fan (2008) and found that the value varies with the location and size of the magnetic structure in radius and latitude. For most reasonable parameter choices, q_J/q_A is between 1 and 2. When comparing with photospheric measurements, we must also remember that considerable changes may take place as magnetic flux passes through the upper 5 % of the solar convection zone. The dynamics of twisted buoyant loops in that region is beginning to be studied in local domains (Cheung *et al.*, 2010).

Of the 131 loops in Cycle 1, only 13 had current-derived twist parameters $[q_J]$ within an order of magnitude of the critical value Q_A. One explanation may be that convective upflows assisting the rise of these loops reduce the drag that they experience, thus making them less susceptible to disruption as they rise and therefore less dependent on twist for coherence. Whatever the cause, we do not see a critical value of twist beyond which loops are unable to traverse our domain.

Additionally, we can look at the latitudinal tilts of the buoyant loops. We calculated these tilts by computing the center of each loop at all longitudes where the center is within 0.02 $R_\odot$ of its peak position and then fitting a linear trend to latitudinal locations of the loop center. We define positive tilts to be those with the eastern side of the loop closer to the Equator than the western side, as in Joy's law. Here we did not consider the polarity of the loops, so values are restricted to the interval $[-90°, 90°]$. The distribution of tilts seen in the 131 loops found in Cycle 1 is shown in Figure 8(b), along with the best-fit Gaussian to that distribution, which peaks at 7.3° but is quite broad. This is similar to observations of tilts in sunspots where the trend towards Joy's law is part of broad distribution in tilt angles (Li and Ulrich, 2012).

5. Magnetic Cycles with Buoyant Loops

Case S3 achieves three magnetic-activity cycles with reversals in global magnetic polarity. If we define the cycle period as the time between changes in the sign of the antisymmetric components of the toroidal field at low latitudes, as in Brown *et al.* (2011), then Cycles 1 and 2 have periods of 278 and 269 days, respectively. Cycle 3 had not ended at the present end of the simulation, but has been simulated for 228 days. The coexistence of cyclic magnetic activity and buoyant loops provides an opportunity to probe the relationship between axisymmetric fields, which are commonly used in 2D dynamo models (see review by Charbonneau, 2010), and the buoyant transport of magnetic flux.

We have chosen to conduct our analysis primarily using Cycle 1 since the process of finding and characterizing buoyant loops is too data-intensive to be carried out conveniently for all three cycles. Figure 9 shows a time–latitude plot of the mean toroidal field (averaged in longitude and in radius over the lower convection zone from 0.72 to 0.84 $R_\odot$), as well as the location in time and latitude of the 131 buoyant loops detected in Cycle 1. It is evident from this representation that the loops do not arise uniformly in time. Although loops tend

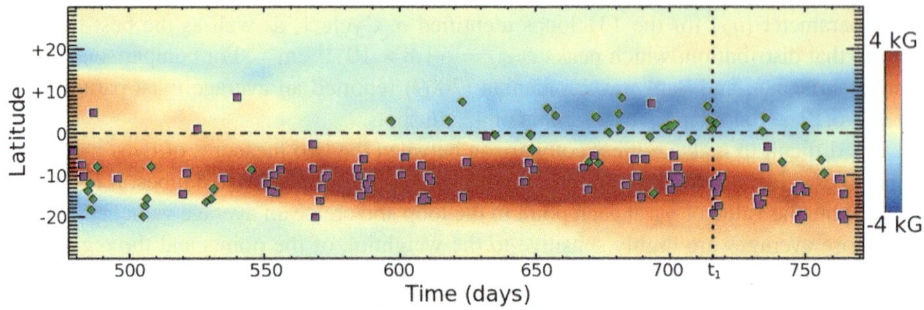

Figure 9 Time–latitude display of the toroidal magnetic field averaged in longitude and radius during the peak of cycle 1. Overplotted symbols indicate the time and latitude of 131 buoyant loops in the style of a synoptic map, with positive polarity loops shown as pink squares and negative polarity loops as green diamonds. Some loops may be present from the previous cycle, particularly prior to Day 550 in the southern hemisphere. Time t_1 at which the snapshots in Figures 1 and 3 are taken is indicated by the dotted line.

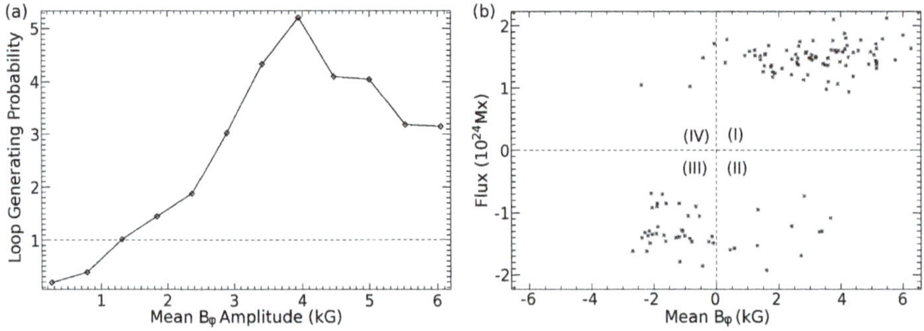

Figure 10 (a) Relative probability that a region with given mean B_ϕ will produce a buoyant loop compared to the production rate of buoyant loops in Cycle 1 averaged over all events. For Cycle 1 the average occurrence rate for loops was 7.6×10^{-3} day^{-1} degree^{-1}. We normalized all probabilities by this rate, therefore the dashed line represents the average loop-production rate. Note that nearly 60 % of the times and latitudes considered have mean field strengths of less than 1.5 kG. (b) Total magnetic flux as a function of axisymmetric toroidal magnetic field averaged in radius at the latitude and time of each of the 131 buoyant loops in Figure 9. While most loops are associated with mean magnetic fields of the correct sense, there are 15 loops in quadrants II and IV that arise from wreath segments which are canceled in the longitudinal averaging procedure by large or stronger wreath segments at the same time and latitude of the opposite polarity.

to appear at times and latitudes when the mean toroidal fields are strong, they can also appear at times and latitudes with relatively weak mean fields. There are even examples in which loops have the opposite polarity to the longitudinally-averaged mean fields at that time and latitude. This is consistent with the non-axisymmetric nature of the wreaths shown in Figures 1(c)–(d) where smaller-scale segments of intense toroidal field can be masked in the longitudinal average by larger segments of the opposite polarity.

5.1. Relation of Loop Emergence and Mean Field Strength

In many mean-field models it is assumed that buoyant magnetic flux (which can be used as a proxy for the sunspot number) at a given latitude and time is proportional to the axisymmetric toroidal field strength at that location and time at the generation depth. In particular, the Babcock–Leighton model postulates that the buoyant transport of magnetic flux occurs

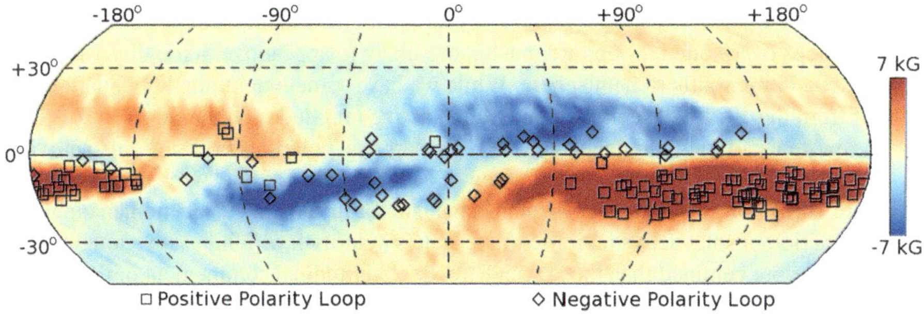

Figure 11 Time-averaged toroidal magnetic field on a spherical surface between $\pm 45°$ of latitude at $0.79\ R_{\odot}$ during Cycle 1. Symbols indicate the rotational phase in longitude of the 131 buoyant loops from Figure 9 at the time they were launched. Squares indicate positive polarity loops while diamonds indicate negative polarity loops. Both the wreaths and loops are confined in longitude. Loops are particularly concentrated in the strong positive wreath segment in the southern hemisphere.

whenever the axisymmetric magnetic field exceeds some threshold value (*e.g.* Durney, 1995; Chatterjee, Nandy, and Choudhuri, 2004). Here we can test this assumption by looking at the probability that a region with a given axisymmetric field strength will produce a buoyant loop. Figure 10 shows the relative probability that a region with a given mean field strength will produce a buoyant loop. Over Cycle 1, the average production rate of buoyant loops is roughly one loop every two days within 30° of the Equator. Regions with $\langle B_\phi \rangle \leq 1.5$ kG cover about 60 % of the time–latitude domain and produce loops at or below the average rate. The generation probability per unit time and latitude rises to five times the mean rate for regions with $\langle B_\phi \rangle \approx 3.9$ kG. Interestingly, the generation probability then falls for the regions of the strongest $\langle B_\phi \rangle$. Indeed, the strongest regions of axisymmetric field are only about three times more likely to produce buoyant loops than the average production rate. The relatively small sample size invites further study on this topic, as only 5 % of the domain is covered by fields above 4.2 kG, which fall in the last four bins. However, the implication that axisymmetric toroidal fields above some threshold value are less likely to produce buoyant loops may have significant implications for mean-field models of the solar dynamo.

To further explore the correlation between the axisymmetric field strength and the amount of buoyant magnetic flux, we can look for correlations between the amount of flux in a given buoyant loop and the axisymmetric fields at the time and latitude of its launch. Figure 10(b) shows the magnetic flux in each of the 131 buoyant loops from Cycle 1 as a function of the average value of axisymmetric toroidal field in the lower convection zone at the time of launch. Out of 131 loops, 15 were launched when the axisymmetric B_ϕ was of the opposite sense. Interestingly, Stenflo and Kosovichev (2012) report that roughly 5 % of moderate to large active regions violate Hale's polarity law.

5.2. Preferential Longitudes for Loop Creation

The longitudinal concentration of sunspots into so-called active longitudes has been observed for the past several solar cycles (Henney and Harvey, 2002). These active longitudes provide observational evidence that the creation of buoyant magnetic structures is not a purely axisymmetric process. Magnetic wreaths in case S3 tend to be confined in longitude, as was shown in Figure 1(c). These wreath segments are generally between 90° and 270° in longitude. Loops tend to be generated in these wreath segments, and thus are more likely to appear in those longitudinal patches than other longitudes. Figure 11 shows the time-averaged value of B_ϕ at 0.80 $R_\odot$ over Cycle 1 with the longitudinal position of the 131 buoyant loops overplotted. Loops are much more likely to appear over a roughly 180° patch in longitude in the southern hemisphere. While we have some longitudinal modulation, it is still far from the 10° to 20° confinement seen in active longitudes on the Sun.

The existence of longitudinal patches of both magnetic polarities in case S3 also provides a potential explanation for the small fraction of active regions of the "wrong" polarity seen in Figure 10(b). This provides a possible mechanism for the analogous phenomena in which a small fraction of solar active regions violate Hale's polarity law. While active longitudes in the Sun are more confined than those seen here, the longitudinal confinement of the wreaths in case S3 may provide a possible pathway toward understanding active longitudes.

6. Summary and Reflections: Buoyant Loops in Convective Dynamos

This article has explored the first global convective-dynamo simulation to achieve buoyant magnetic loops that transport coherent magnetic structures through the convection zone. These buoyant structures arise from large-scale magnetic wreaths, which have been previously described in both persistent (Brown *et al.*, 2010) and cyclic states (Brown *et al.*, 2011; Nelson *et al.*, 2013). In this work we have focused on case S3, which possesses large-scale magnetic wreaths that undergo cycles of magnetic activity and produce many buoyant magnetic loops. Case S3 was able to achieve buoyant loops due to the use of a dynamic Smagorinsky SGS model, which greatly reduced diffusive processes in the simulation.

Although case S3 has a rotation rate greater than the current Sun, the dynamics achieved may be applicable to solar dynamo action. The most salient non-dimensional parameter for the creation of toroidal wreaths is the Rossby number, which considers the local vorticity [ω] and rotation rate [Ω] as $\mathrm{Ro} = \omega/2\Omega$. In case S3 the Rossby number at mid-convection zone is 0.581, indicating that the convection is rotationally constrained, as is also expected in the bulk of the solar convection zone.

Much of the work on buoyant magnetic flux has generally regarded convection as a purely disruptive process. In our dynamo studies here, convection plays a key role in both the creation of the strong, coherent magnetic fields and the advection of magnetic flux radially outward. Turbulent intermittency provides an effective mechanism for the amplification of magnetic fields to energy densities well above equipartition with the resolved flows (Nelson *et al.*, 2013). Convection also assists in the transport process by the upflows helping to advect the loops. Without convection, buoyant transport of magnetic flux is generally regarded as a low-wavenumber instability on axisymmetric fields. With convection, buoyant loops are formed on convective length scales as the result of non-axisymmetric processes. The loops realized in case S3 are not large-scale instabilities of axisymmetric flux tubes, but rather they result from turbulently amplified coherent structures becoming buoyant and being advected by convective upflows. Similar upward advection of magnetic structures by convection has been seen when considering the impact of convection on flux tubes (Weber, Fan, and Miesch, 2011, 2012) or specified magnetic structures (Jouve and Brun, 2009).

When we consider moderate numbers of buoyant loops over an activity cycle, we find a number of trends in their collective behavior. In all of these trends, it is important to note that our statistical sample of 158 loops is significant, but still relatively small. First, loops in our simulation clearly show a hemispheric polarity preference analogous to Hale's polarity law for solar active regions, although case S3 shows a slightly higher rate of violations to this trend compared to the Sun. Second, the buoyant loops tend to show latitudinal tilts similar to Joy's law for solar active regions. As in the Sun, a wide variety of tilt angles are observed, though the average tilt angle places the leading edge of the buoyant loop closer to the Equator than the trailing edge. Third, the buoyant loops tend to show a degree of twist similar to the twist inferred from photospheric measurements of vector magnetic fields. Again, a wide variety of twist parameters are measured centered about a relatively low negative mean value. Finally, there are ranges in longitude that demonstrate repeated emergence of magnetic flux. This longitudinal modulation in the creation of magnetic loops is reminiscent of active longitudes observed in the Sun, but on broader longitudinal ranges than active longitudes in the Sun.

Buoyant transport of magnetic fields is a key ingredient in many models of the solar dynamo. Mean-field models often use parameterizations to represent this buoyant transport. We have considered connections between the axisymmetric toroidal fields in case S3 and the magnetic flux in the buoyant loops. We find that total flux in a given buoyant loop is only weakly dependent on the strength of the mean field from which that buoyant loop was generated. Additionally, we find that the probability that a buoyant loop will be generated in regions of relatively weak mean fields is significant, and that the strongest mean fields may be less likely to generate buoyant loops than regions of moderate axisymmetric fields.

This simulation is a first step towards connecting convective-dynamo models and flux emergence in the Sun and Sun-like stars. As we consider the role of turbulent convection, we find clear indications that it plays important roles in the dynamo that generates buoyant magnetic loops and the transport of those loops. This simulation invites continued effort towards linking convective-dynamo models and simulations of flux emergence.

Acknowledgements We thank Kyle Augusten, Chris Chronopolous, and Yuhong Fan for discussions and advice. This research is partly supported by NASA through Heliophysics Theory Program grants NNX08AI57G and NNX11AJ36G. BPB is supported through NSF Astronomy and Astrophysics postdoctoral fellowship AST 09-02004. CMSO is supported by NSF grant PHY 08-21899. MSM is also supported by NASA SR&T grant NNH09AK14I. NCAR is sponsored by the National Science Foundation. ASB is partly supported by both the Programmes Nationaux Soleil-Terre and Physique Stellaire of CNRS/INSU (France), and by the STARS2 grant 207430 from the European Research Council. The simulations were carried out with NSF TeraGrid and XSEDE support of Ranger at TACC, and Kraken at NICS, and with NASA HECC support of Pleiades.

References

Babcock, H.: 1961, *Astrophys. J.* **133**, 572. doi:10.1086/147060.

Baliunas, S.L., Donahue, R.A., Soon, W.H., Horne, J.H., Frazer, J., Woodard-Eklund, L., *et al.*: 1995, *Astrophys. J.* **438**, 269. doi:10.1086/175072.

Brown, B.P., Browning, M.K., Brun, A.S., Miesch, M.S., Toomre, J.: 2008, *Astrophys. J.* **689**, 1354. doi:10.1086/592397.

Brown, B.P., Browning, M.K., Brun, A.S., Miesch, M.S., Toomre, J.: 2010, *Astrophys. J.* **711**, 424. doi:10.1088/0004-637X/711/1/424.

Brown, B.P., Miesch, M.S., Browning, M.K., Brun, A.S., Toomre, J.: 2011, *Astrophys. J.* **731**, 69. doi:10.1088/0004-637X/731/1/69.

Browning, M.K., Miesch, M.S., Brun, A.S., Toomre, J.: 2006, *Astrophys. J. Lett.* **648**, L157. doi:10.1086/507869.

Brun, A.S., Miesch, M.S., Toomre, J.: 2004, *Astrophys. J.* **614**, 1073. doi:10.1086/423835.

Caligari, P., Moreno-Insertis, F., Schüssler, M.: 1995, *Astrophys. J.* **441**, 886. doi:10.1086/175410.

Charbonneau, P.: 2010, *Living Rev. Solar Phys.* **7**, 3. http://www.livingreviews.org/lrsp-2010-3.

Chatterjee, P., Nandy, D., Choudhuri, A.R.: 2004, *Astron. Astrophys.* **427**, 1019. doi:10.1051/0004-6361.

Cheung, M.C.M., Rempel, M., Title, A.M., Schüssler, M.: 2010, *Astrophys. J.* **720**, 233. doi:10.1088/0004-637X/720/1/233.
Cline, K.S., Brummell, N.H., Cattaneo, F.: 2003, *Astrophys. J.* **588**, 630. doi:10.1086/373894.
Clune, T.C., Elliott, J.R., Miesch, M.S., Toomre, J.: 1999, *Parallel Comput.* **25**, 361. doi:10.1016/S0167-8191(99)00009-5.
DeRosa, M.L., Brun, A.S., Hoeksema, J.T.: 2012, *Astrophys. J.* **757**, 96. doi:10.1088/0004-637X/757/1/96.
Durney, B.: 1995, *Solar Phys.* **160**, 213. doi:10.1007/BF00732805.
Fan, Y.: 2001, *Astrophys. J.* **546**, 509. doi:10.1086/318222.
Fan, Y.: 2008, *Astrophys. J.* **676**, 680. doi:10.1086/527317.
Fan, Y.: 2009, *Living Rev. Solar Phys.* **6**, 4. http://www.livingreviews.org/lrsp-2009-4.
Gaulme, P., Deheuvels, S., Weiss, W.W., Mosser, B., Moutou, C., Bruntt, H., *et al.*: 2010, *Astron. Astrophys.* **524**, A47. doi:10.1051/0004-6361/201014142.
Ghizaru, M., Charbonneau, P., Smolarkiewicz, P.K.: 2010, *Astrophys. J. Lett.* **715**, L133. doi:10.1088/2041-8205/715/2/L133.
Guerrero, G., Käpylä, P.J.: 2011, *Astron. Astrophys.* **533**, A40. doi:10.1051/0004-6361/201116749.
Hempelmann, A., Schmitt, J.H.M.M., Stępień, K.: 1996, *Astron. Astrophys.* **305**, 284.
Henney, C.J., Harvey, J.W.: 2002, *Solar Phys.* **207**, 199. doi:10.1023/A:1016265629455.
Jouve, L., Brun, A.S.: 2009, *Astrophys. J.* **701**, 1300. doi:10.1088/0004-637X/701/2/1300.
Krause, F., Raedler, K.-H.: 1980, *Mean-Field Magnetohydrodynamics and Dynamo Theory*, Pergamon Press, Oxford, 271.
Leighton, R.: 1964, *Astrophys. J.* **140**, 1547.
Li, J., Ulrich, R.K.: 2012, *Astrophys. J.* **758**, 29. doi:10.1088/0004-637X/758/2/115.
Llama, J., Jardine, M., Mackay, D.H., Fares, R.: 2012, *Mon. Not. Roy. Astron. Soc. Lett.* **422**, L72. doi:10.1111/j.1745-3933.2012.01239.x.
Meibom, S., Barnes, S., Latham, D.W., Batalha, N., Borucki, W.J., Koch, D.G., *et al.*: 2011, *Astrophys. J. Lett.* **733**, L9. doi:10.1088/2041-8205/733/1/L9.
Miesch, M.S.: 2005, *Living Rev. Solar Phys.* **2**, 1. http://www.livingreviews.org/lrsp-2005-1.
Miesch, M.S., Brown, B.P.: 2012, *Astrophys. J. Lett.* **746**, L26. doi:10.1088/2041-8205/746/2/L26.
Miesch, M.S., Brun, A.S., Toomre, J.: 2006, *Astrophys. J.* **641**, 618. doi:10.1086/499621.
Moffatt, H.K.: 1978, *Magnetic Field Generation in Electrically Conducting Fluids*, Cambridge University Press, Cambridge, 353.
Moreno-Insertis, F., Emonet, T.: 1996, *Astrophys. J. Lett.* **472**, L53. doi:10.1086/310360.
Morgenthaler, A., Petit, P., Saar, S., Solanki, S.K., Morin, J., Marsden, S.C., *et al.*: 2012, *Astron. Astrophys.* **540**, A138. doi:10.1051/0004-6361/201118139.
Nelson, N.J., Brown, B.P., Brun, A.S., Miesch, M.S., Toomre, J.: 2011, *Astrophys. J. Lett.* **739**, L38. doi:10.1088/2041-8205/739/2/L38.
Nelson, N.J., Brown, B.P., Brun, A.S., Miesch, M.S., Toomre, J.: 2013, *Astrophys. J.* **762**, 73. doi:10.1088/0004-637X/762/2/73.
Oláh, K., Kolláth, Z., Granzer, T., Strassmeier, K.G., Lanza, A.F., Järvinen, S., *et al.*: 2009, *Astron. Astrophys.* **501**, 703. doi:10.1051/0004-6361/200811304.
Parker, E.N.: 1955, *Astrophys. J.* **122**, 293. doi:10.1086/146087.
Petit, P., Dintrans, B., Solanki, S.K., Donati, J.-F., Aurire, M., Lignires, F., *et al.*: 2008, *Mon. Not. Roy. Astron. Soc.* **388**, 80. doi:10.1111/j.1365-2966.2008.13411.x.
Pevtsov, A.A., Canfield, R.C., Metcalf, T.R.: 1995, *Astrophys. J. Lett.* **440**, L109. doi:10.1086/187773.
Pevtsov, A.A., Maleev, D., Longcope, D.: 2003, *Astrophys. J. Lett.* **593**, 1217. doi:10.1086/376733.
Pizzolato, N., Maggio, A., Micela, G., Sciortino, S., Ventura, P.: 2003, *Astron. Astrophys.* **157**, 147. doi:10.1051/0004-6361.
Pope, S.B.: 2000, *Turbulent Flows*, Cambridge University Press, Cambridge.
Racine, E., Charbonneau, P., Ghizaru, M., Bouchat, A., Smolarkiewicz, P.K.: 2011, *Astrophys. J.* **735**, 46. doi:10.1088/0004-637X/735/1/46.
Saar, S.H., Brandenburg, A.: 1999, *Astrophys. J.* **524**, 295. doi:10.1086/307794.
Stenflo, J.O., Kosovichev, A.G.: 2012, *Astrophys. J.* **745**, 129. doi:10.1088/0004-637X/745/2/129.
Tiwari, S.K., Venkatakrishnan, P., Sankarasubramanian, K.: 2009, *Astrophys. J. Lett.* **702**, L133. doi:10.1088/0004-637X/702/2/L133.
Vasil, G.M., Brummell, N.H.: 2009, *Astrophys. J.* **690**, 783. doi:10.1088/0004-637X/690/1/783.
Weber, M.A., Fan, Y., Miesch, M.S.: 2011, *Astrophys. J.* **741**, 11. doi:10.1088/0004-637X/741/1/11.
Weber, M.A., Fan, Y., Miesch, M.S.: 2012, *Solar Phys.* doi:10.1007/s11207-012-0093-7.
Wright, N.J., Drake, J.J., Mamajek, E.E., Henry, G.W.: 2011, *Astrophys. J.* **743**, 48. doi:10.1088/0004-637X/743/1/48.

DOI 10.1007/978-1-4939-1182-0_3
Reprinted from *Solar Physics* Journal, DOI 10.1007/s11207-012-0185-4

Helioseismic Holography of an Artificial Submerged Sound Speed Perturbation and Implications for the Detection of Pre-emergence Signatures of Active Regions

D.C. Braun

Received: 24 August 2012 / Accepted: 29 October 2012 / Published online: 27 November 2012

Abstract We use a publicly available numerical wave-propagation simulation of Hartlep *et al.* (*Solar Phys.* **268**, 321, 2011) to test the ability of helioseismic holography to detect signatures of a compact, fully submerged, 5 % sound-speed perturbation placed at a depth of 50 Mm within a solar model. We find that helioseismic holography employed in a nominal "lateral-vantage" or "deep-focus" geometry employing quadrants of an annular pupil can detect and characterize the perturbation. A number of tests of the methodology, including the use of a plane-parallel approximation, the definition of travel-time shifts, the use of different phase-speed filters, and changes to the pupils, are also performed. It is found that travel-time shifts made using Gabor-wavelet fitting are essentially identical to those derived from the phase of the Fourier transform of the cross-covariance functions. The errors in travel-time shifts caused by the plane-parallel approximation can be minimized to less than a second for the depths and fields of view considered here. Based on the measured strength of the mean travel-time signal of the perturbation, no substantial improvement in sensitivity is produced by varying the analysis procedure from the nominal methodology in conformance with expectations. The measured travel-time shifts are essentially unchanged by varying the profile of the phase-speed filter or omitting the filter entirely. The method remains maximally sensitive when applied with pupils that are wide quadrants, as opposed to narrower quadrants or with pupils composed of smaller arcs. We discuss the significance of these results for the recent controversy regarding suspected pre-emergence signatures of active regions.

Keywords Helioseismology, observations

1. Introduction

For almost two decades, helioseismic methods have been employed to search for evidence of magnetic flux rising through the convection zone (Braun, 1995; Chang, Chou,

Solar Origins of Space Weather and Space Climate
Guest Editors: I. González Hernández, R. Komm, and A. Pevtsov

D.C. Braun (✉)
NWRA, CoRA Office, 3380 Mitchell Ln, Boulder, CO 80301, USA
e-mail: dbraun@nwra.com

and Sun, 1999; Jensen *et al.*, 2001; Zharkov and Thompson, 2008; Kosovichev, 2009; Hartlep *et al.*, 2011; Ilonidis, Zhao, and Kosovichev, 2011; Leka *et al.*, 2012; Birch *et al.*, 2012). Submerged magnetic fields may produce travel-time anomalies due to changes in the wave speed caused by the magnetic field or by the presence of flows and perturbations to the thermal structure associated with the magnetic field (Birch, Braun, and Fan, 2010). If positively identified, such signatures could play an important role in space-weather forecasting, and lead to a physical understanding of the emergence process, which is a key component of the solar-activity cycle. Recent detection of p-mode travel-time anomalies prior to the emergence of several large active regions, obtained with time–distance methods, have been reported (Ilonidis, Zhao, and Kosovichev 2011, 2012b), although no significant travel-time anomalies were subsequently measured from an independent analysis using helioseismic holography (Braun, 2012). Ilonidis, Zhao, and Kosovichev (2012a) suggest that this discrepancy may be due to differences in sensitivity between the methods employed.

Numerical simulations have provided artificial data through which helioseismic analysis and modeling can be tested (Jensen *et al.*, 2003; Benson, Stein, and Nordlund, 2006; Hanasoge *et al.*, 2006; Parchevsky and Kosovichev, 2007; Zhao *et al.*, 2007; Braun *et al.*, 2007; Cameron, Gizon, and Duvall, 2008; Parchevsky and Kosovichev, 2009; Crouch *et al.*, 2010; Cameron *et al.*, 2011; Birch *et al.*, 2011; Hartlep *et al.*, 2011; Braun *et al.*, 2012). Many of these simulations include near-surface flows, sound-speed perturbations, or magnetic structures typical of active regions or supergranulation. Simulations that propagate waves through completely submerged perturbations are rarer (Hartlep *et al.*, 2011), but are critical for testing and developing helioseismic methods that are sensitive to detect active regions prior to their emergence on the surface. In this work, we use one of the simulations of Hartlep *et al.* (2011) to test the sensitivity of helioseismic holography comparatively to subsurface sound-speed perturbations under a variety of applications.

2. Simulation

Hartlep *et al.* (2011) constructed a number of simulations containing p-modes propagating through a spherical domain containing localized perturbations of the sound speed about the standard solar Model S (Christensen-Dalsgaard *et al.*, 1996). No flows or magnetic fields are included. The solar model is convectively stabilized by a neglect of the entropy gradient of the background model, which lowers the acoustic cutoff frequency. The mode amplitudes above 3.5 mHz are thus reduced in amplitude. In addition, the simulation is only populated with p-modes with angular degree [ℓ] between 0 and 170. The simulations span about 17 hours of solar time. A number of simulations using the same code are publicly available and include a variety of sound-speed perturbations at different depths. In this work, we employ the simulation with a peak 5 % sound-speed reduction at a depth of 50 Mm and with a horizontal size of 45 Mm (see Figure 1). The simulated velocity field is provided in arbitrary units and represented in heliographic coordinates, with 512 pixels in longitude and 256 pixels in latitude, and a cadence of one minute. The simulation is stored in a FITS file (sun.stanford.edu/~thartlep/Site/Artificial_Data/Entries/2012/3/21_Subsurface_sound_speed_perturbations.html).

Our primary emphasis is on testing the ability of helioseismic holography to detect p-mode travel-time signatures of the prescribed perturbation within the simulation, and to measure the *relative* sensitivity of the results (in both signal strength and background noise) to changes of methodology. In contrast, direct comparison of measured and expected travel times requires the computation and application of sensitivity functions, which is not attempted here. A prediction of the travel-time shift expected from a given sound-speed perturbation is a non-trivial exercise, but a rough estimate is useful. We estimate the travel-time

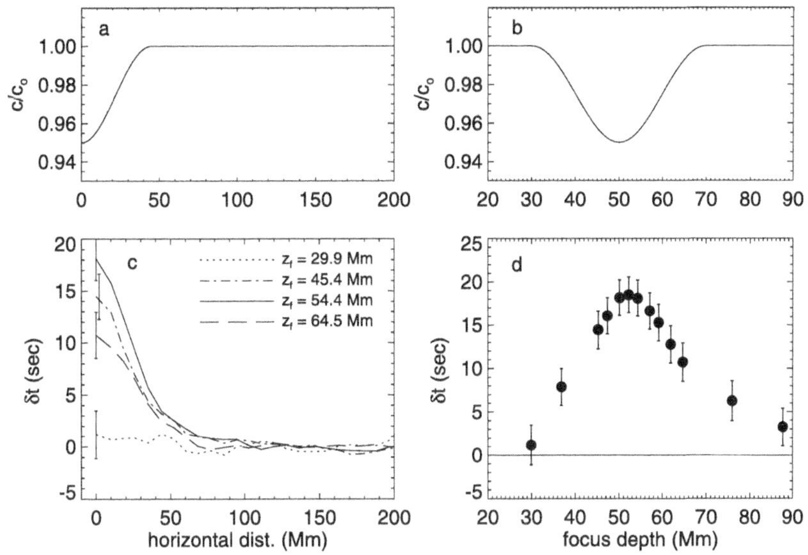

Figure 1 The sound-speed ratio $[c/c_0]$ in the simulation, where c is the perturbed sound speed and c_0 is the background sound speed of Model S (top panels), and the travel-time measurements (bottom panels) made with helioseismic holography using a "nominal" methodology (see text). (a) The variation with horizontal distance from the center of the circularly symmetric sound-speed ratio at a depth of 50 Mm below the surface of the simulation. (b) The variation with depth of the sound-speed ratio at the center of the perturbation. (c) The variation with horizontal distance of the azimuthally averaged mean travel-time shift measured using lateral-vantage helioseismic holography applied to the simulation at focus depths of 29.9 Mm (dotted line), 45.4 Mm (dash–dotted line), 54.4 Mm (solid line), and 64.5 Mm (long dashed line). The travel-time shifts are averaged over 0.7°-wide annuli centered on the location of the perturbation. (d) The travel-time shift at the center of the perturbation as a function of focus depth. The error bars in panels (c) and (d) indicate the standard deviation of the realization noise determined from a region away from the perturbation (see text).

shift in the geometric-optics limit as the path integral of the fractional sound-speed perturbation (Equation (1) of Hartlep *et al.*, 2011) weighted by the inverse of the background sound speed in Model S. For convenience, the path is chosen as purely horizontal through the center of the perturbation. This procedure yields a travel-time increase of 23 seconds.

3. The Nominal Procedure and Results

Helioseismic holography (hereafter HH) is described extensively elsewhere (Lindsey and Braun, 1997, 2000, 2004; Chang *et al.*, 1997). For our purposes, it is useful to enumerate the data-analysis steps taken to define the "nominal," or baseline, procedure. This provides the context for investigating the sensitivity of the results to changes in methodology (discussed in Section 4).

The basic idea is to apply Green's functions to the solar oscillation field at the surface of the Sun (or in this case, a simulation) to estimate the amplitudes of incoming and outgoing waves at targets (or "focal points") at or below the surface. In the "lateral-vantage" or "deep-focus" configuration of HH (Lindsey and Braun, 2004; Braun, Birch, and Lindsey, 2004; Braun and Birch, 2008a), travel-time perturbations are extracted from the cross-covariances between these "ingression" and "egression" amplitudes with a focus below the surface (Figure 2).

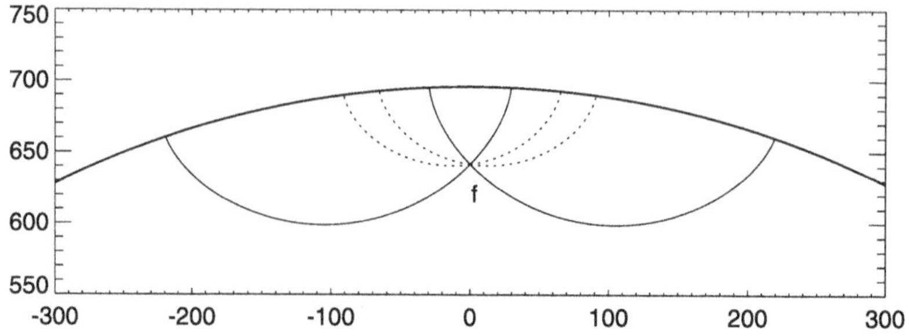

Figure 2 Ray paths for p-modes converging on a focal point $[f]$ that is 54.4 Mm below the surface of the spherical domain of the simulation (denoted by the thick solid line). The thinner solid lines denote rays spanning $\pm 45°$ from the horizontal direction. This is the nominal range of impact angles for lateral-vantage HH. In Section 4.4 we perform HH with smaller ranges of impact angles. The dashed lines indicate the ray paths for impact angles spanning $\pm 7.5°$ from the horizontal direction. The scale is in megameters.

To establish some common notation, we define the three-dimensional (3D) Fourier transform in time $[t]$ and two spatial dimensions $[x, y]$ of a function $A(x, y, t)$ as $\hat{A}(k_x, k_y, \omega)$, where k_x and k_y are the horizontal wavenumber components and ω is the temporal frequency. We define the Fourier transform in only the temporal dimension of A as $\tilde{A}(x, y, \omega)$. Thus, the steps involved in the data analysis are as follows:

i) The simulated surface velocity data, provided in heliographic coordinates, are remapped onto a Postel projection $\Psi(x, y, t)$. The nominal spacing of the Postel grid is $\delta x = \delta y = 8.54$ Mm (0.7°), which is the original spacing of the velocity data in heliographic coordinates. The central tangent point $(x, y) = (0, 0)$ is defined as 34.2 Mm (2.8°) south of the location of the perturbation.
ii) The 3D Fourier transform $[\hat{\Psi}(k_x, k_y, \omega)]$ of the Postel-projected data is computed in both spatial dimensions and in time. In the temporal-frequency domain, the data within a bandpass of 2.5 and 5.5 mHz are extracted for further analysis. The simulation contains very little p-mode power above 3.5 mHz.
iii) A phase-speed filter is applied to $\hat{\Psi}$. The nominal method employs filters that are Gaussian in the phase speed $[w \equiv \omega/k$ (where $k^2 = k_x^2 + k_y^2$)$]$ for each depth with peak phase speeds $[w_0]$ and widths $[\delta w]$ specified in Table 1. In Section 4.3 we examine the sensitivity of the results to variations in the form of the filter.
iv) A set of depths is chosen (Table 1) and Green's functions for both diverging $[G_+^P]$ and converging $[G_-^P]$ waves are computed in the same Postel-projected grid as the data (Lindsey and Braun, 2000). The Green's functions are multiplied by spatial masks defining a given pupil $[P]$. The nominal set of pupils represent quadrants (or "arcs") of annuli extending outward in four directions and are denoted E, W, N, and S. The annulus widths are determined by ray theory from the paths of acoustic modes diverging from the subsurface focus point and spanning a range of "impact angles" $\pm 45°$ from the horizontal direction (see Figure 2). In Section 4.4 we explore the sensitivity of the results to narrower ranges of impact angles, and in Section 4.5 we employ different azimuthal extents of the pupil arcs.
v) For each pupil quadrant $[P]$, the ingression $[H_-^P]$ and egression $[H_+^P]$ amplitudes are estimated by convolutions of the data cube $[\Psi]$ with G_-^P and G_+^P, respectively, in both time and the two spatial coordinates. This is performed using a plane-parallel approximation by the simple product of $\hat{G}_\pm^P$ and $\hat{\Psi}$ (Lindsey and Braun, 2000) in the 3D

Table 1 Pupil sizes, modes, and filter parameters.

z_f [Mm]	Pupil Radii [Mm]	ℓ at 3 mHz	w_0 [$\mathrm{km\,s^{-1}}$]	δw [$\mathrm{km\,s^{-1}}$]
29.9	16.0 – 128	124 – 175	74	37
37.0	19.5 – 159	108 – 153	87	43
45.4	24.4 – 190	95 – 134	96	49
47.6	25.8 – 195	92 – 130	101	50
49.9	27.1 – 209	89 – 126	105	52
52.4	28.5 – 216	86 – 122	108	54
54.4	29.2 – 224	84 – 119	111	55
57.1	31.3 – 230	82 – 115	114	57
59.2	32.0 – 237	79 – 113	117	58
62.1	34.1 – 251	77 – 109	119	60
64.5	36.2 – 254	75 – 106	122	62
76.1	41.8 – 292	67 – 95	140	69
87.9	48.0 – 327	60 – 85	153	77

Fourier domain. The validity and consequences of this approximation are explored in Section 4.1.

vi) The cross-covariance functions between ingression and egression amplitudes corresponding to opposite quadrants (*e.g.*, E and W, N and S) are computed. The four resulting cross-covariance functions are summed.

vii) Mean travel-time maps are determined from the sum of the four cross-covariance functions. The nominal method uses the "phase method" (Braun and Lindsey, 2000). In Section 4.2 we compare the phase method with results from fits of the cross-covariances to Gabor wavelets.

viii) Maps of the mean travel-time shifts are determined from the residual of the travel-time maps after subtracting a two-dimensional (2D) polynomial fit to a "quiet Sun" area excluding the perturbation. As shown in Section 4.1 this procedure helps to remove the effects of the plane-parallel approximation used in step v).

Table 1 shows the pupil ranges for each selected focus depth z_f, determined from ray theory. Also listed are the range of mode degrees [ℓ] at 3 mHz, sampled by the pupil, and the parameters for the Gaussian phase-speed filter (see Section 4.3) at each depth. The highest value of ℓ at each depth represents waves propagating horizontally through the focal point, while the lowest value indicates modes which propagate at impact angles of $\pm 45°$ from the horizontal direction (see Figure 2).

Figure 3 shows maps of the travel-time shifts for a sample of focus depths. The perturbation is clearly seen as an increase in travel-time shift with a maximum of between 15 and 20 seconds at the expected horizontal position. Figure 1c shows the azimuthal averages of the travel-time shifts for several focus depths while Figure 1d shows the variation of the travel-time shift at the center of the perturbation (hereafter "peak travel-time shift") with focus depth. It is clear that the horizontal and vertical dependences of the travel-time shifts reasonably characterize the shape of the perturbation.

We measure a background realization noise [σ] as the standard deviation of the mean travel-time shifts within an annulus spanning distances 111 – 195 Mm from the center of the Postel projection. For the "nominal" maps shown in Figure 3, σ is about 2.1 seconds and does not vary substantially with depth. We find that the background noise for maps made at

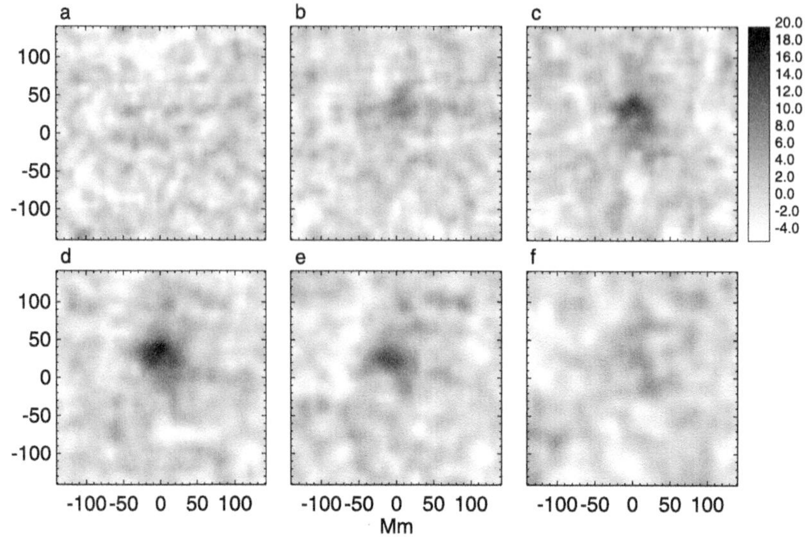

Figure 3 Maps of the mean travel-time shift using the nominal methodology of lateral-vantage HH as applied to the Hartlep *et al.* simulation and for focus depths of (a) 29.9 Mm, (b) 37.0 Mm, (c) 45.4 Mm, (d) 54.4 Mm, (e) 64.5 Mm, and (f) 76.1 Mm. The grayscale indicates the travel-time shift in seconds.

different focus depths is correlated. For example, there is a correlation coefficient (measured after excluding a region around the perturbation) of 0.96 between maps made at 54.4 and 52.4 Mm, and a correlation of 0.56 between maps at 54.4 and 47.6 Mm.

4. Tests of the Methodology

4.1. Tests of the Plane-Parallel Approximation

In Section 3, step v, a convolution in time and horizontal spatial coordinates between the Green's functions $G_{\pm}^{P}$ and the data Ψ is computed in the Fourier domain under the assumption that the functions $G_{\pm}^{P}$ are invariant with respect to translation in the Postel coordinate frame (this assumption has been termed the "plane-parallel" approximation: Lindsey and Braun, 2000; Braun and Birch, 2008b). The use of this approximation is highly desirable, since it decreases computing time and resources by several orders of magnitude. For example, without its use, separate Green's functions for each target pixel would have to be computed, stored, accessed, and operated on with a 3D multiplication by the datacube in the computations.

A major result of this approximation is the introduction of a systematic bias in the mean travel-time shift, which is a function of the horizontal distance between the focus and the central tangent point of the Postel projection. The reason for this bias is straightforward: In the Postel (also known as azimuthal-equidistant) projection, distances measured along any line intersecting the central tangent point (hereafter simply called the "center") are accurate, but distances between all other points differ from their true great-circle values. Thus, a locus of constant phase of waves propagating either away from or towards the center is warped in the projected plane into an ellipse with the semi-minor axis aligned towards the center (Figure 4). These wavefronts do not match the assumed circular wavefronts (and pupils)

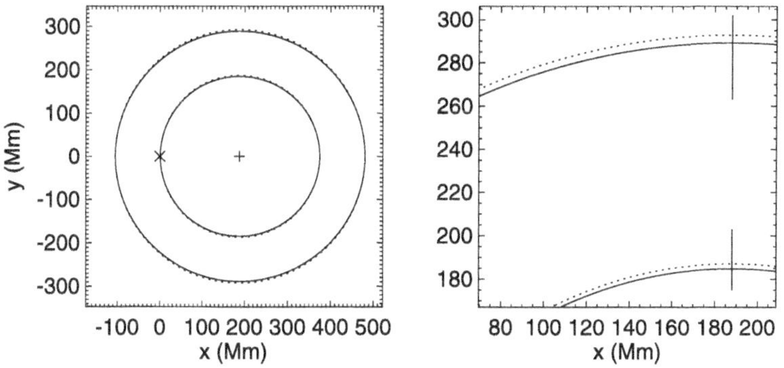

Figure 4 Examples of the distortion in wavefronts at the outer pupil boundaries centered on a target focus point (denoted by the plus sign) placed 188 Mm to the right of the center of the Postel projection, which is denoted by the × symbol. The larger and smaller dotted circles show wavefronts at the outer pupil boundaries, as projected onto the Postel frame, for focus depths of 76.1 and 45.4 Mm, respectively. The solid circles show circular wavefronts as assumed in the plane-parallel approximation. The dotted and solid wavefronts coincide along the x-axis but deviate at other places, with the maximum deviation occurring at the top and bottom. The deviations are difficult to discern by eye in the left panel. The right panel shows a magnified version of the upper part of the left panel. The vertical line segments in the right panel indicate the length of the horizontal wavelengths of modes that propagate from the focus depth to the outer pupil boundaries.

of the computed Green's functions. For the depths and pupil parameters listed in Table 1 the distortion in distance is small compared to the horizontal wavelengths of the modes. For example, in Figure 4 are drawn wavefronts at the outer pupil boundaries (where the distortion is greatest) corresponding to focus positions placed 188 Mm (15.5°) to the right of the center and at depths of 45.4 and 76.1 Mm. The maximum distortion of the wavefronts for these depths is 2.3 and 3.6 Mm, respectively, and these values are small compared to the horizontal wavelengths (30 and 40 Mm) of the modes considered. However, the distortion in projected distances results in observable spurious mean travel-time variations that vary with the azimuthal angle of propagation from the focus as well as the distance between the focus and the wavefront. At the outer edge of the pupils these spurious shifts can be as large as 20 – 30 seconds. However, the net travel-time shift as assessed over the entire pupil is typically less than ten seconds over tangent-point distances below 200 Mm (*e.g.* see Figure 5).

To correct for this spatially varying bias, we fit and subtract a 2D polynomial to the raw mean travel-time maps (Section 3, step viii). A circular mask excluding the perturbation is applied before the polynomial fit. Figure 5 shows cuts through a mean travel-time map with and without this correction.

Since all of the distortions introduced by the plane-parallel approximation increase with tangent-point distance, it is worthwhile to test the approximation by computing travel-time shift maps with varying positions of the tangent point. A similar test was performed by Braun (2012) on solar observations, but comparing only measurements of realization noise. The simulation here provides a larger, isolated, signal that provides a complementary target for this type of test. Figure 6 shows that maps made using tangent points spaced 200 Mm apart, after correction for the bias discussed above, have residual differences on the order of a second. For smaller distances of approximately 20 – 30 Mm these residuals are well below a tenth of a second.

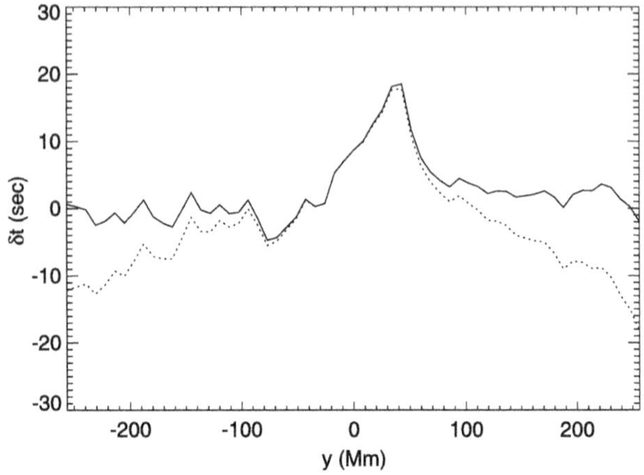

Figure 5 Comparison of mean travel-time shifts with and without a polynomial subtraction designed to remove the effects of the plane-parallel approximation. The dotted (solid) line is a vertical cut (at $x = 0$) of the uncorrected (corrected) mean travel-time map at a focus depth of 54.4 Mm.

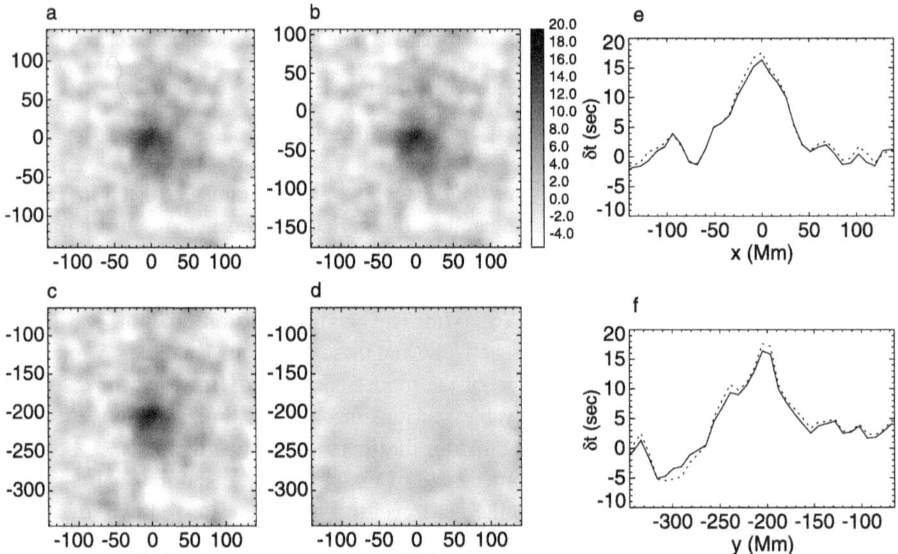

Figure 6 Maps of the mean travel-time shifts at a focus depth of 54.4 Mm and with the tangent point (center) of the Postel frame placed at the following locations: (a) centered on the perturbation, (b) 34 Mm to the north of the perturbation, and (c) 205 Mm to the north of the perturbation. (d) The difference between the maps shown in (c) and (a). The rightmost plots show horizontal (e) and vertical (f) cuts through the center of the perturbation of map (a) shown as solid lines and map (c) shown as dotted lines.

4.2. Comparisons of Travel-Time Measurements

As we note in Section 3 step vii, the mean travel times are determined from the sum of the cross-covariance functions. In the nominal procedure, there are four cross-covariances of the form

$$\tilde{C}^{\mathrm{EW}}(\mathbf{r}, z_{\mathrm{f}}, \omega) = \tilde{H}_{+}^{\mathrm{E}}(\mathbf{r}, z_{\mathrm{f}}, \omega)\tilde{H}_{-}^{W*}(\mathbf{r}, z_{\mathrm{f}}, \omega), \quad (1)$$

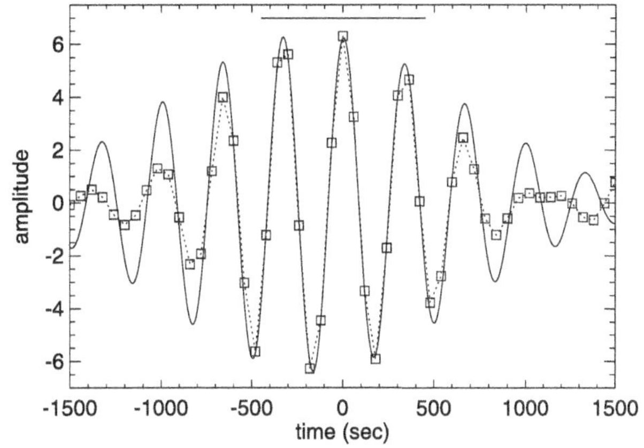

Figure 7 The squares connected by a dotted line show a cross-covariance function between the ingression and egression amplitudes, summed over the four opposite-quadrant pairs, for a single spatial location and a focus depth of 54.4 Mm. The solid curve represents a fit to the cross-covariance function, sampled over a 14-minute window denoted by the horizontal line at the top, to a Gabor wavelet (Equation (4)). The amplitude is in arbitrary units.

where the asterisk denotes the complex conjugate, $\mathbf{r} = (x, y)$, and we have also included the dependence on focus depth z_{f}. The temporal Fourier transform of the sum

$$\tilde{C} = \tilde{C}^{\mathrm{EW}} + \tilde{C}^{\mathrm{NS}} + \tilde{C}^{\mathrm{WE}} + \tilde{C}^{\mathrm{SN}}, \tag{2}$$

is used in the "phase method" (Braun and Lindsey, 2000; Braun and Birch, 2008b) to compute the mean travel time through

$$\tau_{\mathrm{pm}}(\mathbf{r}, z_{\mathrm{f}}) = \arg\bigl(\bigl\langle\tilde{C}(\mathbf{r}, z_{\mathrm{f}}, \omega)\bigr\rangle_{\Delta\omega}\bigr)/\omega_0, \tag{3}$$

where the brackets indicate the average over the bandwidth $\Delta\omega$, and ω_0 is the mean frequency. The desired travel-time shift $[\delta t]$ is obtained from τ_{pm} by subtracting a 2D polynomial fit to a quiet Sun region (step viii). A typical summed cross-covariance function, transformed back to the temporal domain, is shown in Figure 7.

An alternative method for extracting travel-time shifts is to fit the cross-covariance function to a Gabor wavelet:

$$g = A\cos\bigl(\omega_0[t - \tau_{\mathrm{gf}}]\bigr)\exp\biggl(-\frac{1}{2}\biggl[\frac{t - \tau_{\mathrm{en}}}{\sigma}\biggr]^2\biggr) \tag{4}$$

where A, σ, and τ_{en} are the amplitude, width, and position of a Gaussian envelope, ω_0 is the mean frequency, and τ_{gf} is the (phase) travel time, which is used instead of τ_{pm} to determine the travel-time shift δt. We have applied MPFIT routines (Markwardt, 2009) to perform a nonlinear least-squares fitting of the summed cross-covariance functions to Gabor wavelets for a focus depth of 54.4 Mm. The initial guesses of τ_{gf} in the fits were based on the peak closest to $t = 0$ of the cross-covariance function. Figure 7 shows an example of the fit of a single cross-covariance. Figure 8 shows that there is remarkable agreement between the mean travel-time shifts as determined from the phase method and the Gabor fits. We note that fine tuning the initial guesses based on the peaks of the cross-covariance functions to the left (or right) of the central peak yields phase times $[\tau_{\mathrm{gf}}]$ that agree to within a fraction of a second of the times obtained using the central peak minus (or plus) the period $[2\pi/\omega_0]$. Thus, due to statistical fluctuations in the mean frequency, maps made using fits to these peaks are noisier than maps made using the central peak.

4.3. Sensitivity to Different Phase-Speed Filters

In Section 3 step iii, a phase-speed filter is applied to the Fourier transform $\hat{\Psi}$ of the datacube. Phase-speed filters are widely used in both time–distance helioseismology (Duvall

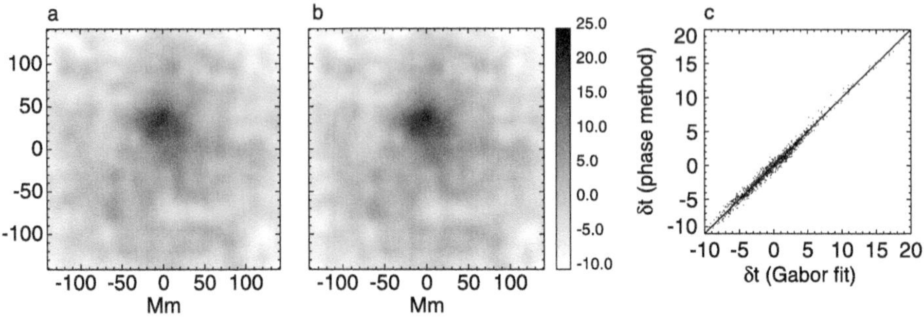

Figure 8 Comparisons of maps of mean travel-time shifts for a focus depth of 54.4 Mm obtained (a) using the nominal method including the phase method for extracting travel times from the cross-covariance functions and (b) using fits to Gabor wavelets to the same cross-covariance functions in the temporal domain. No corrections for the bias introduced by the plane-parallel approximation have been performed here; rather, a simple mean has been subtracted from each map. (c) A scatter plot of the two maps, compared to a line with unit slope.

et al., 1997; Couvidat and Birch, 2006) and helioseismic holography (Braun and Birch, 2006). The nominal procedure for lateral-vantage HH uses Gaussian filters

$$\phi = \exp\left(-\frac{1}{2}\left[\frac{w - w_0}{\delta w}\right]^2\right) \tag{5}$$

with a peak phase speed $[w_0]$ and width $[\delta w]$ such that the square of the filter has values of one and one-half at the highest and lowest wavenumbers, respectively, at 3 mHz as listed in Table 1. The use of phase-speed filters reduces the noise contributed by convective (non-wave) motions as well as from p-modes outside the range of desired phase speeds. Recently, Ilonidis, Zhao, and Kosovichev (2012a, 2012b) claim that different types of filters can affect the measured strength of subsurface signatures of emerging active regions. We have compared results using the nominal Gaussian filter, results using a "flat-top" filter similar to that employed by Ilonidis, Zhao, and Kosovichev (2012a), and results using no phase-speed filter. Figures 9 and 10 show comparable peak travel-time shifts in the simulated perturbation between the three cases, although the flat-top filter may be somewhat less sensitive to the variation with depth of the perturbation. This is also confirmed by computing the correlation coefficient between maps for different depths. For example, travel-time shifts at focus depths of 54.4 and 64.5 have a correlation of 0.45 using the Gaussian filter, but 0.64 using the flat-top filter. These correlation coefficients were computed with the perturbation masked out, so they measure correlations in the background realization noise. Consistently higher correlation coefficients for all of the flat-top filtered results over this depth range (45 – 65 Mm) are observed. In general, the use of either filter produces somewhat less noise (as determined from the standard deviation of the realization noise outside of the perturbation) than using no filter, as expected (see Figure 10d).

A restriction in the simulation to mode power below $\ell = 170$ means that the tests performed here are not sensitive to variations in the filter properties below $w = 70\ \mathrm{km\,s^{-1}}$. Nonetheless, our general findings are consistent with expectations based on experience analyzing solar data for lateral-vantage holography performed for similar focus depths.

4.4. Sensitivity to Different Quadrant Widths

We explored the effect of changes to the range of impact angles of p-modes interacting with the perturbation, by decreasing the pupil width from the nominal values in Table 1. New

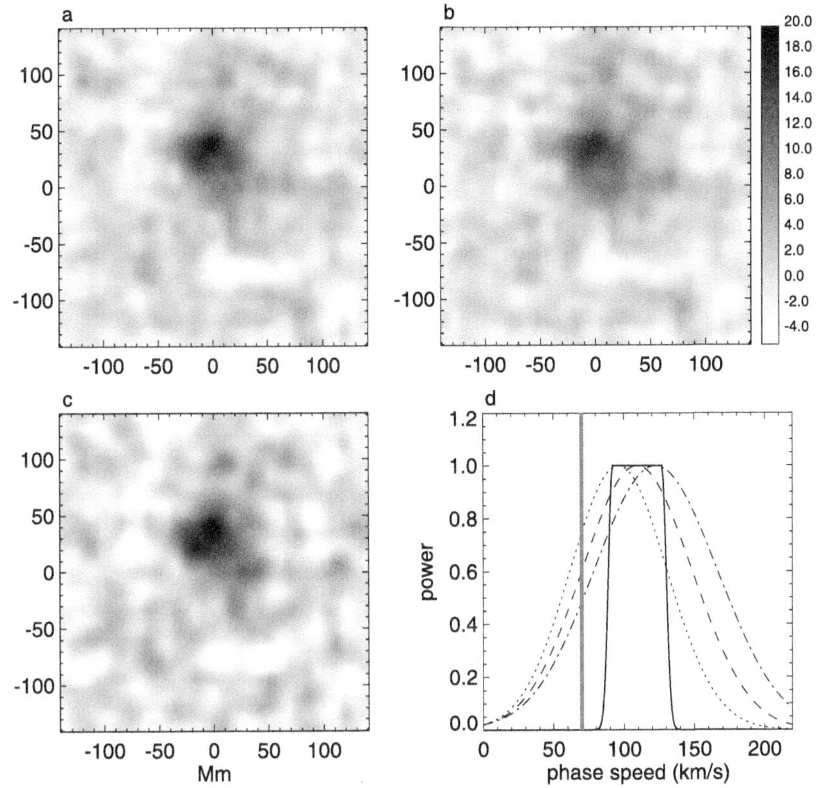

Figure 9 Mean travel-time shift maps made for a focus depth of 54.4 Mm and different phase-speed filtering: (a) a Gaussian phase-speed filter, (b) a flat-top filter, and (c) no phase-speed filtering. (d) The square of the filter function for several filters used: dotted, dashed, and dot–dashed lines indicate the nominal Gaussian filters corresponding to focus depths of 45, 54.4, and 64.5 Mm, respectively. The solid line shows the flat-top filter used in this study. There is no p-mode power in the simulation to the left of the vertical gray line. Thus, the tests here are not sensitive to differences between the filters at these low phase speeds.

pupil widths were computed using ray theory for impact angle extrema of $\pm 25°$, $\pm 15°$, and $\pm 7.5°$ at the focus depth of 54.4 Mm. Figure 2 shows rays corresponding to impact angles of the nominal $\pm 45°$ and the smallest range, $\pm 7.5°$, considered.

Figure 11 shows that there is no substantial change in the strength of the perturbation as the impact angle is changed, within the uncertainty specified by the background realization noise. This result is expected, since the travel-time shifts due to a simple sound-speed perturbation should not depend on impact angle. There is a slight increase of realization noise, which also appears to take on a more fine-scale oscillatory pattern, as the pupil quadrant widths are decreased (Figure 11b). This is likely a diffraction (side-lobe) artifact due to the narrow pupil. The widths of the pupils for these angles ($\pm 7.5°$) are smaller than the horizontal wavelength of the modes. To resolve this fine structure, the travel-time shift maps shown in Figure 11 were made with a grid spacing of half of the nominal value, by applying a Fourier interpolation of the original data.

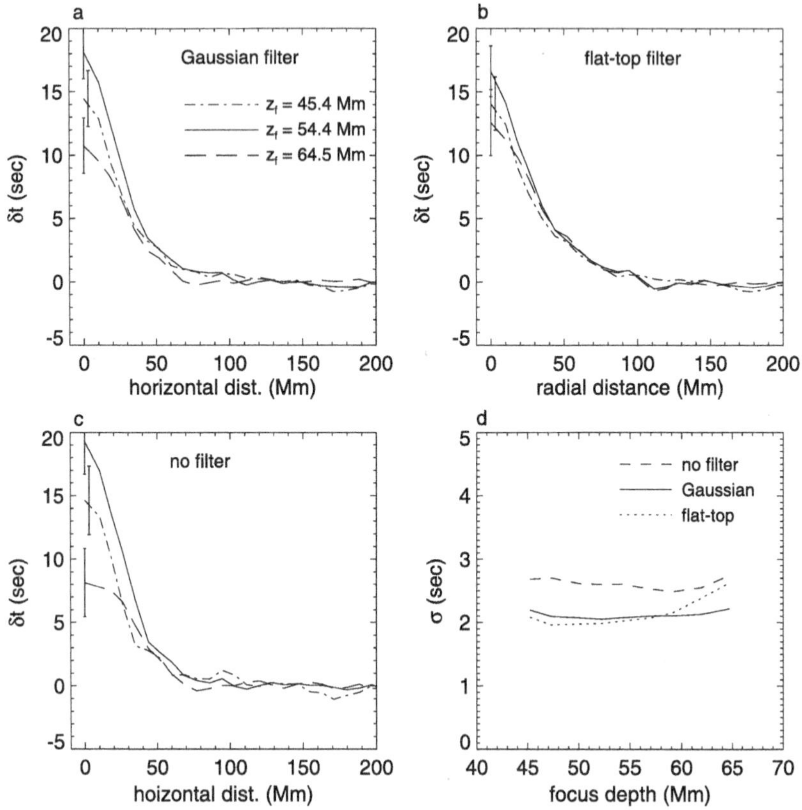

Figure 10 Mean travel-time shifts for different focus depths, averaged in annuli centered on the perturbation (as in Figure 1c), using three different types of phase-speed filtering: (a) the nominal Gaussian filters, (b) the flat-top filter, and (c) no filtering. The depths are 45.3 (dash–dotted line), 54.4 (solid line), and 64.5 Mm (long dashed line). Error bars represent the standard deviation [σ] of the background realization noise in a region surrounding the perturbation (see text). The variation of σ with focus depth is shown in panel (d) for the three cases: Gaussian (solid line), flat-top (dotted line), and no filter (dashed line).

4.5. Sensitivity to Pupil Arc Size

The advantage of using four quadrants to compute the ingression/egression cross-covariances derives primarily from the utility in making measurements sensitive to flows as well as perturbations producing mean (horizontal-direction-averaged) travel-time shifts (Gizon and Birch, 2005). Ilonidis, Zhao, and Kosovichev (2012a, 2012b) have proposed several refinements, for application to time–distance (hereafter TD) methods, for the detection of subsurface signatures of emerging active regions. These include: i) dividing the annulus into a greater number of opposing arc pairs (*i.e.*, 6, 8, 10, 12, and 14 arcs), ii) making multiple measurements with different angular orientations of each set of arcs, and iii) combining all of the TD cross-covariances made with the different arcs and their orientations before the determination of the travel times. There are four different orientations used for each arc configuration in this scheme, as each set of arcs is rotated one-quarter of the angular extent of an arc.

We explore similar procedures for HH using the simulation of Hartlep *et al.* (2011). The results here complement the tests made for HH on Doppler observations obtained with the

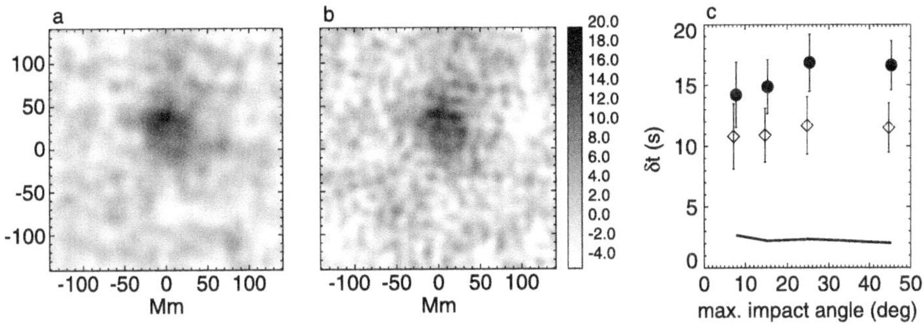

Figure 11 Mean travel-time shift maps made for a focus depth of 54.4 Mm and with different ranges of impact angle: (a) $\pm 45^\circ$ and (b) $\pm 7.5^\circ$. Both maps were made using the same flat-top filter shown in Figure 9. (c) Measurements of the mean travel-time shift in the perturbation against the maximum (absolute) impact angle. The filled circles show the peak shift; the diamonds show the average shift within a 25 Mm radius. Error bars denote the standard deviation of the realization noise, which is also plotted as a solid line.

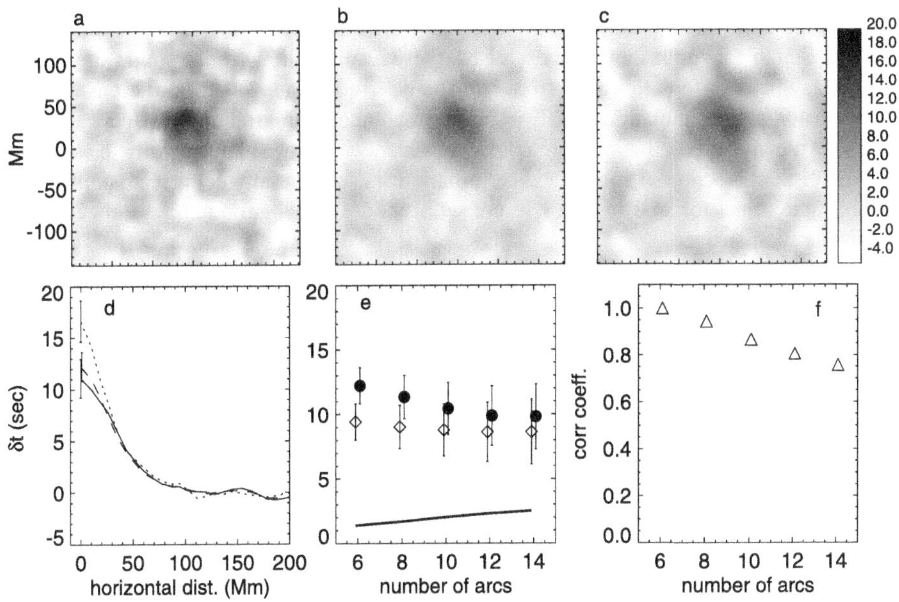

Figure 12 Mean travel-time shifts at a focus depth of 54.4 Mm from HH using different pupil geometries: (a) the nominal method using a fixed set of four quadrant pupils, (b) using six arcs and four orientations, and (c) the combination of 6, 8, 10, 12, and 14 arc configurations with four orientations of each configuration. The flat-top filter shown in Figure 9 is used. (d) Azimuthal averages of the travel-time shift over annuli centered on the perturbation for the three maps shown in the three top panels: quadrants (dotted line), six arcs (dashed line), and combined 6 – 14 arcs (solid line). (e) Measurements of the mean travel-time shift in the perturbation and the background realization noise σ against the number of pupil arcs used. The filled circles show the peak shift, and the diamonds show the average shift within a 25 Mm radius. Error bars denote the standard deviation of the realization noise, which is also plotted as a solid line. (f) The correlation coefficient between travel-time shift maps made using six arcs and the other arc configurations.

MDI instrument by Braun (2012). Figure 12 shows some of our results for the measurements on the simulated perturbation. In general, the use of six arcs produces a weaker (by about 25 %) travel-time signature in the perturbation than using quadrants. A slight trend of a decreasing signal strength with the number of arcs from 6 to 14 is also observed (Figure 12e), although the net decrease is within the background noise. The realization noise increases with the number of arcs used from about 1.4 seconds for the four orientations of the six-arc set to 2.5 seconds for the four orientations of the 14-arc set. These can be compared with the 2-second noise measured using the nominal quadrant method. These results are consistent with the increase in noise using smaller arcs observed by Braun (2012) using MDI data. The map made from combining all cross-covariances from all pupil-arc configurations and orientations has a realization noise of 1.9 seconds, which is essentially identical to the nominal (quadrant) map. It is significant that maps made with different arc lengths are highly correlated with each other (Figure 12f).

5. Conclusions

In summary, we find that helioseismic holography, as performed in the nominal lateral-vantage configuration and using the plane-parallel approximation, is in conformance with expectations well suited for detecting and characterizing subsurface sound-speed perturbations of the kind included in the simulation of Hartlep *et al.* (2011) at depths of at least 50 Mm. Suitable caution should be exercised: these results follow from a single simulation, which may have different physics from real solar perturbations. Other limitations, such as the inclusion in the simulation of only a subset (in both temporal frequencies and wavenumbers) of known solar oscillations, are noted. However, we believe that generally the viability of HH for the detection of subsurface perturbations is substantially confirmed, particularly its ability to select for analysis the relevant set of modes passing through a localized target below the solar surface.

Furthermore, mindful of the caveats mentioned above, we find no evidence that the sensitivity of the procedure, as assessed by the mean travel-time shift at the expected position of the perturbation, is enhanced by the use of the flat-top filter or different pupils, as suggested in the critique by Ilonidis, Zhao, and Kosovichev (2012a) of the results of Braun (2012). Specifically, the holographic signatures are influenced little by the detailed profile of the phase-speed filter, and very little more by the lack of any such filter. We also find that holography remains maximally sensitive when applied with spatially extended pupils, as opposed to restricting or partitioning them. The main effect of partitioning the pupil to smaller arcs is, if anything, a reduction of the signature and the appearance of diffraction effects.

Gabor-wavelet fitting can be applied to helioseismic holography as it is with other time–distance techniques, and so this should not be regarded as a discriminating qualification against it. In the case of the simulation, the results are essentially identical to those of the phase-method and in conformance with expected travel-time shifts given the size and amplitude of the perturbation.

Our tests do not attempt to replicate the time–distance procedures applied by Ilonidis, Zhao, and Kosovichev (2011); thus, we draw no conclusion about the sensitivity of their own measurements to the changes in methodology that they advocate. In attempting to understand the discrepancies of the results between Ilonidis, Zhao, and Kosovichev (2011) and Braun (2012), we can reasonably infer that negative holography results suggest that the suspected perturbation is different than the simple sound-speed perturbation simulated by Hartlep *et al.* (2011). Furthermore, it seems that the use of the plane-parallel approximation can be ruled out as a contributing factor to the negative results of Braun (2012).

It is possible that the physics of the suspected signatures are such that, unlike a simple sound-speed perturbation, the use of narrow pupils or different filters may be critical. The signatures may also produce complicated changes to the cross-covariance functions, perhaps due to unknown effects of magnetic fields (Ilonidis, Zhao, and Kosovichev, 2012a). Further tests need to be performed on the relevant data. In our opinion, it is possible that the signatures of Ilonidis, Zhao, and Kosovichev (2011) may represent noise. We return to the point made by Braun (2012) suggesting the need for blind "hare-and-hound" tests as a minimal condition for the signatures of the signals found by Ilonidis, Zhao, and Kosovichev (2011) to be established as pre-emergence signatures of deeply submerged magnetic fields. Tests with simulated data on artificial perturbations such as those reported here provide the critical context under which similar analyses of solar observations may be understood. In general, the results presented here provide confidence in helioseismic holography as a useful method for probing submerged perturbations (Leka *et al.*, 2012; Birch *et al.*, 2012).

Acknowledgements This work is supported by the NASA Heliophysics program through contract NNH12CF23C, by the NASA SDO Science Center project through contract NNH09CE41C, and by the Solar Terrestrial program of the National Science Foundation through grant AGS-1127327. We thank Charlie Lindsey and an anonymous referee for their useful comments. We are grateful to Thomas Hartlep for providing the data used in this study.

References

Benson, D., Stein, R., Nordlund, Å.: 2006, Supergranulation scale convection simulations. In: Leibacher, J., Stein, R.F., Uitenbroek, H. (eds.) *Solar MHD Theory and Observations: A High Spatial Resolution Perspective* **354**, Astron. Soc. Pac., San Francisco, 92 – 96.

Birch, A.C., Braun, D.C., Fan, Y.: 2010, An estimate of the detectability of rising flux tubes. *Astrophys. J.* **723**, L190 – L194.

Birch, A.C., Parchevsky, K., Braun, D.C., Kosovichev, A.: 2011, Hare and hounds tests of helioseismic holography. *Solar Phys.* **272**, 11 – 28.

Birch, A.C., Braun, D.C., Leka, K.D., Barnes, G., Javornik, B.: 2012, Helioseismology of pre-emerging active regions II: average emergence properties. *Astrophys. J.* accepted.

Braun, D.C.: 1995, Sunspot seismology: new observations and prospects. In: Ulrich, R.K., Rhodes, E.J. Jr., Dappen, W. (eds.) *GONG 1994. Helio- and Astro-Seismology from the Earth and Space* **76**, Astron. Soc. Pac., San Francisco, 250.

Braun, D.C.: 2012, Comment on "Detection of emerging sunspot regions in the solar interior". *Science* **336**, 296. doi:10.1126/science.1215425.

Braun, D.C., Birch, A.C.: 2006, Observed frequency variations of solar p-mode travel times as evidence for surface effects in sunspot seismology. *Astrophys. J. Lett.* **647**, L187 – L190.

Braun, D.C., Birch, A.C.: 2008a, Prospects for the detection of the deep solar meridional circulation. *Astrophys. J. Lett.* **689**, L161 – L165.

Braun, D.C., Birch, A.C.: 2008b, Surface-focused seismic holography of sunspots: I. Observations. *Solar Phys.* **251**, 267 – 289. ADS:2008SoPh..251..267B, doi:10.1007/s11207-008-9152-5.

Braun, D.C., Birch, A.C., Lindsey, C.: 2004, Local helioseismology of near-surface flows. In: Danesy, D. (ed.) *SOHO 14 Helio- and Asteroseismology: Towards a Golden Future* **SP-559**, ESA, Noordwijk, 337 – 340.

Braun, D.C., Lindsey, C.: 2000, Phase-sensitive holography of solar activity. *Solar Phys.* **192**, 307 – 319. ADS:2000SoPh..192..307B, doi:10.1023/A:1005239216840.

Braun, D.C., Birch, A.C., Benson, D., Stein, R.F., Nordlund, A.: 2007, Helioseismic holography of simulated solar convection and prospects for the detection of small-scale subsurface flows. *Astrophys. J.* **669**, 1395 – 1405.

Braun, D.C., Birch, A.C., Rempel, M., Duvall, T.L. Jr.: 2012, Helioseismology of a realistic magnetoconvective sunspot simulation. *Astrophys. J.* **744**, 77 – 86.

Cameron, R., Gizon, L., Duvall, T.L. Jr.: 2008, Helioseismology of sunspots: confronting observations with three-dimensional MHD simulations of wave propagation. *Solar Phys.* **251**, 291 – 308. ADS:2008SoPh..251..291C, doi:10.1007/s11207-008-9148-1.

Cameron, R.H., Gizon, L., Schunker, H., Pietarila, A.: 2011, Constructing semi-empirical sunspot models for helioseismology. *Solar Phys.* **268**, 293 – 308. ADS:2011SoPh..268..293C, doi:10.1007/s11207-010-9631-3.

Chang, H.-K., Chou, D.-Y., Sun, M.-T.: 1999, In search of emerging magnetic flux underneath the solar surface with acoustic imaging. *Astrophys. J. Lett.* **526**, L53 – L56. doi:10.1086/312366.
Chang, H.-K., Chou, D.-Y., Labonte, B., TON Team: 1997, Ambient acoustic imaging in helioseismology. *Nature* **389**, 825 – 827. doi:10.1038/39822.
Christensen-Dalsgaard, J., Dappen, W., Ajukov, S.V., Anderson, E.R., Antia, H.M., Basu, S., Baturin, V.A., Berthomieu, G., Chaboyer, B., Chitre, S.M., Cox, A.N., Demarque, P., Donatowicz, J., Dziembowski, W.A., Gabriel, M., Gough, D.O., Guenther, D.B., Guzik, J.A., Harvey, J.W., Hill, F., Houdek, G., Iglesias, C.A., Kosovichev, A.G., Leibacher, J.W., Morel, P., Proffitt, C.R., Provost, J., Reiter, J., Rhodes, E.J. Jr., Rogers, F.J., Roxburgh, I.W., Thompson, M.J., Ulrich, R.K.: 1996, The current state of solar modeling. *Science* **272**, 1286 – 1292.
Couvidat, S., Birch, A.C.: 2006, Optimal Gaussian phase-speed filters in time-distance helioseismology. *Solar Phys.* **237**, 229 – 243. ADS:2006SoPh..237..229C, doi:10.1007/s11207-006-0209-z.
Crouch, A.D., Birch, A.C., Braun, D.C., Clack, C.T.M.: 2010, Helioseismic probing of the subsurface structure of sunspots. In: Choudhary, D.P., Strassmeier, K.G. (eds.) *The Physics of Sun and Star Spots, IAU Symp.* **273**, Cambridge Univ. Press, Cambridge, 384 – 388. doi:10.1017/S1743921311015602.
Duvall, T.L. Jr., Kosovichev, A.G., Scherrer, P.H., Bogart, R.S., Bush, R.I., de Forest, C., Hoeksema, J.T., Schou, J., Saba, J.L.R., Tarbell, T.D., Title, A.M., Wolfson, C.J., Milford, P.N.: 1997, Time-distance helioseismology with the MDI instrument: initial results. *Solar Phys.* **170**, 63 – 73. ADS:1997SoPh..170...63D, doi:10.1023/A:1004907220393.
Gizon, L., Birch, A.C.: 2005, Local helioseismology. *Living Rev. Solar Phys.* **2**. http://www.livingreviews.org/lrsp-2005-6.
Hanasoge, S.M., Larsen, R.M., Duvall, T.L. Jr., De Rosa, M.L., Hurlburt, N.E., Schou, J., Roth, M., Christensen-Dalsgaard, J., Lele, S.K.: 2006, Computational acoustics in spherical geometry: steps toward validating helioseismology. *Astrophys. J.* **648**, 1268 – 1275.
Hartlep, T., Kosovichev, A.G., Zhao, J., Mansour, N.N.: 2011, Signatures of emerging subsurface structures in acoustic power maps of the Sun. *Solar Phys.* **268**, 321 – 327.
Ilonidis, S., Zhao, J., Kosovichev, A.: 2011, Detection of emerging sunspot regions in the solar interior. *Science* **333**, 993 – 996. doi:10.1126/science.1206253.
Ilonidis, S., Zhao, J., Kosovichev, A.: 2012a, Response to "Comment on detection of emerging sunspot regions in the solar interior". *Science* **336**, 296. doi:10.1126/science.1215539.
Ilonidis, S., Zhao, J., Kosovichev, A.G.: 2012b, Helioseismic detection of emerging magnetic flux. In: Shibahashi, H., Takata, M., Lynas-Gray, A.E. (eds.) *Progress in Solar/Stellar Physics with Helio- and Asteroseismology* **462**, Astron. Soc. Pac., San Francisco, 283.
Jensen, J.M., Duvall, T.L. Jr., Jacobsen, B.H., Christensen-Dalsgaard, J.: 2001, Imaging an emerging active region with helioseismic tomography. *Astrophys. J. Lett.* **553**, L193 – L196.
Jensen, J.M., Olsen, K.B., Duvall, T.L. Jr., Jacobsen, B.H.: 2003, Test of helioseismic time-distance inversion using 3-D finite-difference wavefield modeling. In: Sawaya-Lacoste, H. (ed.) *GONG+ 2002. Local and Global Helioseismology: The Present and Future* **SP-517**, ESA, Noordwijk, 319 – 320.
Kosovichev, A.G.: 2009, Photospheric and subphotospheric dynamics of emerging magnetic flux. *Space Sci. Rev.* **144**, 175 – 195. doi:10.1007/s11214-009-9487-8.
Leka, K.D., Barnes, G., Birch, A.C., Gonzalez-Hernandez, I., Dunn, T., Javornik, B., Braun, D.C.: 2012, Helioseismology of pre-emerging active regions I: overview, data, and target selection criteria. *Astrophys. J.* accepted.
Lindsey, C., Braun, D.C.: 1997, Helioseismic holography. *Astrophys. J.* **485**, 895 – 903.
Lindsey, C., Braun, D.C.: 2000, Basic principles of solar acoustic holography. *Solar Phys.* **192**, 261 – 284. ADS:2000SoPh..192..261L, doi:10.1023/A:1005227200911.
Lindsey, C., Braun, D.C.: 2004, Principles of seismic holography for diagnostics of the shallow subphotosphere. *Astrophys. J. Suppl.* **155**, 209 – 225.
Markwardt, C.B.: 2009, Non-linear least-squares fitting in IDL with MPFIT. In: Bohlender, D.A., Durand, D., Dowler, P. (eds.) *Astronomical Data Analysis Software and Systems XVIII* **411**, Astron. Soc. Pac., San Francisco, 251.
Parchevsky, K.V., Kosovichev, A.G.: 2007, Three-dimensional numerical simulations of the acoustic wave field in the upper convection zone of the Sun. *Astrophys. J.* **666**, 547 – 558. doi:10.1086/520108.
Parchevsky, K.V., Kosovichev, A.G.: 2009, Numerical simulation of excitation and propagation of helioseismic MHD waves: effects of inclined magnetic field. *Astrophys. J.* **694**, 573 – 581. doi:10.1088/0004-637X/694/1/573.
Zhao, J., Georgobiani, D., Kosovichev, A.G., Benson, D., Stein, R.F., Nordlund, Å.: 2007, Validation of time-distance helioseismology by use of realistic simulations of solar convection. *Astrophys. J.* **659**, 848 – 857.
Zharkov, S., Thompson, M.J.: 2008, Time distance analysis of the emerging active region NOAA 10790. *Solar Phys.* **251**, 369 – 380. ADS:2008SoPh..251..369Z, doi:10.1007/s11207-008-9239-z.

DOI 10.1007/978-1-4939-1182-0_4
Reprinted from *Solar Physics* Journal, DOI 10.1007/s11207-012-0218-z

Active Regions with Superpenumbral Whirls and Their Subsurface Kinetic Helicity

R. Komm · S. Gosain · A. Pevtsov

Received: 18 September 2012 / Accepted: 17 December 2012 / Published online: 10 January 2013

Abstract We search for a signature of helicity flow from the solar interior to the photosphere and chromosphere. For this purpose, we study two active regions, NOAA 11084 and 11092, that show a regular pattern of superpenumbral whirls in chromospheric and coronal images. These two regions are good candidates for comparing magnetic/current helicity with subsurface kinetic helicity because the patterns persist throughout the disk passage of both regions. We use photospheric vector magnetograms from SOLIS/VSM and SDO/HMI to determine a magnetic helicity proxy, the spatially averaged signed shear angle (SASSA). The SASSA parameter produces consistent results leading to positive values for NOAA 11084 and negative ones for NOAA 11092 consistent with the clockwise and counter-clockwise orientation of the whirls. We then derive the properties of the subsurface flows associated with these active regions. We measure subsurface flows using a ring-diagram analysis of GONG high-resolution Doppler data and derive their kinetic helicity, h_z. Since the patterns persist throughout the disk passage, we analyze synoptic maps of the subsurface kinetic helicity density. The sign of the subsurface kinetic helicity is negative for NOAA 11084 and positive for NOAA 11092; the sign of the kinetic helicity is thus anticorrelated with that of the SASSA parameter. As a control experiment, we study the subsurface flows of six active regions without a persistent whirl pattern. Four of the six regions show a mixture of positive and negative kinetic helicity resulting in small average values, while two regions are clearly dominated by kinetic helicity of one sign or the other, as in the case of regions with whirls. The regions without whirls follow overall the same hemispheric rule in their kinetic helicity as in their current helicity with positive values in the southern and negative values in the northern hemisphere.

Keywords Active regions, magnetic fields · Helicity, current · Helioseismology, observations · Velocity fields, interior

Solar Origins of Space Weather and Space Climate
Guest Editors: I. González Hernández, R. Komm, and A. Pevtsov

R. Komm (✉) · S. Gosain · A. Pevtsov
National Solar Observatory, Tucson, AZ 85719, USA
e-mail: komm@nso.edu

1. Introduction

We search for a signature of helicity flow from the solar interior to the photosphere and chromosphere. For this purpose, we study two active regions, NOAA 11084 and NOAA 11092, that show a regular pattern of superpenumbral whirls in chromospheric and coronal images. The two regions are good candidates for such a study not only because the sense of helicity is clearly visible but also because the patterns persist throughout the disk passage of both regions. A study of superpenumbral filaments showed that only about 27 % of sunspots have all their superpenumbral filaments twisted in the same direction (Pevtsov, Balasubramaniam, and Rogers, 2003). In a majority of sunspots, superpenumbral filaments of both clockwise (CW) and counter-clockwise (CCW) curvature were found. The latter suggests that superpenumbral whirls are not hydrodynamic features like hurricanes on Earth, and that their curvature may be associated with the patterns of electric currents flowing in sunspots. Such patterns of opposite currents are observed in many sunspots (*e.g.* Pevtsov, Canfield, and Metcalf, 1994). For active regions with whirls, the sign of helicity derived at photospheric heights has a one-to-one correspondence with the sense of chirality observed at chromospheric heights (Tiwari *et al.*, 2008).

As suggested by several recent studies, helicity may play an important role in a broad range of solar phenomena from the dynamo to flares and coronal mass ejections (Rüdiger, Pipin, and Belvedère, 2001; Pevtsov, 2008; Kazachenko *et al.*, 2012). Excess of helicity in coronal magnetic structures can lead to their instability and eruption (Low, 1996). The twist (kinetic helicity) of subsurface flows might serve as a proxy for the twist (magnetic helicity) of magnetic flux tubes below the solar surface because either the flows have to respond to the changes in helicity of magnetic fields or the fields are being pushed and twisted by subsurface turbulent flows. Active regions associated with subphotospheric patterns of strong kinetic helicity were found to be more flare productive (Mason *et al.*, 2006; Reinard *et al.*, 2010; Komm *et al.*, 2011).

The subsurface kinetic helicity and its relationship with magnetic helicity has been studied before using data from the *Global Oscillation Network Group* (GONG) and the *Michelson Doppler Imager* (MDI) onboard the *Solar and Heliospheric Observatory* (SOHO). A time–distance analysis of a rotating sunspot (Zhao and Kosovichev, 2003) shows that its subsurface kinetic helicity is comparable to current helicity estimates (Pevtsov, Canfield, and Metcalf, 1995). A study of the product of divergence and curl of horizontal flows, as a proxy of the vertical kinetic helicity density, has shown that its sign is mainly negative in the northern hemisphere and positive in the southern one (Komm *et al.*, 2007). The hemispheric rule of the kinetic helicity is most likely due to the Coriolis force (Brun, Miesch, and Toomre, 2004; Egorov, Rüdiger, and Ziegler, 2004). Since the current helicity shows on average the same hemispheric relation (Pevtsov, Canfield, and Metcalf, 1995; Pevtsov, Canfield, and Latushko, 2001; Hagino and Sakurai, 2005; Zhang, 2006), it appears that current and kinetic helicity might be closely related. However, a recent study of kinetic and magnetic helicity of active regions found that while these quantities follow the same hemispheric rule, there seems to be no significant correlation between kinetic and magnetic helicity (Maurya, Ambastha, and Reddy, 2011).

To further investigate this question, we study two sunspots with strong coronal whirls. The selected sunspots are simple round structures and do not exhibit major flaring activity. We first determine the magnetic helicity of these regions using photospheric vector magnetograms obtained from the *Vector Spectromagnetograph* (VSM) instrument of the *Synoptic Optical Long-term Investigations of the Sun* (SOLIS) synoptic facility (Keller, Harvey, and Giampapa, 2003) and the *Helioseismic and Magnetic Imager* (HMI) instrument

(Schou *et al.*, 2012) onboard the *Solar Dynamics Observatory* (SDO) spacecraft (Pesnell, Thompson, and Chamberlin, 2012). As a helicity proxy, we calculate the spatially averaged signed shear angle (SASSA), a simple measure of magnetic twist (Tiwari, Venkatakrishnan, and Sankarasubramanian, 2009), which is less sensitive to the presence of polarimetric noise than the mean twist parameter α or the current helicity density (Gosain, Tiwari, and Venkatakrishnan, 2010). We compare the sign of the photospheric twist, as inferred from the SASSA parameter with the chromospheric pattern of whirls. Further, we monitor the sign of SASSA for the two sunspots during their disk passage using continuous series of vector magnetograms from SOLIS and HMI to see if the sign of photospheric twist (sign of SASSA) persists in the same way as the whirls are maintained in the chromospheric and coronal images.

We then compare the magnetic helicity proxy with the properties of the subsurface flows associated with these active regions. Since the pattern of whirls persists throughout the disk passage and we are interested in the average behavior, it is sufficient to analyze synoptic maps of subsurface flows. We apply a ring-diagram analysis to GONG high-resolution Doppler data to determine daily subsurface flows and then calculate synoptic maps of subsurface kinetic helicity from the measured velocity maps (Komm *et al.*, 2004; Komm, 2007). We derive the kinetic helicity of each region and compare the subsurface values with the magnetic helicity proxies. As a control experiment, we study the subsurface flows of six active regions of similar size, strength, and flare activity that do not show a pattern of whirls in the chromosphere or corona. From this comparison, we try to find characteristics that distinguish between regions with and without whirls.

2. Data and Analysis

2.1. Active Regions and Magnetic Helicity

We study two active regions, NOAA 11084 and 11092, that show a regular pattern of superpenumbral whirls in chromospheric and coronal images (Table 1). Figure 1 shows images obtained by the *Atmospheric Imaging Assembly* (AIA: Lemen *et al.*, 2012) instrument onboard SDO. The He II 304 Å images show chromospheric and transition region structures, while the Fe IX 171 Å images show coronal loops. The "whirly" nature of the fibrils and loops of the two sunspots is clearly visible. From the twistedness of the coronal structures, one expects the chirality of NOAA 11084 to be positive and that of NOAA 11092 to be negative. With the first region located in the southern hemisphere and the second one in the northern hemisphere, the two regions follow the sign convention expected from the empirical hemispheric rule for helicity (Pevtsov, Canfield, and Metcalf, 1995). Judging from their location in latitude, both regions belong to the Solar Cycle 24. By their polarity orientation, NOAA 11084 was of non-Hale polarity, while NOAA 11092 was of Hale polarity.

We are using photospheric vector magnetograms obtained from SOLIS/VSM data (http://solis.nso.edu) to measure the helicity of these active regions (for details regarding SOLIS/VSM data see Balasubramaniam and Pevtsov, 2011; Pietarila *et al.*, 2013). The SOLIS/VSM instrument takes four full Stokes profiles of the Fe I 630.15 – 630.25 nm line pair for each pixel along a 2048 pixel-long spectrograph slit. The pixel size is about 1 arcsecond for data used in this study. To construct a vector magnetogram of the full solar disk, the image of the Sun is scanned in the declination with steps of 1 arcsecond. The total time for recording a single full-disk magnetogram is about 20 minutes. The Stokes profiles are inverted in the framework of the Milne–Eddington (ME) model of a stellar atmosphere

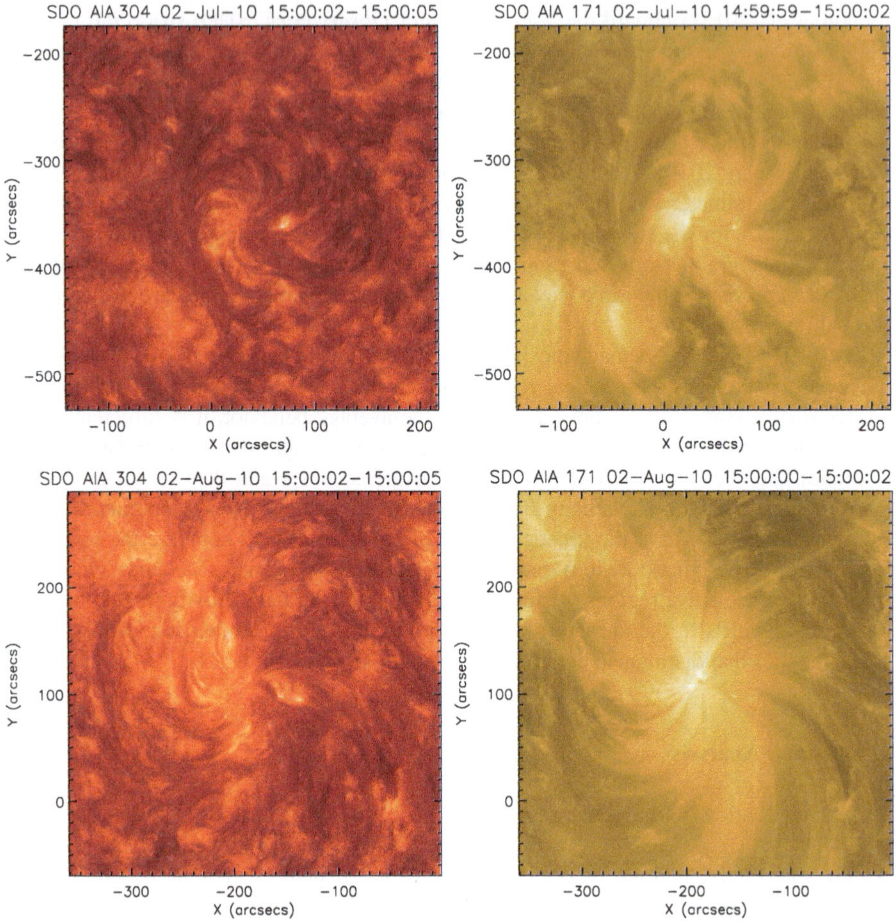

Figure 1 Chromospheric and transition region structures (left) and coronal loops (right) of two active regions obtained by the SDO/AIA instrument in He II 304 Å (left) and Fe IX 171 Å (right) (http://aia.lmsal.com/). Top: NOAA 11084 is a small region in the southern hemisphere with a clockwise (CW) whirl. Bottom: NOAA 11092 is a medium region in the northern hemisphere with a counter-clockwise (CCW) whirl. Both sunspots have negative polarity and the curvature of the whirls is defined by following the field lines.

(Skumanich and Lites, 1987). For illustration purposes, we use quick-look magnetograms derived by applying the integral method (Ronan, Mickey, and Orrall, 1987). After all pixels are inverted, the azimuth ambiguity in horizontal fields is resolved using the non-potential field computation (NPFC) method (Georgoulis, 2005).

We also use vector magnetograms obtained with the HMI instrument onboard SDO. The SDO/HMI instrument produces full-disk vector magnetograms with 0.5 arcsec pixels every 12 minutes using polarization measurement at six wavelengths along the Fe I 617.3 nm spectral line. These spectropolarimetric profiles are inverted by the Very Fast Inversion of the Stokes Algorithm (VFISV) code (Borrero *et al.*, 2011) under ME approximation. The resulting maps of magnetic and thermodynamic parameters are available on a daily basis from the Joint Science Operations Center or JSOC (http://jsoc.stanford.edu). The 180° azimuthal field ambiguity is resolved using the acute angle method (Metcalf *et al.*, 2006).

Table 1 The active regions with and without whirls characterized by their NOAA number, the date of central meridian crossing, the Carrington Rotation (CR), the latitude and longitude in the ring-diagram grid (in degree), the magnetic activity index (MAI in G) calculated over a ring-diagram patch, the total corrected area of the group in millionths of the solar hemisphere (www.swpc.noaa.gov), the number of flares (B class, C class), the flare index (FI in 10^{-3} erg s^{-1} cm^{-2}), and the total flare energy index (TFE in erg cm^{-2}).

NOAA #	Date	CR	Lat.	Long.	MAI	Area	Flares	FI	TFE
11084	02/07/10	2098	−22.5	142.5	31	101	1, 0	0.10	0.013
11092	03/08/10	2099	15.0	82.5	47	204	11, 1	9.13	10.451
11106	16/09/10	2101	−22.5	217.5	69	83	22, 0	7.73	3.305
11126	18/11/10	2103	−30.0	105.0	36	54	8, 0	3.59	0.795
11109	27/09/10	2101	22.5	75.0	57	282	38, 5	22.00	7.429
11124	13/11/10	2103	15.0	172.5	37	133	18, 1	9.65	4.480
11127	21/11/10	2103	22.5	60.0	70	64	4, 0	0.75	0.459
11131	07/12/10	2104	30.0	210.0	92	322	7, 0	1.88	0.266

Magnetic helicity arises from twisting and kinking of an individual flux tube and by linking and knotting of flux tubes (Berger and Field, 1984). Two commonly used magnetic helicity proxies are the vertical component of the current helicity density h_c:

$$\langle h_c \rangle = \langle J_z \cdot B_z \rangle = \left\langle \left(\frac{\partial B_y}{\partial x} - \frac{\partial B_x}{\partial y} \right) \cdot B_z \right\rangle, \tag{1}$$

where J_z is the vertical current density, B_z is the vertical magnetic field, and B_x and B_y are the horizontal magnetic field components (Bao and Zhang, 1998; Hagino and Sakurai, 2005) and the mean twist parameter α_z

$$\langle \alpha_z \rangle = \langle J_z / B_z \rangle, \tag{2}$$

where the angled braces indicate a spatial average over all pixels (Pevtsov, Canfield, and Metcalf, 1994, 1995). We employ instead the spatially averaged signed shear angle (SASSA):

$$\mathrm{SASSA} = \hat{\Psi} = \left\langle \operatorname{atan}\left(\frac{B_{y\mathrm{o}} B_{x\mathrm{p}} - B_{y\mathrm{p}} B_{x\mathrm{o}}}{B_{x\mathrm{o}} B_{x\mathrm{p}} + B_{y\mathrm{o}} B_{y\mathrm{p}}} \right) \right\rangle, \tag{3}$$

where $B_{x\mathrm{o}}$, $B_{y\mathrm{o}}$ and $B_{x\mathrm{p}}$, $B_{y\mathrm{p}}$ are the observed and potential transverse field components (Tiwari, Venkatakrishnan, and Sankarasubramanian, 2009). The twist of the sunspot magnetic field is well represented by the sign of SASSA. The SASSA parameter has been shown to be a robust quantity if the photospheric field is not force-free and to be less sensitive to polarimetric noise as compared to the mean twist parameter (Tiwari *et al.*, 2009; Gosain, Tiwari, and Venkatakrishnan, 2010). To compute the SASSA parameter, we transform the vector field in heliographic coordinates and deproject the magnetograms to solar disk center (Hagyard, 1987; Gary and Hagyard, 1990).

In addition, we analyze six regions without whirls that appear more potential in chromospheric images (Table 1). These six regions are reasonably similar to the ones with whirls in size, strength, and flare activity. A quick examination of SOLIS magnetograms shows that all active regions belong to Cycle 24 (judging from the polarity orientation and latitude). Two regions, NOAA 11084 and 11126, do not follow the Hale polarity law. One region, NOAA 11106, has a positive leading polarity, all other regions have a negative leading polarity. The two regions with whirls and most regions without whirls are stable throughout their disk passage, only NOAA 11124 is growing rapidly during its disk passage. The six regions without whirls have similar areas compared to the regions with whirls; the size of a region is not a distinguishing parameter. The eight regions show not much flare activity; the largest flares produced are C-class flares, as indicated in Table 1. For a more quantitative representation of the flare activity of active regions, we employ the flare index (FI) and the total flare energy (TFE) index. The flare index is computed as the sum of all X-ray flare peak fluxes (for example, a C-class flare corresponds to 10^{-3} erg s^{-1} cm^{-2}). The total flare energy index (TFE) is computed as the sum of all flare peak fluxes multiplied by the duration of the flare and scaled by 0.3 assuming that this is an adequate representation of the integral in time (per unit area).

2.2. Ring-Diagram Analysis and Kinetic Helicity

We analyze observations obtained during Carrington Rotations 2098 and 2099 (16 June 2010 – 7 August 2010) that include active regions NOAA 11084 and 11092. We use high-resolution full-disk Doppler data from the GONG network (http://gong.nso.edu/data) to derive the subsurface flows associated with these regions. In addition, we study the subsurface flows of six regions without whirls that are listed in Table 1 analyzing GONG Dopplergrams obtained during four Carrington Rotations CR 2101 – 2104 (5 September 2010 – 22 December 2010) that contain these regions. We determine the horizontal components of solar subsurface flows with a ring-diagram analysis using the dense-pack technique (Haber *et al.*, 2002) adapted to GONG data (Corbard *et al.*, 2003). The full-disk Doppler images are divided into 189 overlapping regions with centers spaced by 7.5° ranging over ±52.5° in latitude and central meridian distance (CMD). Each region is apodized with a circular function reducing the effective diameter to 15° before calculating three-dimensional power spectra. We derive daily flow maps of horizontal velocities from the 189 dense-pack patches. For this study, we combine them to form synoptic flow maps at 16 depths from 0.6 to 16 Mm. We also estimate the vertical-velocity component from the divergence of horizontal flows using mass conservation (Komm *et al.*, 2004; Komm, 2007). To focus on the spatial variation of the flows near active regions, we remove the large-scale trends in latitude of the differential rotation and the meridional flows and calculate residual synoptic flow maps by subtracting a low-order polynomial fit in latitude of the longitudinal average of the flows for each Carrington rotation (Komm *et al.*, 2004, 2005). The ring-diagram grid is comparable to the size of active regions and the resulting horizontal flows are complicated near active regions (Figure 2).

From the subsurface flow maps, we calculate the kinetic helicity density (h_k) which is a scalar that corresponds to changing orientation in space of fluid particles and is thus associated with mixing and turbulence (Lesieur, 1987; Moffatt and Tsinober, 1992). The kinetic helicity density is defined as the scalar product of the velocity ($\boldsymbol{v}$) and vorticity vector ($\boldsymbol{\omega}$):

$$h_k = \boldsymbol{\omega} \cdot \boldsymbol{v} \tag{4}$$

and the vorticity is defined as the curl of the velocity vector ($\boldsymbol{\omega} = \nabla \times \boldsymbol{v}$). We calculate vorticity and helicity from the measured subsurface velocity components (Komm *et al.*, 2004; Komm, 2007). To compare with the magnetic helicity proxies, we calculate the vertical contribution to the helicity scalar (h_z) defined as the product of the vertical velocity (v_z) and the curl of the horizontal velocity components (ω_z) which is the equivalent of the vertical component of the current helicity density h_c (see Equation (1) and replace B with v). We integrate h_z in volume over a suitable range in depth and a horizontal dense-pack area (A_{dp}):

$$H_z = \int_{r_1}^{r_2} h_z A_{dp}\, dr. \tag{5}$$

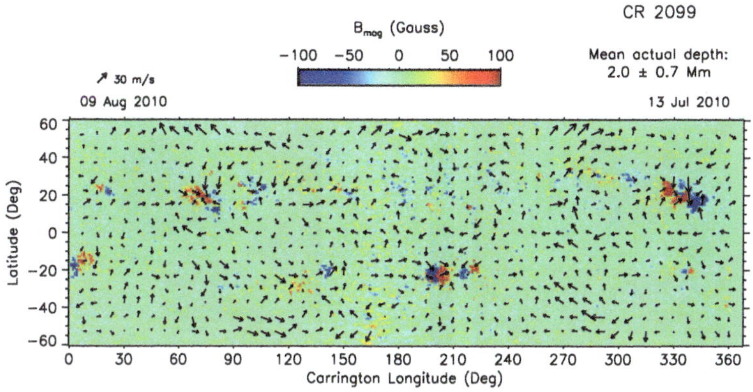

Figure 2 A synoptic map of subsurface flows at a depth of 2 Mm (arrows) for Carrington rotation 2099 superposed on a line-of-sight magnetogram (background) shows that subsurface flows are complicated near active regions, such as NOAA 11092 (82.5° longitude, 15° N latitude) and that the ring-diagram grid is comparable to the size of active regions.

For simplicity, we refer to the quantity, h_z, as kinetic helicity density and to H_z as kinetic helicity throughout this study.

The vertical vorticity component is small compared to the other two vorticity components due to the large size of the analysis areas in the horizontal direction. Fortunately, the whirl pattern persists during the disk passage and we can reduce the noise by using synoptic maps created from averaging over daily maps. Also, for the active regions studied here, the contributions to the kinetic helicity density in the other two directions are less significant than the vertical one with values smaller than two times the errors at all depths and locations.

As a corresponding measure of solar activity, we convert the SOLIS synoptic maps to absolute values (in Gauss) and bin them into circular areas with 15° diameter centered on the same grid as the dense-pack mosaic. In this way, we estimate a value of the unsigned magnetic flux density that matches the dense-pack grid in resolution, called magnetic activity index (MAI, see Basu, Antia, and Bogart, 2004).

3. Results

3.1. Current Helicity Proxies

Figures 3 and 4 show SOLIS/VSM quick-look vector magnetograms for the leading sunspots of the active regions, NOAA 11084 and 11092. These sunspots are nearly circular and change little during their disk passage. We calculate the magnetic helicity proxy for different days during the disk passage of the two regions (Table 2). The values are typically calculated over a region of 10 × 10 arcsec. The standard deviation is included to show the width of the distribution; the standard error of the mean is about a factor of ten (SOLIS/VSM) or 20 (SDO/HMI) smaller. Even though the distribution of SASSA values within each active region is broad, the SASSA parameter is consistently positive for NOAA 11084 and negative for NOAA 11092; the sign of the twist derived from the SASSA parameter does not change during the disk passage of the regions. This agrees with the persistence of twisted coronal structures above these sunspots during their disk passage, as noticed in SDO/AIA chromospheric and coronal images. Both VSM and HMI data sets lead to the same sign for both

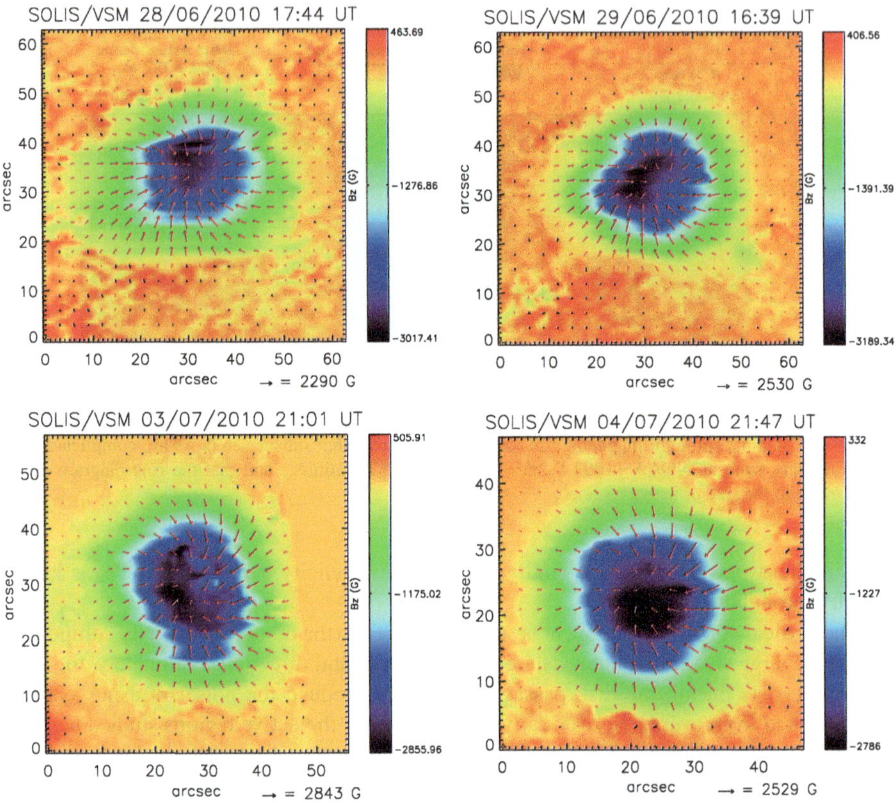

Figure 3 Quick-look vector magnetograms of the leading spot of active region NOAA 11084 for four different days during its disk passage derived from SOLIS/VSM data.

regions. The values cannot be directly compared, since the data have different properties (*e.g.*, resolution and atmospheric seeing). The sense of twist of the two regions as derived from photospheric, chromospheric/transition region and coronal observations appears to be i) the same at different heights, ii) persistent during the disk passage, and iii) consistent with the hemispheric rule (established by Pevtsov, Canfield, and Metcalf, 1995).

For both instruments, the SASSA parameter exhibits significant error bars (defined as one standard deviation of the mean). The amplitude of error bars is similar for HMI and VSM, which suggests that the error is mainly defined by the properties of the magnetic field and not by the atmospheric seeing or the spatial resolution. Large error bars are in a general agreement with the fact that many sunspots show patches of both positive and negative helicity inside the same polarity fields (Pevtsov, Canfield, and Metcalf, 1994).

Next, we calculate the magnetic helicity proxies for the six active regions without whirls (Table 3). The SASSA parameter derived from both data sets lead to the same sign except for one region (NOAA 11109). For regions without whirls, the helicity proxies are overall consistent with the hemispheric rule of negative values in the northern and positive ones in the southern hemisphere. The magnetic helicity proxy leads to similar values for two regions without whirls (NOAA 11124 and 11127) compared to the regions with whirls. The other four regions have smaller SASSA values than the regions with whirls.

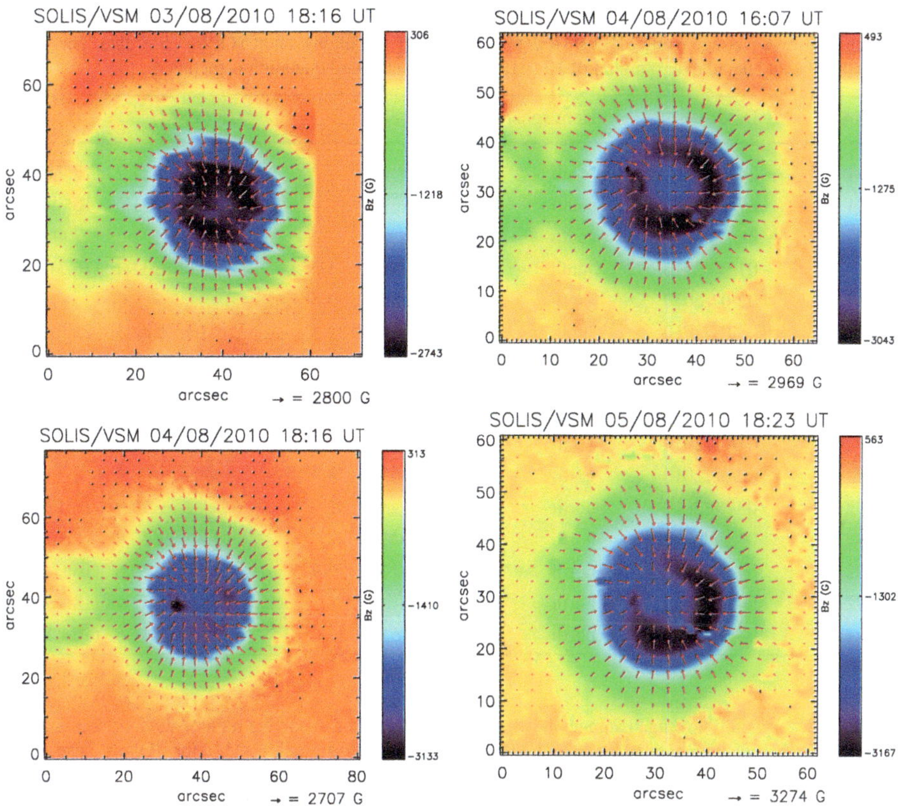

Figure 4 Quick-look vector magnetograms of the leading spot of active region NOAA 11092 for different days during its disk passage derived from SOLIS/VSM data.

3.2. Subsurface Helicity

We now study the subsurface properties of the flows beneath these active regions. Figures 5 and 6 show the kinetic helicity density, h_z, as a function of latitude and depth at a longitude centered on NOAA 11084 and 11092, respectively. Both regions show significant helicity values from the surface to about 8 Mm. The helicity of the subsurface flows associated with NOAA 11084 is clearly negative, while the kinetic helicity of NOAA 11092 is positive. The large values at high latitudes are due to systematics and are not significant, as indicated by the low signal-to-error ratio.

We derive the kinetic helicity, H_z, of these regions by integrating from 2.0 to 8.5 Mm in depth (six grid points) over the area of a dense-pack patch; the volume is about 1.67×10^{23} m^3. Both regions show significant values in this depth range. Since the center of an active region does not necessarily coincide with a dense-pack grid point, we calculate the helicity not only on the most central grid point but also for its neighbors in either direction or a total of 3×3 grid points. By definition, neighboring dense-pack patches overlap and are not independent. To avoid spurious results, we integrate only over grid locations where the helicity density value is at least twice as large as the error at two or more locations in depth. In Table 4, we show the kinetic helicity, $\langle H_z \rangle$, averaged over the grid points with values that are at least twice as large as the errors. To show the spread of the helicity values, we

Table 2 The spatially averaged signed shear angle, $\hat{\Psi}$ (degree), of the active regions with whirls calculated during their disk passage using SDO/HMI data (at 00:00 UT for all days) and SOLIS/VSM data. The values are typically calculated over a region of 10 × 10 arcsec; the standard deviation is included.

NOAA #	Date	Hemisphere	$\hat{\Psi}$ (HMI)	$\hat{\Psi}$ (VSM)
11084	28/06/2010	south		$5.0° \pm 34.7°$ (17:44 UT)
11084	29/06/2010	south		$2.5° \pm 31.6°$ (16:39 UT)
11084	30/06/2010	south		$9.0° \pm 31.2°$ (17:44 UT)
11084	01/07/2010	south	$8.12° \pm 33.2°$	$16.0° \pm 32.4°$ (16:31 UT)
11084	02/07/2010	south	$5.54° \pm 30.3°$	
11084	03/07/2010	south	$5.84° \pm 31.3°$	$7.4° \pm 22.5°$ (21:01 UT)
11084	04/07/2010	south	$3.58° \pm 29.8°$	$6.4° \pm 24.9°$ (21:47 UT)
11084	05/07/2010	south	$2.13° \pm 29.8°$	$3.6° \pm 30.1°$ (17:10 UT)
11092	01/08/2010	north	$-8.88° \pm 31.9°$	
11092	02/08/2010	north	$-9.63° \pm 33.9°$	
11092	03/08/2010	north	$-6.26° \pm 32.1°$	$-1.2° \pm 50.9°$ (16:12 UT)
11092	03/08/2010	north		$-5.3° \pm 26.8°$ (18:16 UT)
11092	04/08/2010	north	$-8.77° \pm 29.9°$	$-0.3° \pm 30.6°$ (16:07 UT)
11092	04/08/2010	north		$-0.9° \pm 32.9°$ (18:16 UT)
11092	05/08/2010	north	$-5.61° \pm 31.6°$	$-0.5° \pm 41.5°$ (16:22 UT)

Table 3 The spatially averaged signed shear angle, $\hat{\Psi}$ (degree), of the active regions without whirls calculated during their disk passage using SDO/HMI and SOLIS/VSM data (at 19:00 UT for all days). The values are typically calculated over a region of 10 × 10 arcsec; the standard deviation is included.

NOAA #	Date	Hemisphere	$\hat{\Psi}$ (HMI)	$\hat{\Psi}$ (VSM)
11106	16/09/2010	south	$0.99° \pm 30.1°$	$4.86° \pm 32.0°$
11126	18/11/2010	south	$-0.70° \pm 39.3°$	$-0.81° \pm 15.3°$
11109	27/09/2010	north	$-0.79° \pm 28.7°$	$2.55° \pm 12.6°$
11124	13/11/2010	north	$-7.84° \pm 37.9°$	$-3.57° \pm 18.4°$
11127	18/11/2010	north	$-9.57° \pm 35.0°$	$-11.7° \pm 47.1°$
11131	07/12/2010	north	$-3.12° \pm 31.6°$	$-7.1° \pm 21.6°$

include the mean value, $\overline{H_z}$, using uniform weights and the standard error of the mean. The table also includes the largest value for each active region, H_{max}, the number of grid points with significant helicity values, N, and the number of all locations with significant positive or negative helicity density values, n_{p} and n_{n}. The subsurface flows of NOAA 11084 are characterized by negative kinetic helicity and the ones of NOAA 11092 are characterized by positive values, as expected from Figures 5 and 6. Table 4 includes the sign of the magnetic helicity represented by the SASSA parameter. We find that the sign of the subsurface kinetic helicity of regions with whirls has negative correlation with that of their current helicity.

Next, we calculate the kinetic helicity values for eight ring days in order to see whether the kinetic helicity changes sign during the disk passage of these regions. The first and last data set, centered on 52.5° CMD east and west, lead to large values and even larger errors; the errors increase with increasing distance from disk center. We thus focus on the six daily

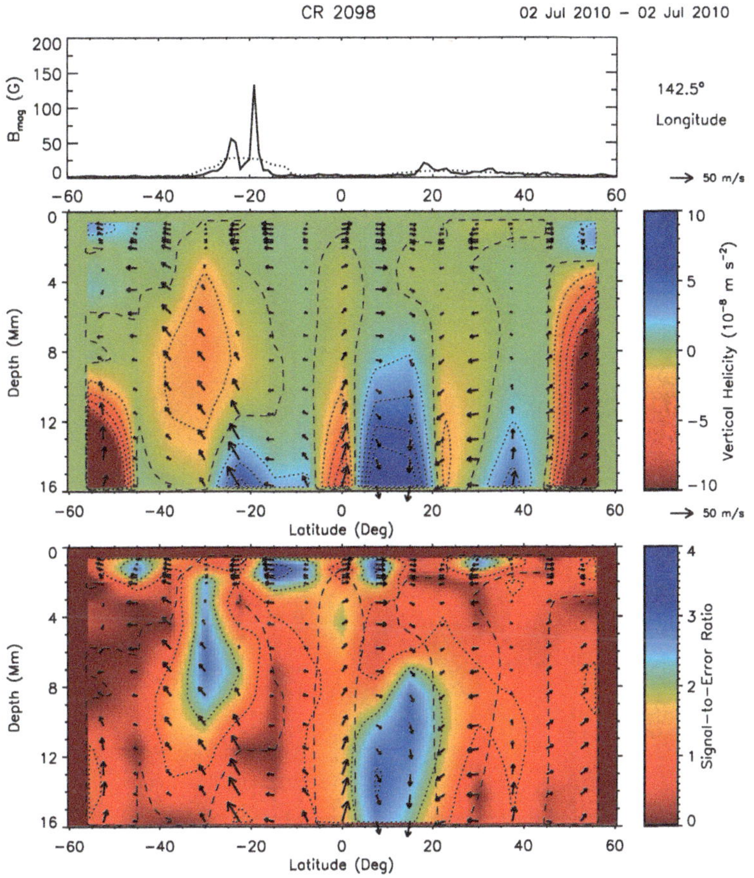

Figure 5 Active region NOAA 11084 with a clockwise (CW) whirl is associated with subsurface locations of predominantly negative kinetic helicity density, as shown in the slice in latitude and depth at 142.5° longitude (top: magnetic activity; middle: kinetic helicity density; bottom: signal-to-error ratio). The flows are indicated by arrows; the vertical ones are enhanced by a factor of six. The values have been smoothed (via a moving window averaging).

sets closer to disk center. For the larger of the two regions, NOAA 11092, there is no sign change in the kinetic helicity during its disk passage. The kinetic helicity values are positive at 7.5° north in all six daily sets and the values are greater than the errors. The daily values confirm the result derived from the synoptic data. For the smaller region, NOAA 11084, it is not that clear. The helicity is negative in the two measurements closest to disk center. For the two daily sets centered at $\pm 7.5°$ CMD, the helicity density values are negative at all depths from 2.0 to 8.5 Mm at 30° south and all values are greater than the errors. But the values are noisier for the other four days. The helicity density values are positive and greater than the errors at 7.1 to 8.5 Mm in the data set centered at 37.5° east, while the errors are larger than the values at all other depths and locations.

As a control experiment, we perform the same analysis for six active regions without whirls. Figures 7 and 8 show the kinetic helicity density as a function of latitude and depth for two such regions. The region in the southern hemisphere (NOAA 11126) is characterized by positive helicity (Figure 7), while the one in the northern hemisphere (NOAA 11124)

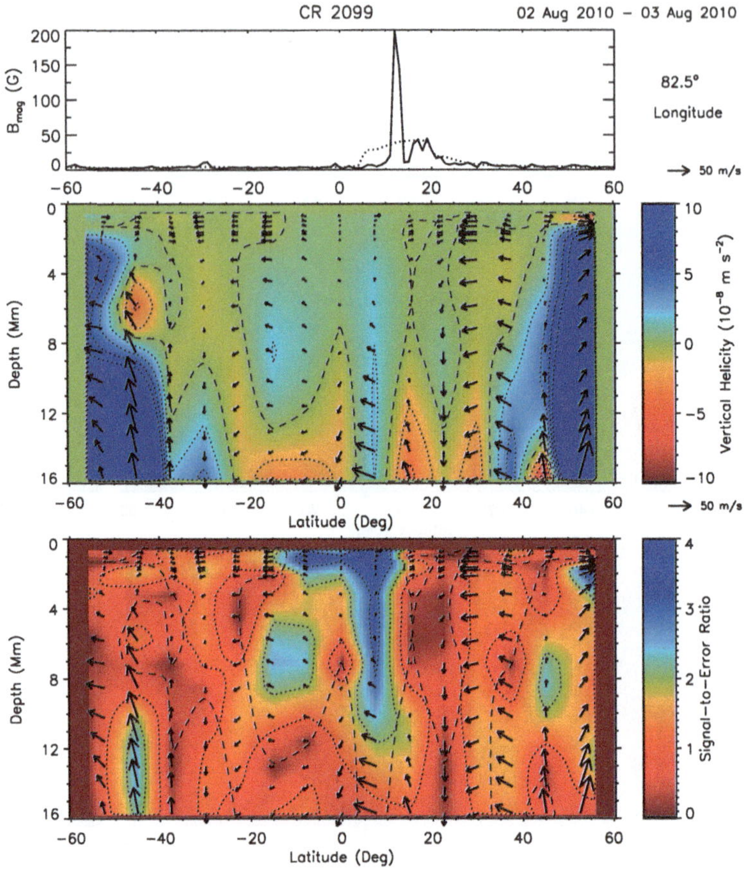

Figure 6 Active region NOAA 11092 with a counter-clockwise (CCW) whirl is associated with subsurface locations of predominantly positive kinetic helicity density, as shown in the slice in latitude and depth at 82.5° longitude (top: magnetic activity; middle: kinetic helicity density; bottom: signal-to-error ratio). The flows are indicated by arrows; the vertical ones are enhanced by a factor of six. The values have been smoothed (via a moving window averaging). (Compare with Figure 5.)

Table 4 The kinetic helicity of active regions with whirls (10^{19} $m^4\,s^{-2}$) integrated from 2.0 to 8.5 Mm over a dense-pack region ($\langle H_z \rangle$: average helicity and error, $\overline{H_z}$: mean value and standard error of the mean, H_{max}: largest value). The number of grid points used (N), the number of locations with positive and negative helicity density (n_p, n_n), and the sign of current helicity proxy $\hat{\Psi}$ are included.

NOAA #	Hemi.	Sign($\hat{\Psi}$)	$\langle H_z \rangle$	$\overline{H_z}$	H_{max}	N	n_p	n_n
11084	south	+	-2.16 ± 0.28	-1.14 ± 0.95	−3.85	4	3	10
11092	north	−	3.52 ± 0.33	3.44	4.05	2	12	0

shows negative helicity values (Figure 8). The other regions without whirls follow the same hemispheric rule with the exception of NOAA 11131 (Table 5). This region is one of four where grid points with large helicity values of opposite sign are present, which results in relatively small values of average kinetic helicity. The regions NOAA 11106 and 11124 are the

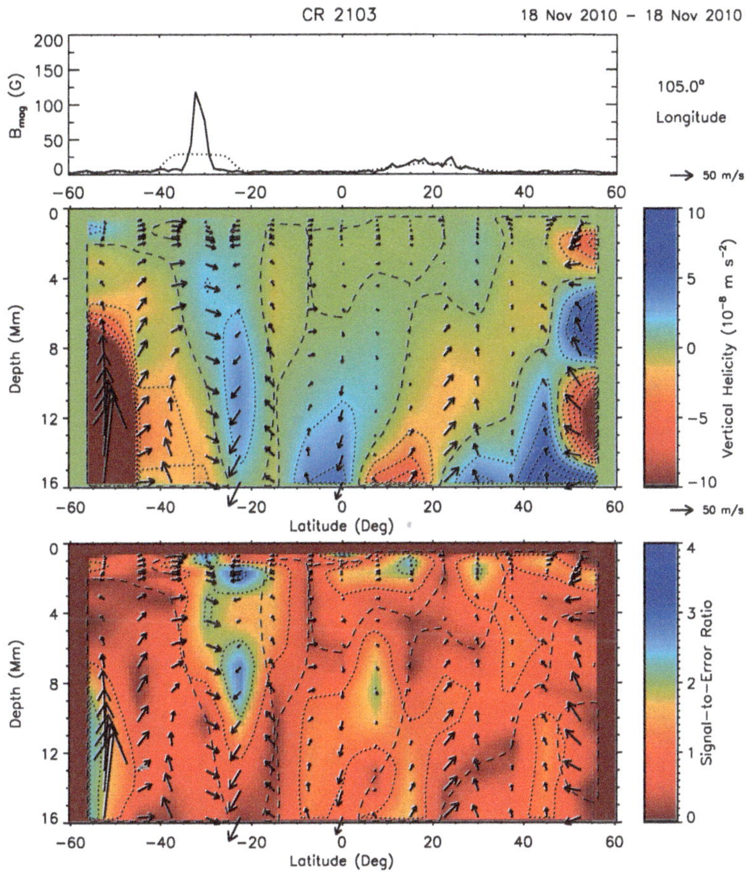

Figure 7 Active region NOAA 11126 (without whirl) is associated with subsurface locations of predominantly positive kinetic helicity density which differs from that of NOAA 11084 (top: magnetic activity; middle: kinetic helicity density; bottom: signal-to-error ratio). The flows are indicated by arrows; the vertical ones are enhanced by a factor of six. The values have been smoothed (via a moving window averaging). (Compare with Figure 5.)

only ones where one sign dominates the helicity values resulting in average values comparable to that of the regions with whirls. Table 5 also includes the sign of the magnetic helicity represented by the SASSA parameter. The sign of kinetic helicity of regions without whirls agrees in three (SOLIS) or four (HMI) of six cases with the sign of the corresponding magnetic helicity. The regions where the sign is different between these parameters are regions where large kinetic helicity values of opposite sign are present. For example, NOAA 11131 has positive kinetic helicity on average, but the largest value is negative (in agreement with the SASSA parameter). The kinetic helicity of regions without whirls exhibits positive correlation with their magnetic helicity.

4. Discussion

We study two active regions, NOAA 11084 and 11092, that show a regular pattern of superpenumbral whirls. The pattern persists throughout the disk passage of both regions. This

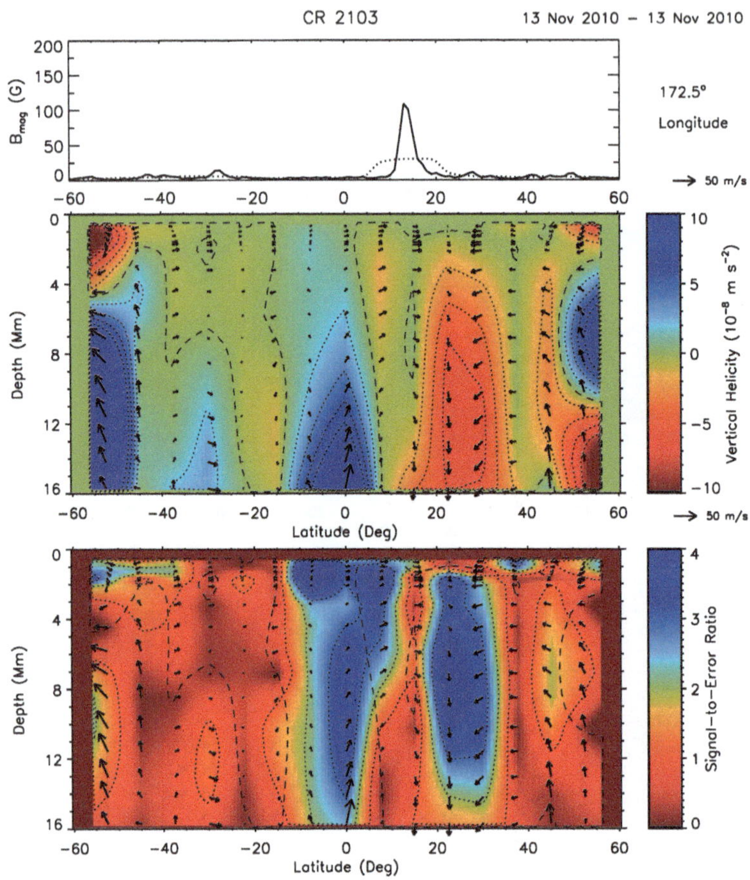

Figure 8 Active region NOAA 11124 (without whirl) is associated with subsurface locations of predominantly negative kinetic helicity density which differs from that of NOAA 11092 (top: magnetic activity; middle: kinetic helicity density; bottom: signal-to-error ratio). The flows are indicated by arrows; the vertical ones are enhanced by a factor of six. The values have been smoothed (via a moving window averaging). (Compare with Figure 6.)

makes them ideal candidates to search for a signature of helicity flow from the solar interior to the photosphere and chromosphere. We use SOLIS/VSM and SDO/HMI vector magnetograms to determine a proxy of the magnetic helicity (SASSA) of these two regions. The SASSA parameter leads to consistent results with positive helicity values for active region NOAA 11084 and negative ones for region NOAA 11092. The sense of twist of the two regions as derived from photospheric, chromospheric/transition region and coronal observations seems to be i) the same at different heights, ii) persistent during the disk passage and iii) consistent with the hemispheric rule (established by Pevtsov, Canfield, and Metcalf, 1995).

The SASSA parameter is an indicator of the non-potentiality of the magnetic field (*i.e.*, angular departure of the transverse magnetic field from the potential transverse field), with a sign attached to it. A large value of SASSA therefore basically indicates large non-potentiality. However, only for round, regular and symmetric sunspots it can be associated with or interpreted as "whirliness" (sense as well as strength of whirls) of the sunspot fibrils.

Table 5 The kinetic helicity of active regions without whirls (10^{19} $m^4\,s^{-2}$) integrated from 2.0 to 8.5 Mm over a dense-pack region ($\langle H_z \rangle$: average helicity and error, $\overline{H_z}$: mean value and standard error of the mean, $H_{\max}$: largest value). The number of grid points used (N), the number of locations with positive and negative helicity density (n_p, n_n), and the sign of current helicity proxy $\hat{\Psi}$ are included. The asterisk (*) indicates regions with large kinetic helicity values of either sign.

NOAA #	Hemi.	Sign($\hat{\Psi}$)	$\langle H_z \rangle$	$\overline{H_z}$	$H_{\max}$	N	n_p	n_n
11106	south	+	7.02 ± 0.57	6.26 ± 1.56	9.21	3	16	0
11126*	south	−	1.16 ± 0.20	0.40 ± 1.42	2.76	3	6	3
11109*	north	−/+	-1.97 ± 0.18	-1.36	-6.24	2	6	5
11124	north	−	-3.74 ± 0.21	-2.44 ± 1.26	-6.00	7	8	29
11127*	north	−	-0.40 ± 0.05	-0.30 ± 1.35	-3.13	4	6	7
11131*	north	−	0.24 ± 0.02	0.69 ± 2.84	-4.95	3	12	6

In other instances such as i) near the polarity inversion line (PIL), where typically the shear is high (Hagyard *et al.*, 1984), and ii) in an evolving/emerging flux region where the shear is high during the emergence phase, and relaxes once the field has fully emerged (Schmieder *et al.*, 1996; Gosain, 2011), we expect the magnitude of SASSA to be high due to high non-potentiality.

We then use GONG high-resolution Doppler data to determine the subsurface flows associated with these regions using a ring-diagram analysis. We derive the subsurface kinetic helicity density, h_z, from the synoptic flow maps and integrate the values over a suitable depth range from 2.0 to 8.5 Mm. We focus here on the vertical contribution to kinetic helicity because it directly corresponds to a commonly used magnetic helicity proxy (the vertical component of the current helicity). We find significant values of subsurface kinetic helicity, H_z, associated with these two regions. The subsurface helicity is negative for NOAA 11084 and positive for NOAA 11092. For the larger one of the two, NOAA 11092, we can establish that the sign of kinetic helicity does not change during its disk passage, which agrees with the persistence of the whirl pattern. For the weaker region NOAA 11084, we cannot reliably determine the daily variation of kinetic helicity; the daily measurements are too noisy. The sign of the average kinetic helicity of the two regions with whirls is opposite that of the corresponding magnetic helicity.

In addition, we study the subsurface flows of six active regions that do not show persistent whirl patterns. These six regions have been selected to be reasonably similar to the ones with whirls with regard to size, strength, and flare activity. Four of these regions contain round and stable sunspots and show small values of the magnetic helicity proxy compared to the regions with whirls. Two "non-whirly" active regions (NOAA 11124 and 11127) have high values of SASSA (Table 3) comparable to that of the "whirly" active regions studied here. In the case of NOAA 11127 there is a highly sheared PIL harboring a filament in close proximity to the leading sunspot. The presence of this highly sheared region might influence the shear magnitude and therefore lead to a large value of SASSA for this region. A subsurface shear flow parallel to the neutral line of a filament has been observed in a previous study (Hindman, Haber, and Toomre, 2006). In the case of NOAA 11124 the active region was in its emerging phase during 10 – 13 November 2010. This is also one of two regions without whirls that show large subsurface kinetic helicity values. So the large magnitude of SASSA could be due to the ongoing emergence of pre-stressed magnetic flux. The other "non-whirly" region with large subsurface kinetic helicity, NOAA 11106, has a small SASSA value. The large subsurface kinetic helicity might be associated with the large decaying region that surrounds the small sunspot of NOAA 11106.

For regions without persistent whirls, the magnetic helicity and subsurface kinetic helicity show a hemispheric preference with mainly negative values in the northern hemisphere and mainly positive values in the southern one. This agrees with the hemispheric rule established for magnetic helicity (Pevtsov, Canfield, and Metcalf, 1995; Pevtsov, Canfield, and Latushko, 2001; Hagino and Sakurai, 2005) and for subsurface flows (Komm *et al.*, 2007). However, four regions without whirls contain locations with kinetic helicity of comparable size but of different sign, which is not too surprising since these are after all regions without persistent helicity pattern of a single sign. The sign of kinetic helicity is well-defined for active regions with whirls and it is opposite to that of the current helicity. This seems to be a key difference between regions with and without whirls.

There are two different mechanisms that describe the interaction between turbulent twisted flows with magnetic activity. In the case of the α effect, right-handed eddies lead to left-handed magnetic mean fields or the opposite sign between kinetic and magnetic helicity (Krause and Rädler, 1981). In the Σ effect, right-handed eddies lead to right-handed twist of flux tubes or the same sign between kinetic and magnetic helicity (Longcope, Fisher, and Pevtsov, 1998; Fisher *et al.*, 1999). The Σ effect seems to be relevant for active regions without whirls. Active regions with whirls are "different". In an attempt to understand the origin of this difference, we have analyzed several measures including size of active regions, magnetic activity index (MAI), flare index (FI), and a newly introduced total flare energy (TFE) index. None of these parameters is distinctively different for "whirly" and "non-whirly" active regions. Thus, it is unclear at this time, why for whirly regions kinetic and magnetic helicities are negatively correlated but the non-whirly regions show a positive correlation between the two helicities.

The activity indices listed in Table 1 were selected for their perceived capability to represent the flare activity of active regions. Thus, for example, the area of an active region or its magnetic flux are traditionally seen as indicators of the total flare potential of an active region. The FI and TFE indices represent the actual flare activity of the regions during their disk passage. One can notice, however, that these indices do not correlate extremely well with each other although they all follow the same general tendency (*e.g.*, smaller in size regions tend to have smaller flare index). The latter suggests that these various indices may represent different aspects of flare activity of solar active regions. A more detailed comparison of these activity indices is outside the scope of this paper. As an additional note, we believe that although TFE and FI are computed in a similar manner, the TFE index may represent the total flare activity of an active region better as it takes into account not only the peak X-ray flare emission, but also the duration of the flares.

These are intriguing results. However, we have only two examples of active regions with whirls and six regions without whirls. We clearly need to analyze a larger sample to be able to draw conclusions that are statistically significant. For this reason, we will search for more active regions that have whirls identifiable in chromospheric and coronal images (Hα, for example). For such active regions with whirls that persist during their disk passage, we can determine the sign of the magnetic helicity from inspecting chromospheric or coronal images and do not have to rely on the availability of vector magnetograms. We will search for such regions during epochs where subsurface flows can be determined from MDI, GONG, or the *Helioseismic and Magnetic Imager* (HMI) instrument onboard SDO. This will allow us to better characterize the relationship between magnetic and kinetic helicity and to identify the defining characteristics of active regions with whirls.

Acknowledgements This work utilizes GONG and SOLIS data obtained by the NSO Integrated Synoptic Program (NISP), managed by the National Solar Observatory, which is operated by the Association of Universities for Research in Astronomy (AURA), Inc. under a cooperative agreement with the National Science Foundation. GONG data were acquired by instruments operated by the Big Bear Solar Observatory, High Altitude Observatory, Learmonth Solar Observatory, Udaipur Solar Observatory, Instituto de Astrofísica de Canarias, and Cerro Tololo Interamerican Observatory. SDO is a NASA mission and the SDO/HMI data used here are courtesy of NASA/SDO and the HMI Science Team. This work was supported by NSF/SHINE Award No. 1062054 to the National Solar Observatory. R.K. was partially supported by NASA grant NNX10AQ69G to Alysha Reinard. We thank the referee for helpful comments.

References

Balasubramaniam, K.S., Pevtsov, A.: 2011, *Society of Photo-Optical Instrumentation Engineers, (SPIE) Conference Series* **8148**.
Bao, S., Zhang, H.: 1998, *Astrophys. J. Lett.* **496**, L43.
Basu, S., Antia, H.M., Bogart, R.S.: 2004, *Astrophys. J.* **610**, 1157.
Berger, M.A., Field, G.B.: 1984, *J. Fluid Mech.* **147**, 133.
Borrero, J.M., Tomczyk, S., Kubo, M., Socas-Navarro, H., Schou, J., Couvidat, S., Bogart, R.: 2011, *Solar Phys.* **273**, 267.
Brun, A.S., Miesch, M.S., Toomre, J.: 2004, *Astrophys. J.* **614**, 1073.
Corbard, T., Toner, C., Hill, F., Hanna, K.D., Haber, D.A., Hindman, B.W., Bogart, R.S.: 2003, In: Sawaya-Lacoste, H. (ed.) *Local and Global Helioseismology: The Present and Future* **SP-517**, ESA, Noordwijk, 255.
Egorov, P., Rüdiger, G., Ziegler, U.: 2004, *Astron. Astrophys.* **425**, 725.
Fisher, G.H., Longcope, D.W., Linton, M.G., Fan, Y., Pevtsov, A.A.: 1999, In: Núñez, M., Ferriz-Mas, A. (eds.) *Stellar Dynamos: Nonlinearity and Chaotic Flows* **CS-178**, Astron. Soc. Pac., San Francisco, 35.
Gary, G.A., Hagyard, M.J.: 1990, *Solar Phys.* **126**, 21.
Georgoulis, M.K.: 2005, *Astrophys. J. Lett.* **629**, L69.
Gosain, S.: 2011, *IAU Symp.* **273**, 347.
Gosain, S., Tiwari, S.K., Venkatakrishnan, P.: 2010, *Astrophys. J.* **720**, 1281.
Haber, D.A., Hindman, B.W., Toomre, J., Bogart, R.S., Larsen, R.M., Hill, F.: 2002, *Astrophys. J.* **570**, 885.
Hagino, M., Sakurai, T.: 2005, *Publ. Astron. Soc. Japan* **57**, 481.
Hagyard, M.J.: 1987, *Solar Phys.* **107**, 239.
Hagyard, M.J., Teuber, D., West, E.A., Smith, J.B.: 1984, *Solar Phys.* **91**, 115.
Hindman, B.W., Haber, D.A., Toomre, J.: 2006, *Astrophys. J.* **653**, 725.
Kazachenko, M.D., Canfield, R.C., Longcope, D.W., Qiu, J.: 2012, *Solar Phys.* **277**, 165.
Keller, C.U., Harvey, J.W., Giampapa, M.S.: 2003, *Proc. SPIE* **4853**, 194.
Komm, R.: 2007, *Astron. Nachr.* **328**, 269.
Komm, R.W., Corbard, T., Durney, B.R., González Hernández, I., Hill, F., Howe, R., Toner, C.: 2004, *Astrophys. J.* **605**, 554.
Komm, R., Howe, R., Hill, F., González Hernández, I., Toner, C., Corbard, T.: 2005, *Astrophys. J.* **631**, 636.
Komm, R., Howe, R., Hill, F., Miesch, M., Haber, D., Hindman, B.: 2007, *Astrophys. J.* **667**, 571.
Komm, R., Ferguson, R., Hill, F., Barnes, G., Leka, K.D.: 2011, *Solar Phys.* **268**, 389. doi:10.1007/s11207-010-9552-1.
Krause, F., Rädler, K.-H.: 1981, *Mean Field Magnetohydrodynamics*, Pergamon, New York.
Lemen, J.R., Title, A.M., Akin, D.J., Boerner, P.F., Chou, C., Drake, J.F., *et al.*: 2012, *Solar Phys.* **275**, 17. doi:10.1007/s11207-011-9776-8.
Lesieur, M.: 1987, *Turbulence in Fluids*, Kluwer, Dordrecht.
Longcope, D.W., Fisher, G.H., Pevtsov, A.A.: 1998, *Astrophys. J.* **507**, 417.
Low, B.C.: 1996, *Solar Phys.* **167**, 217.
Mason, D., Komm, R., Hill, F., Howe, R., Haber, D., Hindman, B.: 2006, *Astrophys. J.* **645**, 1543.
Maurya, R.A., Ambastha, A., Reddy, V.: 2011, *J. Phys. Conf. Ser.* **271**, 012003.
Metcalf, T.R., Leka, K.D., Barnes, G., Lites, B.W., Georgoulis, M.K., Pevtsov, A.A., Balasubramaniam, K.S., Gary, G.A., Jing, J., Li, J., Liu, Y., Wang, H.N., Abramenko, V., Yurchyshyn, V., Moon, Y.-J.: 2006, *Solar Phys.* **237**, 267.
Moffatt, H.K., Tsinober, A.: 1992, *Annu. Rev. Fluid Mech.* **24**, 281.
Pesnell, W.D., Thompson, B.J., Chamberlin, P.C.: 2012, *Solar Phys.* **275**, 3.
Pevtsov, A.A.: 2008, *J. Astrophys. Astron.* **29**, 49.
Pevtsov, A.A., Balasubramaniam, K.S., Rogers, W.J.: 2003, *Adv. Space Res.* **32**, 1905.

Pevtsov, A.A., Canfield, R.C., Latushko, S.M.: 2001, *Astrophys. J. Lett.* **549**, L261.
Pevtsov, A.A., Canfield, R.C., Metcalf, T.R.: 1994, *Astrophys. J. Lett.* **425**, L117.
Pevtsov, A.A., Canfield, R.C., Metcalf, T.R.: 1995, *Astrophys. J. Lett.* **440**, L109.
Pietarila, A., Bertello, L., Harvey, J.W., Pevtsov, A.A.: 2013, *Solar Phys.* **282**, 91. doi:10.1007/s11207-012-0138-y.
Reinard, A.A., Henthorn, J., Komm, R., Hill, F.: 2010, *Astrophys. J. Lett.* **710**, L121.
Ronan, R.S., Mickey, D.L., Orrall, F.Q.: 1987, *Solar Phys.* **113**, 353.
Rüdiger, G., Pipin, V.V., Belvedère, G.: 2001, *Solar Phys.* **198**, 241. doi:10.1023/A:1005217606269.
Schmieder, B., Demoulin, P., Aulanier, G., Golub, L.: 1996, *Astrophys. J.* **467**, 881.
Schou, J., Borrero, J.M., Norton, A.A., Tomczyk, S., Elmore, D., Card, G.L.: 2012, *Solar Phys.* **275**, 327.
Skumanich, A., Lites, B.W.: 1987, *Astrophys. J.* **322**, 473.
Tiwari, S.K., Venkatakrishnan, P., Sankarasubramanian, K.: 2009, *Astrophys. J. Lett.* **702**, L133.
Tiwari, S.K., Joshi, J., Gosain, S., Venkatakrishnan, P.: 2008, In: Hasan, S.S., Gangadhara, R.T., Krishan, V. (eds.) *Turbulence, Dynamos, Accretion Disks, Pulsars and Collective Plasma Processes*, 329.
Tiwari, S.K., Venkatakrishnan, P., Gosain, S., Joshi, J.: 2009, *Astrophys. J.* **700**, 199.
Zhang, M.: 2006, *Astrophys. J. Lett.* **646**, L85.
Zhao, J., Kosovichev, A.G.: 2003, *Astrophys. J.* **592**, 446.

DOI 10.1007/978-1-4939-1182-0_5
Reprinted from *Solar Physics* Journal, DOI 10.1007/s11207-013-0274-z

A Study of Connections Between Solar Flares and Subsurface Flow Fields of Active Regions

Yu Gao · Junwei Zhao · Hongqi Zhang

Received: 16 November 2012 / Accepted: 6 March 2013 / Published online: 4 April 2013

Abstract We investigate the connections between the occurrence of major solar flares and subsurface dynamic properties of active regions. For this analysis, we select five active regions that produced a total of 11 flares with peak X-ray flux intensity higher than M5.0. The subsurface velocity fields are obtained from time–distance helioseismology analysis using SDO/HMI (*Solar Dynamics Observatory/Helioseismic and Magnetic Imager*) Doppler observations, and the X-ray flux intensity is taken from GOES (*Geostationary Operational Environmental Satellites*). It is found that among the eight amplitude bumps in the evolutionary curves of subsurface kinetic helicity, five (62.5%) of them had a flare stronger than M5.0 occurring within 8 hours, either before or after the bumps. Another subsurface parameter is the Normalized Helicity Gradient Variance (NHGV), reflecting kinetic helicity spread in different depth layers; it also shows bumps near the occurrence of these solar flares. Although there is no one-to-one correspondence between the flare and the subsurface properties, these observational phenomena are worth further studies to better understand the flares' subsurface roots, and to investigate whether the subsurface properties can be used for major flare forecasts.

Keywords Sun: photosphere · Sun: helicity · Sun: flare

Solar Origins of Space Weather and Space Climate
Guest Editors: I. González Hernández, R. Komm, and A. Pevtsov

Y. Gao (✉) · H. Zhang
National Astronomical Observatories of China, Chinese Academy of Sciences, Beijing, P.R. China
e-mail: gy@bao.ac.cn

H. Zhang
e-mail: hzhang@bao.ac.cn

J. Zhao
W.W. Hansen Experimental Physics Laboratory, Stanford University, Stanford, CA 94305-4085, USA
e-mail: junwei@sun.stanford.edu

1. Introduction

Helical features of observed solar magnetic fields have been widely investigated in the previous two decades. Helicity is a physical quantity closely related to the mechanism of solar dynamo, and in this regard, hemispheric preponderance in magnetic (or current) helicity distribution was found and confirmed by different observational groups (Seehafer, 1990; Pevtsov, Canfield, and Metcalf, 1995; Bao and Zhang, 1998; Hagino and Sakurai, 2004; Zhang *et al.*, 2010). On the other hand, the helicity of the magnetic field is supposedly connected with the storage and transfer of energy inside active regions. Pevtsov, Canfield, and Metcalf (1995) and Bao, Ai, and Zhang (2001) found that active regions that did not follow the hemispheric sign rule of helicity tend to be more active than the others in the flare productivity, but this characteristic was not supported by the case study performed by Sakurai and Hagino (2003).

It was found that significant helicity accumulation occurred preceding some major flares (Park *et al.*, 2008). During flares, active regions release magnetic free energy accumulated when magnetic non-potentiality increases. A related flare triggering mechanism is the rapid change of magnetic helicity or annihilation of magnetic helicity (Moon *et al.*, 2002; Kusano *et al.*, 2004).

Another flare triggering mechanism may be related to the plasma motion in solar active regions. As the statistical studies indicate, kinetic helicity is an important parameter that may cause instability and thus magnetic reconnection. The horizontal components of subsurface flows obtained from the ring-diagram analysis, which is a local helioseismology technique, have been used by many authors to investigate the connections between subsurface dynamics and solar eruptive events. An analysis of AR 10486 showed systematic variations in synoptic and daily maps of kinetic helicity that might be the subsurface signatures of the flare events that would occur in this active region later (Komm *et al.*, 2004). A relation was found between the maximum values of the unsigned kinetic helicity density, which was calculated from the subsurface flows associated with each active region, and the total flare X-ray intensity of the active regions (Komm *et al.*, 2005). Mason *et al.* (2006) reported that the maximum unsigned zonal and meridional vorticity components of active regions were correlated with the total flare intensity. More recently, Reinard *et al.* (2010) developed a new parameter, normalized helicity gradient variance, to investigate the connections of subsurface dynamics with solar flares, and found that this parameter was expected to increase two to three days in advance of flare occurrences.

Now, the *Helioseismic and Magnetic Imager* (Scherrer *et al.*, 2012; Schou *et al.*, 2012) onboard the NASA mission *Solar Dynamics Observatory* (SDO/HMI) provides continuous observations of the Sun, and time–distance data-analysis pipeline provides continuous subsurface flow fields for different depths using the HMI data (Zhao *et al.*, 2012). The subsurface flow maps inside active regions give an unprecedented opportunity to study connections between the subsurface dynamics and solar flares with a better spatial resolution and a better temporal cadence. Recently, employing these newly available data, Gao, Zhao, and Zhang (2012) studied the relationship between photospheric current helicity and subsurface kinetic helicity in active regions, and found a quite high correlation between the temporal evolutions of these two different quantities. However, that study did not address whether there was a connection between subsurface properties and occurrences of solar flares. In this paper, we focus on this topic and investigate whether there is such a connection by analyzing a few selected flare-productive active regions. We introduce data acquisition and definition of parameters in Section 2, present our results in Section 3, discuss our results and conclude in Section 4.

Table 1 Detailed information of flares that are analyzed in this study.

Flare	Date	Start time	Peak time	Disk location	Active region No.
M6.6	2011.02.13	17:28 UT	17:38 UT	S20°, E04°	AR 11158
X2.2	2011.02.15	01:44 UT	01:55 UT	S21°, W21°	AR 11158
X1.5	2011.03.09	23:13 UT	23:23 UT	N08°, W09°	AR 11166
M6.0	2011.08.03	13:17 UT	13:48 UT	N16°, W30°	AR 11261
M9.3	2011.08.04	03:41 UT	03:57 UT	N19°, W36°	AR 11261
M5.3	2011.09.06	01:35 UT	01:50 UT	N14°, W07°	AR 11283
X2.1	2011.09.06	22:12 UT	22:20 UT	N14°, W18°	AR 11283
X1.8	2011.09.07	22:32 UT	22:38 UT	N14°, W28°	AR 11283
M6.7	2011.09.08	15:37 UT	15:46 UT	N14°, W40°	AR 11283
X1.9	2011.09.24	09:21 UT	09:40 UT	N12°, E60°	AR 11302
M7.4	2011.09.25	04:31 UT	04:50 UT	N11°, E47°	AR 11302

2. Data Acquisition and Parameters

2.1. Data Acquisition

The line-of-sight magnetic field maps used in our analysis are from HMI (Schou *et al.*, 2012), and the subsurface velocity maps are from the HMI time–distance data-analysis pipeline (Zhao *et al.*, 2012). The subsurface velocity maps are typically derived from an 8-hr period, and cover an area of roughly $30° \times 30°$ with active regions located near the center of the area, with a depth coverage from near the surface to about 20 Mm. The spatial resolution of these subsurface velocity maps is $0.06°$ pixel^{-1}. While each flow map is obtained from an 8-hr period, the temporal step used in our analysis is 4 hours. Therefore, each flow map is not completely independent from the maps before and after it, as there is data overlapping between them. X-ray flux density used in this analysis is from GOES (*Geostationary Operational Environment Satellite*) X-ray data lists provided by Space Weather Prediction Center.

For this study, we select five active regions, which produced a total of 11 flares with peak X-ray flux intensity higher than M5.0. The detailed information as regards these flares and active regions is listed in Table 1. One randomly selected sample image showing the magnetic field, overplotted by the subsurface horizontal flow field, is shown for each of five active regions in Figure 1.

2.2. Definition of Parameters

Kinetic helicity is defined as

$$H_k = \mathbf{v} \cdot (\nabla \times \mathbf{v}), \tag{1}$$

where **v** is three-dimensional velocity field with three components. Sometimes it is useful to study weighted kinetic helicity, defined as kinetic helicity dividing the square of local speed:

$$\alpha^k = \mathbf{v} \cdot (\nabla \times \mathbf{v})/|\mathbf{v}|^2. \tag{2}$$

In this study, we only investigate the vertical component of α^k, and that is

$$\alpha_z^k = v_z(\partial v_x/\partial y - \partial v_y/\partial x)/\left(v_x^2 + v_y^2 + v_z^2\right). \tag{3}$$

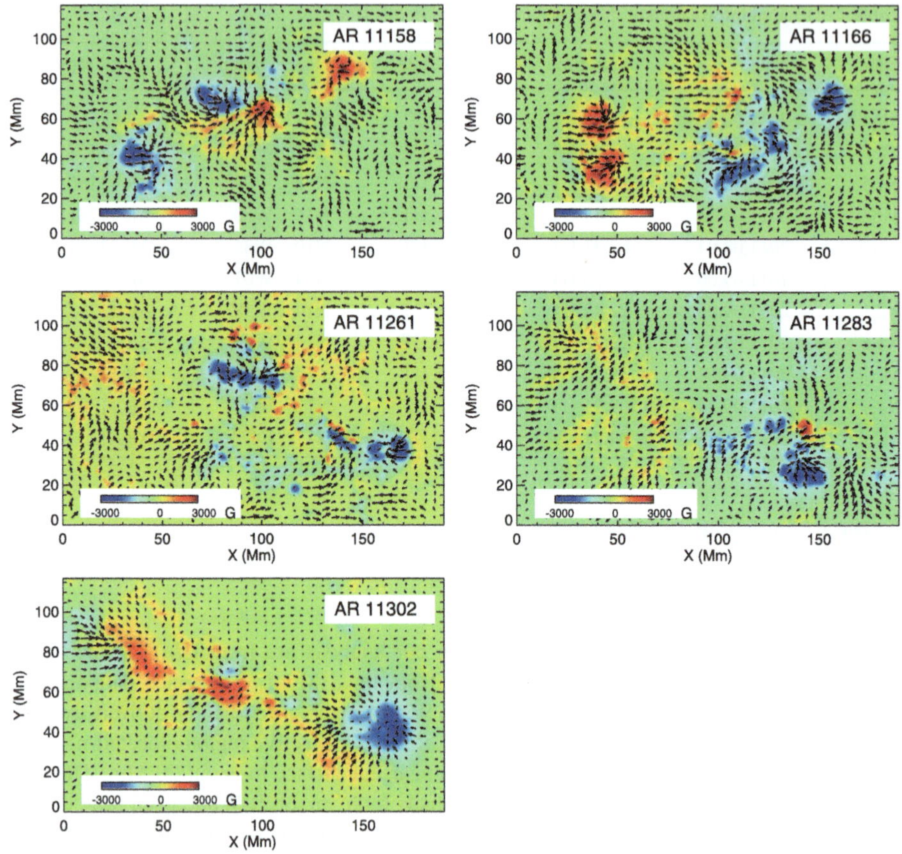

Figure 1 A randomly selected sample of magnetic field (background image), overplotted by horizontal subsurface flow field (arrows) at the depth of 0 – 1 Mm, for each of the five studied active regions. The longest arrow in each panel represents the strongest horizontal flow speed: 327 m s^{-1}, 381 m s^{-1}, 381 m s^{-1}, 326 m s^{-1}, and 624 m s^{-1}, respectively.

When analyzing this parameter, we only use the velocities at the depth of 0 – 1 Mm. It was already demonstrated that the flow field at this depth was similar to the photospheric flow field derived from combining the correlation tracking and minimization of magnetic induction equations (Liu, Zhao, and Schuck, 2012).

Another parameter we investigate in this study is similar to the Normalized Helicity Gradient Variance (NHGV) defined by Reinard *et al.* (2010). This parameter is obtained by following these steps: differences in helicity at each depth from one time period and the period next to it are first computed:

$$\Delta H_z(t) = H_z(t) - H_z(t-1), \tag{4}$$

where z represents the depths of 0 – 1, 1 – 3, 3 – 5, 5 – 7, 7 – 9, and 9 – 11 Mm, respectively, as listed in Zhao *et al.* (2012). Then the spread of the temporal helicity change with depth is determined by summing the changes in ΔH_z with depths:

$$\Delta H(t) = \sum_z \big(\Delta H_z(t) - \Delta H_{z+1}(t)\big), \tag{5}$$

where z takes only the first, third, and fifth values. The change in helicity with depth is summed as

$$H(t) = \sum_z \big(H_z(t) - H_{z+1}(t)\big), \tag{6}$$

which is expected to capture the overall spread of helicity values. Finally, NHGV is defined as a multiplication of $\Delta H(t)$ and $H(t)$, *i.e.*,

$$\mathrm{NHGV} = \Delta H(t) H(t). \tag{7}$$

Slightly different from the approach taken by Reinard *et al.* (2010), we compute this parameter only in the regions where $|B_z| > 50$ G and normalize it by the average values in the regions where $|B_z| < 50$ G. The selection of this criterion threshold, 50 G, is arbitrary, but it is a common practice in similar studies, *e.g.*, Bao and Zhang (1998). When computing both parameters defined above, $\langle \alpha_z^k \rangle$ and NHGV, we used a Cartesian coordinate system. Since the computation is only limited to active regions and their surrounding areas, there is no significant difference between the Cartesian and spherical coordinates for such computations based on our past experience.

3. Results

3.1. Temporal Variation of α_z^k and X-Ray Flux Intensity

We compute the density of the z-component of the subsurface kinetic helicity $\langle \alpha_z^k \rangle$ averaged from areas where $|B_z| > 50$ G. We also compute the $\langle \alpha_z^k \rangle$ for each magnetic-field polarity separately by averaging areas where $B_z > 50$ G and $B_z < -50$ G (note that depending on the hemispheric location of the active regions, the leading polarity of the active region can be positive or negative). In particular, when we compute $\langle \alpha_z^k \rangle$, we only select areas that show prominent kinetic helicity, *i.e.*, where $|\alpha_z^k| > \langle |\alpha_z^k| \rangle + 1\sigma$.

Figure 2 shows an example of evolutionary curves of the X-ray flux intensity and the subsurface kinetic helicities calculated for both the leading and the following polarities of NOAA AR 11158. The curves of $\langle \alpha_z^k \rangle$ show some strong variations, and we define where the $\langle \alpha_z^k \rangle$ amplitude is 2σ above the mean value as an amplitude bump. It can be found there is a bump in $\langle \alpha_z^k \rangle$ in the leading polarity about seven hours before the start of the X2.2 flare. While the $\langle \alpha_z^k \rangle$ in the following polarity also shows some amplitude bumps, the connection of these bumps with the powerful X-class flare is not obvious.

Figure 3 shows the evolutionary curves of X-ray flux taken around the flare events that occurred in the other four active regions, together with the evolutionary curves of subsurface kinetic helicity $\langle \alpha_z^k \rangle$ computed from the leading polarities of these active regions. Since our analyses show that the subsurface $\langle \alpha_z^k \rangle$ in the following polarities does not exhibit clear connection with the flare occurrences, we do not display results from the following polarities of these active regions. In these four active regions there are a few amplitude bumps in $\langle \alpha_z^k \rangle$, and some of them are close in time with the powerful flares. Combining the results shown in Figures 2 and 3, we summarize as follows: among the eight $\langle \alpha_z^k \rangle$ amplitude bumps, five of them (62.5 %) happened within eight hours, either before or after, of a flare stronger than M5.0; among the 11 flares we studied, five of them (45.5 %) occurred no more than eight hours apart, again either before or after, from an $\langle \alpha_z^k \rangle$ amplitude bump. Note that eight hours is the time duration of the data used to compute the time–distance subsurface flow maps. There is no clear one-to-one correspondence between the helicity bump and the occurrence

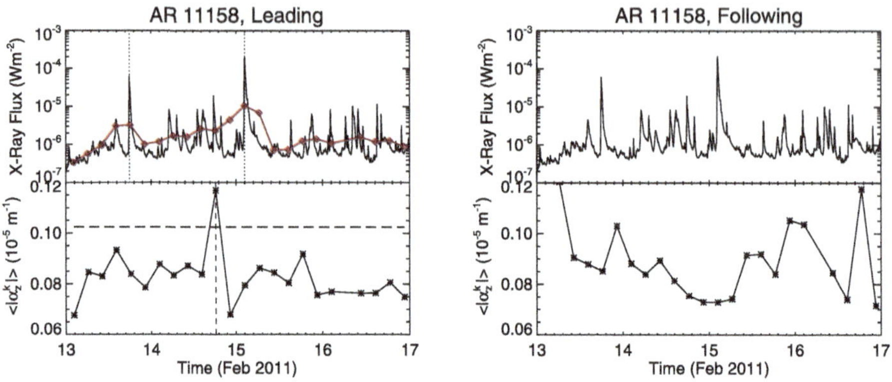

Figure 2 Evolution of the X-ray flux intensity and the subsurface kinetic helicity for the leading polarity (left panel) and the following polarity (right panel) of NOAA AR 11158. The leading polarity is positive for this active region. The X-ray flux intensity is averaged every eight hours with a four-hour step (shown as red curve in the left panel) so as to better match in temporal periods the data points of the subsurface kinetic helicity. The horizontal dashed line in the left lower panel indicates 2σ above the mean, and values above this dashed line are considered as a kinetic helicity bump.

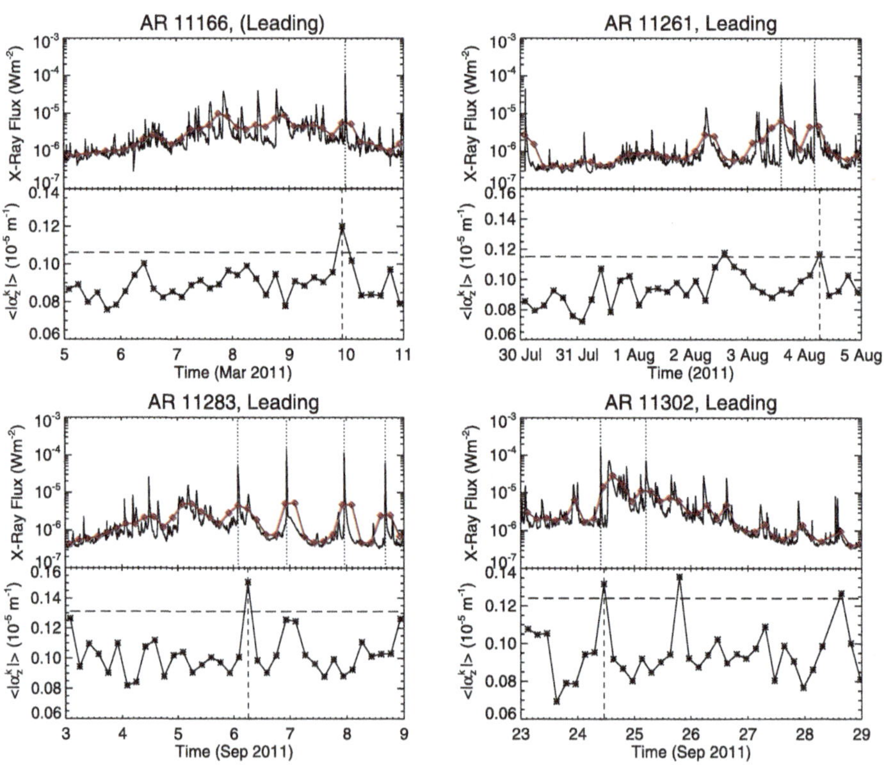

Figure 3 Same as the left panel of Figure 2, but for the leading polarities in the other four active regions.

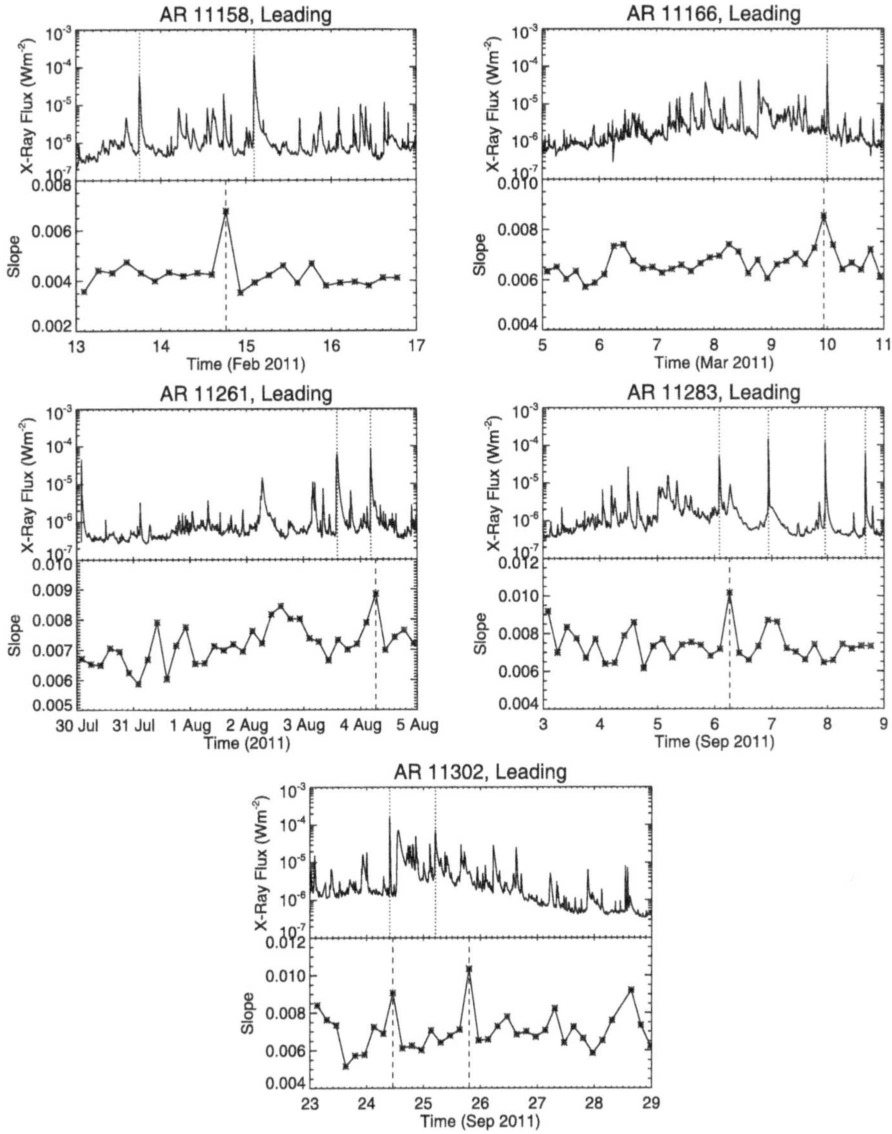

Figure 4 Evolutionary curves of X-ray flux intensity and the fitted slopes of the subsurface kinetic helicity distribution for all the five active regions.

of a flare. The averaged X-ray flux intensity, shown as red curves in Figures 2 and 3, seems to show a better correlation with the $\langle\alpha_z^k\rangle$ variations, but the statistical significance is not high.

Furthermore, we establish that the bumps of the subsurface $\langle\alpha_z^k\rangle$ are a true variation in the profile of kinetic helicity rather than a random enhancement caused by some singular points. For each map of the α_z^k obtained from one certain time period, we divide all the values of kinetic helicity into ten equal segments covering the minimum value to the maximum. Since each map of α_z^k may have different minimum and maximum values, these ten segments

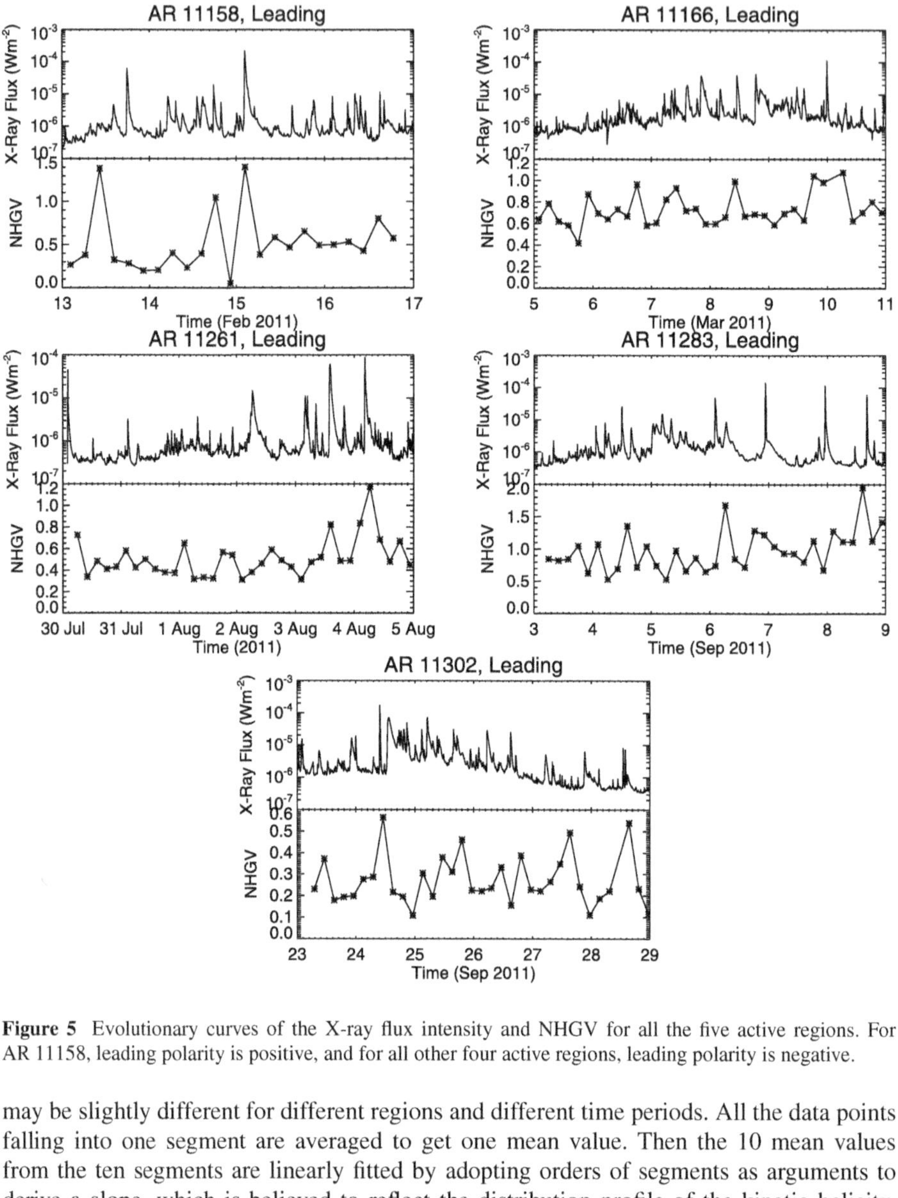

Figure 5 Evolutionary curves of the X-ray flux intensity and NHGV for all the five active regions. For AR 11158, leading polarity is positive, and for all other four active regions, leading polarity is negative.

may be slightly different for different regions and different time periods. All the data points falling into one segment are averaged to get one mean value. Then the 10 mean values from the ten segments are linearly fitted by adopting orders of segments as arguments to derive a slope, which is believed to reflect the distribution profile of the kinetic helicity. Figure 4 shows all the fitted slopes for the different time periods of different active regions. Comparing the evolutionary curves of these fitted slopes with the variations in $\langle \alpha_z^k \rangle$ shown in Figures 2 and 3, one is able to find the amplitude bumps seen in Figure 4 match nicely with those seen in Figures 2 and 3. The purpose of this is just to demonstrate that the bumps in the subsurface $\langle \alpha_z^k \rangle$ seen in Figures 2 and 3 are reliable signals.

3.2. Temporal Variation of NHGV and X-Ray Flux Intensity

We perform a similar analysis over the variations of NHGV together with the X-ray flux intensity, and the results over all the five active regions are shown in Figure 5. Once again,

NHGV is only shown for the leading polarity, as there is no clear correspondence between the two quantities in the following polarity. Similar to what is found in Section 3.1, for roughly half of the major flares listed in Table 1, an amplitude bump in NHGV can be found within eight hours of the flare occurrence, either before or after. It is also true that near about half of the large NHGV bumps, powerful flare events occurred. At the moment, we adopt this parameter for a preliminary co-investigation for the result obtained for the $\langle\alpha_z^k\rangle$. Quantitative diagnosis of the connection between these two parameters needs further study.

4. Discussion

By analyzing subsurface velocity fields of some flare-productive active regions, together with the X-ray flux intensity, we have found that in about half of the cases the subsurface kinetic helicity shows bumps within eight hours either before or after the flare onset. These bumps in subsurface kinetic helicity are reliable, as can be demonstrated by an alternative approach of analyzing the slopes of distributions of the subsurface kinetic helicity. Meanwhile, our analysis of NHGV, a parameter that reflects kinetic helicity spread in different depth layers of active regions, also shows that major flares are often associated with strong variations in NHGV.

Despite that the $\langle\alpha_z^k\rangle$ are displayed with a four-hour cadence in Figures 2 and 4, the actual temporal resolution of these data is eight hours, quite poor compared to the rapid development of flare events. Therefore, the bumps found in both $\langle\alpha_z^k\rangle$ and NHGV within 8-hrs range of flare occurrence may be considered as occurring at about the same time as the flares. It is not very clear to us whether the strong changes in subsurface kinetic helicity and NHGV lead to or help the flare occurrence, or whether these changes in subsurface properties are caused by the occurrences of powerful flares high above the photosphere. However, considering the high mass density and the low plasma β of the subsurface interior, we tend to believe the former rather than the latter.

Previous studies using the subsurface velocity field, derived from the ring-diagram analysis, have found promising connections between subsurface dynamics and flare events (*e.g.* Komm *et al.*, 2005; Mason *et al.*, 2006; Reinard *et al.*, 2010). Both the spatial resolution and the temporal cadence of the velocity fields in these ring-diagram analyses were poorer than the time–distance velocity fields that are used in this study. Although our analysis reveals some promising connections between the subsurface dynamical properties and the flare events, it also shows some ambiguity between them. A clear subsurface indicator of an upcoming major flare may not be there, and most likely, such a subsurface indicator, if there is one, only exists statistically. This prompts us to perform a statistical study once observations of more flare events in more active regions become available from HMI observations. On the other hand, regardless of whether the subsurface properties can be used as flare precursors, the subsurface anomalies that occur close to the flare onset are interesting phenomena, worth further investigation.

Acknowledgements SDO is a NASA mission, and HMI project is supported by NASA contract NAS5-02139 to Stanford University. This work is partially supported by the National Natural Science Foundation of China under the grants 11028307, 10921303, 11103037, 11173033, 11221063 and 41174153, and by Chinese Academy of Sciences under grant KJCX2-EW-T07. We thank the anonymous referee for his/her comments and suggestions that helped to improve the quality of this paper.

References

Bao, S., Zhang, H.: 1998, *Astrophys. J.* **496**, L43.

Bao, S., Ai, G., Zhang, H.: 2001, In: Brekke, P., Fleck, B., Gurman, J.B. (eds.) *Recent Insights into the Physics of the Sun and Heliosphere—Highlights from SOHO and Other Space Mission, IAU Symp.* **203**, Astron. Soc. Pac., 247.

Gao, Y., Zhao, J., Zhang, H.: 2012, *Astrophys. J. Lett.* **761**, L9.

Hagino, M., Sakurai, T.: 2004, *Publ. Astron. Soc. Pac.* **56**, 831.

Komm, R.W., Howe, R., González Hernández, I., Hill, F., Haber, D., Hindman, B., Corbard, T.: 2004, In: Danesy, D. (ed.) *SOHO 14 Helio- and Asteroseismology: Towards a Golden Future, ESA* **SP-559**, ESA, Noordwijk, 520.

Komm, R., Howe, R., Hill, F., González Hernández, I., Toner, C.: 2005, *Astrophys. J.* **630**, 1184.

Kusano, K., Maeshiro, T., Yokoyama, T., Sakurai, T.: 2004, *Astrophys. J.* **610**, 537.

Liu, Y., Zhao, J., Schuck, P.W.: 2012, *Solar Phys.* doi:10.1007/s11207-012-0089-3.

Mason, D., Komm, R., Hill, F., Howe, R., Haber, D., Hindman, B.: 2006, *Astrophys. J.* **645**, 1543.

Moon, Y.J., Chae, J., Choe, G.S., Wang, H.M., Park, Y.D., Yun, H.S., Yurchyshyn, V., Goode, P.R.: 2002, *Astrophys. J.* **574**, 1066.

Park, S., Lee, J., Choe, G.S., Chae, J., Jeong, H., Yang, G., Jing, J., Wang, H.: 2008, *Astrophys. J.* **686**, 1397.

Pevtsov, A.A., Canfield, R.C., Metcalf, T.R.: 1995, *Astrophys. J.* **440**, 109.

Reinard, A.A., Henthorn, J., Komm, R., Hill, F.: 2010, *Astrophys. J. Lett.* **710**, L121.

Sakurai, T., Hagino, M.: 2003, *Adv. Space Res.* **32**, 1943.

Scherrer, P.H., Schou, J., Bush, R.I., Kosovichev, A.G., Bogart, R.S., Hoeksema, J.T., Liu, Y., Duvall, T.L., Jr., Zhao, J., Title, A.M., *et al.*: 2012, *Solar Phys.* **275**, 207.

Schou, J., Scherrer, P.H., Bush, R.I., Wachter, R., Couvidat, S., Rabello-Soares, M.C., Bogart, R.S., Hoeksema, J.T., Liu, Y., Duvall, T.L., Jr., *et al.*: 2012, *Solar Phys.* **275**, 229.

Seehafer, N.: 1990, *Solar Phys.* **125**, 219.

Zhang, H., Sakurai, T., Pevtsov, A., Gao, Y., Xu, H., Sokoloff, D., Kuzanyan, K.: 2010, *Mon. Not. Roy. Astron. Soc.* **402**, L30.

Zhao, J., Couvidat, S., Bogart, R.S., Parchevsky, K.V., Birch, A.C., Duvall, T.L. Jr., Beck, J.G., Kosovichev, A.G., Scherrer, P.H.: 2012, *Solar Phys.* **275**, 375.

DOI 10.1007/978-1-4939-1182-0_6
Reprinted from *Solar Physics* Journal, DOI 10.1007/s11207-013-0339-z

A Full-Sun Magnetic Index from Helioseismology Inferences

I. González Hernández · M. Díaz Alfaro · K. Jain · W.K. Tobiska · D.C. Braun · F. Hill · F. Pérez Hernández

Received: 4 October 2012 / Accepted: 28 May 2013 / Published online: 9 July 2013

Abstract Solar magnetic indices are used to model the solar irradiance and ultimately to forecast it. However, the observation of such indices is generally limited to the Earth-facing hemisphere of the Sun. Seismic maps of the far side of the Sun have proven their capability to locate and track medium–large active regions at the non-visible hemisphere. We present here the possibility of using the average signal from these seismic far-side maps, combined with similarly calculated near-side maps, as a proxy to the full-Sun magnetic activity.

Keywords Helioseismology · Magnetic fields · Irradiance forecast

1. Introduction

Photospheric features of solar activity account for a large portion of the total solar irradiance (TSI) variation, with a superimposed modulation due to the solar cycle (Fröhlich, 1993). Magnetic indices related to photospheric activity, such as the Mount Wilson Plage Strength Index (MPSI) and the Mount Wilson Sunspot Index (MWSI: Parker, Ulrich, and Pap, 1998),

Solar Origins of Space Weather and Space Climate
Guest Editors: I. González Hernández, R. Komm, and A. Pevtsov

I. González Hernández (✉) · K. Jain · F. Hill
National Solar Observatory, Tucson, AZ 85719, USA
e-mail: irenegh@nso.edu

M. Díaz Alfaro · F. Pérez Hernández
Instituto de Astrofísica de Canarias, Tenerife 38205, Spain

W.K. Tobiska
Space Environment Technologies, Los Angeles, CA, USA

D.C. Braun
NorthWest Research Associates, Boulder, CO 80301, USA

F. Pérez Hernández
Universidad de la Laguna, Tenerife 38200, Spain

as well as magnetic activity proxies such as the 10.7 cm radio flux ($F_{10.7}$) (Covington, 1969; Kundu, 1965), the Mg II core-to-wing ratio (Viereck *et al.*, 2001) and the Lyman-alpha (Lyα) intensity (Woods *et al.*, 2000) have been traditionally used as input to the modeling and prediction tools for TSI (Tobiska, 2002; Jain and Hasan, 2004) and ultraviolet and extreme ultraviolet (UV/EUV) irradiance (Viereck *et al.*, 2001; Dudok de Wit *et al.*, 2008). However, at each particular point in time, these indices contain information of the magnetic activity only for the solar hemisphere facing the Earth.

The *Solar Wind Anisotropies* (SWAN) instrument, onboard SOHO, has been producing maps of backscattered solar (Lyα) radiation from the interplanetary medium since 1996 (Bertaux *et al.*, 1997). These data are the basis for forecasting several solar indices: $F_{10.7}$, Mg II, and the Lyα intensity (http://swan.projet.latmos.ipsl.fr), proving the potential of using far-side information.

For the last ten years, local helioseismology methods have provided a way to map medium-to-large active regions on the non-visible hemisphere of the Sun using information carried by waves that propagate all the way from the far side to the near side, where these waves are observed (Lindsey and Braun, 2000b; Zhao, 2007). The waves that pass through areas of strong magnetic fields experience a phase shift (Braun *et al.*, 1992) that is measurable when compared with a model representing waves that propagate between two points in the quiet photosphere. Maps of the difference between the model and the measured phase shift present large perturbations in those areas of concentrated magnetic field, such as active regions. Seismic maps calculated using this technique are currently used as another space weather forecasting tool and are available at http://gong.nso.edu/data/farside/ and http://sdo.gsfc.nasa.gov/data.

In this article, we introduce the idea of calculating a magnetic index that accounts for the whole solar surface, the Total Solar Seismic Magnetic Index (TSSMI) from the seismic signature of active regions. The TSSMI is an integrated value over both the observed and the non-visible hemispheres calculated by averaging the seismic signal (phase shift or travel-time difference) produced by surface magnetic activity, namely sunspots and plages, on the waves. This index is a proxy to the photospheric magnetic activity over the whole Sun and could be used to research long-term variations of the Sun without the limitations of the near-side-only observations. Furthermore, the far-side component of this index (Far Side Seismic Magnetic Index, FSSMI) can be used as input to solar irradiance forecast models.

2. Data Analysis

To compute the seismic index, we use near-side and far-side maps for the period of January 2002 to December 2005. The maps are calculated using the phase-sensitive seismic-holography technique as described by Lindsey and Braun (2000a) and Braun and Lindsey (2001). The technique is based on the comparison between the travel path determined for observed waves (as velocity field perturbations in the solar photosphere) compared with a model of the quiet Sun (Green's functions) for each location of the solar surface. When waves travel though an area of concentrated magnetic fields, they accelerate (Braun *et al.*, 1992), producing a phase shift with respect to the model. Hence, the seismic-holography maps show areas of large phase shift, which are interpreted as active regions.

Each far-side synoptic map is computed from a one-day series of 1440 *Global Oscillation Network Group* (GONG) Dopplergrams with a cadence of 60 s. If there are missing observations, the gaps are zero-filled. The Dopplergrams are taken in the photospheric line Ni I λ 6768 Å and have a spatial resolution of ≈ 5 arcsec (Harvey *et al.*, 1996). Each Dopplergram is Postel projected into a 200×200-pixel map. The maps are then stacked into a

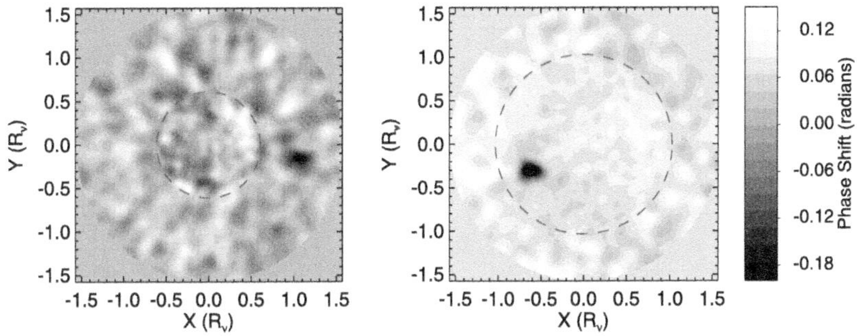

Figure 1 Left panel: Postel projection of the far-side map of the phase shift for 5 September 2005 showing the strong seismic signature of active region NOAA 10808 two days before it appeared at the east limb. Right panel: Postel projection of the helioseismic near-side map showing the same active region on 11 September 2005. The dotted line represents the center of the merged region which is different for the near- and far-side maps.

1440-frame datacube to which the technique as described by Lindsey and Braun (2000a) is applied. The resulting far-side image is itself a Postel projection. To reduce the errors in the calibration, only maps calculated from series with a clear-weather duty cycle greater than 80 % have been considered.

Because of geometrical limitations of the near-side observation area (pupil) combined with variable sensitivity with disk position, helioseismic-holography maps of the far side have been traditionally calculated in two parts: the central area, using the correlation between waves that bounce once at the surface (2-skip) before reaching the observation point, and the peripheral area, using the correlation between waves that do not bounce (1-skip) and bounce twice (3-skip) (Braun and Lindsey, 2001). During this research, we have followed this schema when calculating the far-side maps.

Different seismic imaging techniques have been used to create 3D structural and dynamic maps of active regions in the near side of the Sun for some time (*e.g.* Chou, 2000; Ilonidis, Zhao, and Kosovichev, 2012). They use shorter ray paths that provide more resolution and a plane-parallel approximation to reduce the computer load, but this approach allows only for small areas to be analyzed. For this work, we have created full-Sun maps by extending the same seismic-holography technique that computes the far-side to cover also the near-side. To keep consistency with the far-side calculations, we use waves following long ray paths and spherical geometry computations. Because of the geometrical limitations of the observational pupils, the near-side has been calculated also in two parts: the central part using 1×1-skip ray-path correlations, and the peripheral area, using 1×3-skip correlations. Mapping the observed side with a similar schema allows us to merge the front- and far-side inferences into a single full-Sun map that will be the basis for the calculation of the seismic magnetic index.

Figure 1 shows two seismic maps in Postel projection. The left panel presents the Postel projection of the far-side map of phase shift for 5 September 2005, showing the strong seismic signature of active region NOAA 10808 two days before it appeared at the east limb. The right panel is the corresponding Postel projection of the helioseismic near-side map, showing the same active region on 11 September 2005. Dark areas in the map correspond to negative phase signatures introduced by magnetic regions. The apparent difference in the latitudinal location of the active region in both maps is due to the B_0 angle (the angle of inclination of the solar equator on the ecliptic) being extreme during that period, which appears as a displacement in the y-coordinate in these Postel projected maps centered on

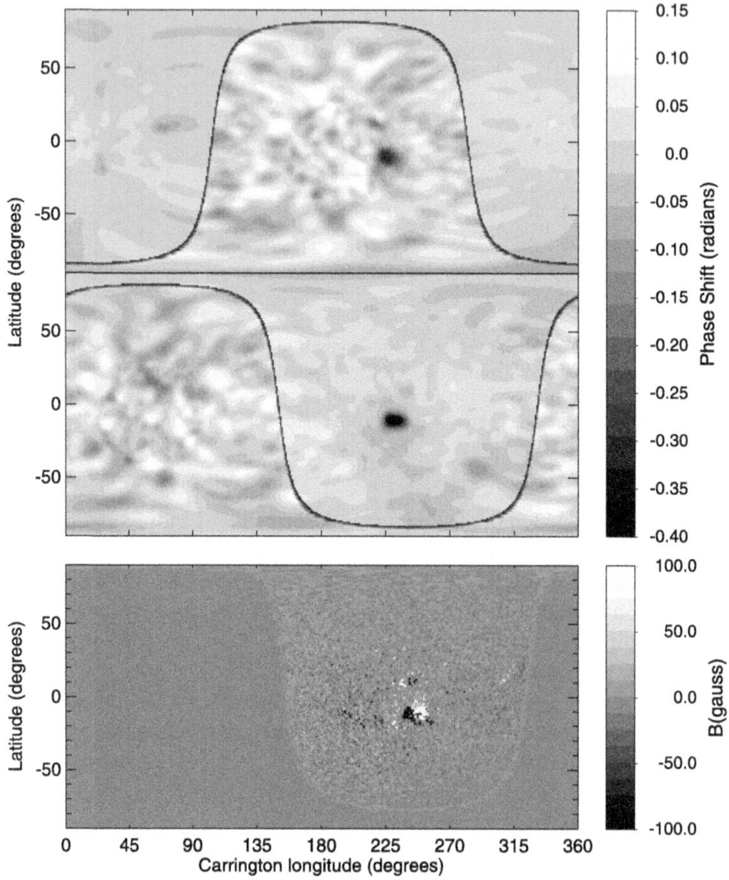

Figure 2 Longitude–latitude projection of the combined front–far side calculation into full-Sun maps. Top: 3 September 2005 map showing active region NOAA 10808. Middle: 13 September 2005 map of the same active region in the front side. Bottom: Longitude–latitude projection of the MDI magnetogram for 13 September 2005, showing the same active region as observed by this instrument onboard SOHO.

disk center. Because we use similar waves for the analysis, the resolution on both maps is similar. However, since the near-side map consist largely of 1×1-skip correlations, the noise is smaller. This is due to the dispersion that waves undergo when bouncing at the surface: the less bounces, the less noisy the results are. The dashed lines represent the separation between the central and the peripheral areas calculated using different wave combinations for each of the maps.

The near-side and far-side separated maps are then re-projected into a longitude–latitude grid and combined into a single map of the full Sun. Examples of full-Sun longitude–latitude projected maps are presented in Figure 2. Far-side and near-side areas are delimited by a dark line. The figures correspond to maps of the same region as in Figure 1 (NOAA 10808) but on different days, 3 September 2005 (top) and 13 September 2005 (middle). The bottom panel is a longitude-latitude projection of the *Michelson Doppler Imager* (MDI: Scherrer *et al.*, 1995) magnetogram for 13 September 2005, showing the same active region as observed by this instrument onboard SOHO.

3. A Near-Side Seismic Magnetic Index (NSSMI)

For validation purposes, we start calculating a near-side seismic index using only the near-side seismic maps. That way we can compare with the standard indices from the observable side of the Sun. The NSSMI is an average over the seismic signal (phase shift) of the near-side maps. We have arbitrarily ignored the positive phase shift in the maps and integrated only over negative values, associated with magnetic features. In addition, we integrate the value of the maps only from -40 to $+40$ degrees latitude, to avoid the noisier latitudes. The seismic far-side maps have been shown to be affected by a solar cycle variation, similar to the superimposed cycle modulation found in TSI measurements. In the case of the seismic maps, this cycle modulation has been associated to either global or localized structural changes in the solar convection zone (González Hernández, Scherrer, and Hill, 2009). To remove this variation each individual map (both near and far side) is corrected by removing a 60-day trailing average.

Figure 3 shows simultaneous values of the composite Lyα index from the Laboratory for Atmospheric and Space Physics (lasp.colorado.edu/lisird/tss), the calculated Mount Wilson sunspot (MWSI) and plage (MPSI) indices (www.astro.ucla.edu), the observed $F_{10.7}$ (www.ngdc.noaa.gov) and Mg II (www.swpc.noaa.gov/ftpmenu/sbuv.html) and the calculated NSSMI from the seismic-holography maps. The sequence spans from January 2002 to December 2005. Figure 4 presents a running mean of 10 days for both the seismic index and the standard solar indices to aid the comparison. A large increase in all of them can be seen associated with the large active regions that produced the Halloween flares at the end of 2003. The different indices have a marked correlation. The Pearson correlation coefficients between each of the traditional ones and the newly calculated seismic index can be found in Table 1 for both the one-day values and the ten-day smoothed series. The corresponding scatter plots are presented in Figure 5.

4. The Total Solar Seismic Magnetic Index (TSSMI)

To complete the full-Sun magnetic index, we follow the same procedure as described in the previous section but using the far-side helioseismic maps to calculate the far-side seismic magnetic index (FSSMI). Finally, we combine the two of them into the total solar seismic magnetic index (TSSMI). Figure 6 presents the three series, near-, far-side, and full-Sun indices.

The 27-day modulation due to the solar rotation is seen in both the near-side index and the far-side one. Two particular scenarios have been highlighted in the figure. Scenario A shows an increase of the integrated phase shift first on the front side index and then in the far side. The solid and dashed lines are plotted 14 days apart, to account for about half solar rotation. Scenario B is the reverse case, when we first have an increase in the far side that then moves to the front side. These are typical cases of active regions emerging in one of the hemispheres (front or far) and moving to the other one. In Scenario A, the active regions live long enough to come back to the front side a second time. The time lag between the near-side index and the far-side one is determined by the solar rotation, as the active regions move from the front to the far side of the Sun and *vice versa*. However, since active regions emerge and decay in a semi-random manner on both the front and far side of the Sun, trying to find a correlation between the two data sets to infer a time lag is not trivial, since it will be positive for certain periods, negative for others, and a complex mix for periods of high activity. The 27-day modulation still shows a residual in the full-Sun index, which is most probably due

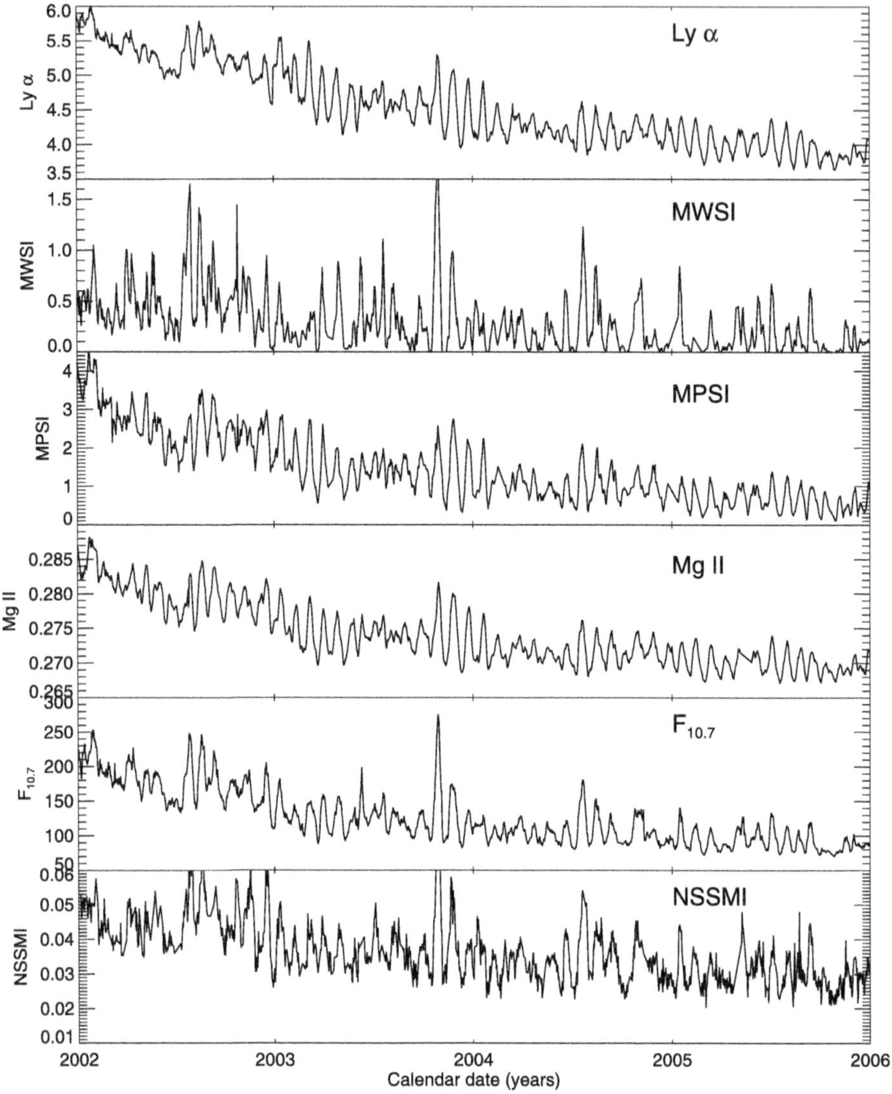

Figure 3 Simultaneous observations/calculations of the Lyα composite index, the Mount Wilson sunspot and plage indices, Mg II, $F_{10.7}$ and the near-side seismic magnetic index. The sequence spans from January 2002 to December 2005. The sign of the seismic magnetic index has been inverted to aid the comparison. The units of the different indices are: solar flux units for $F_{10.7}$, 10^{11} photons $cm^{-2}\,s^{-1}$ for Lyα, Gauss for the MPSI and MWSI and radians for the NSSMI.

to a combination of the location-dependent sensitivity of the helioseismic maps and the fact that the front- and far-side indices have been calculated using a different combination of waves, which introduces a phase difference between the two inferences.

Although the far-side maps are noisier than the near-side ones, the ten-day smooth indices obtained from both series seem to present a similar behavior and background, but with a positive phase shift for the FSSMI. For the purpose of this article, we have simply added

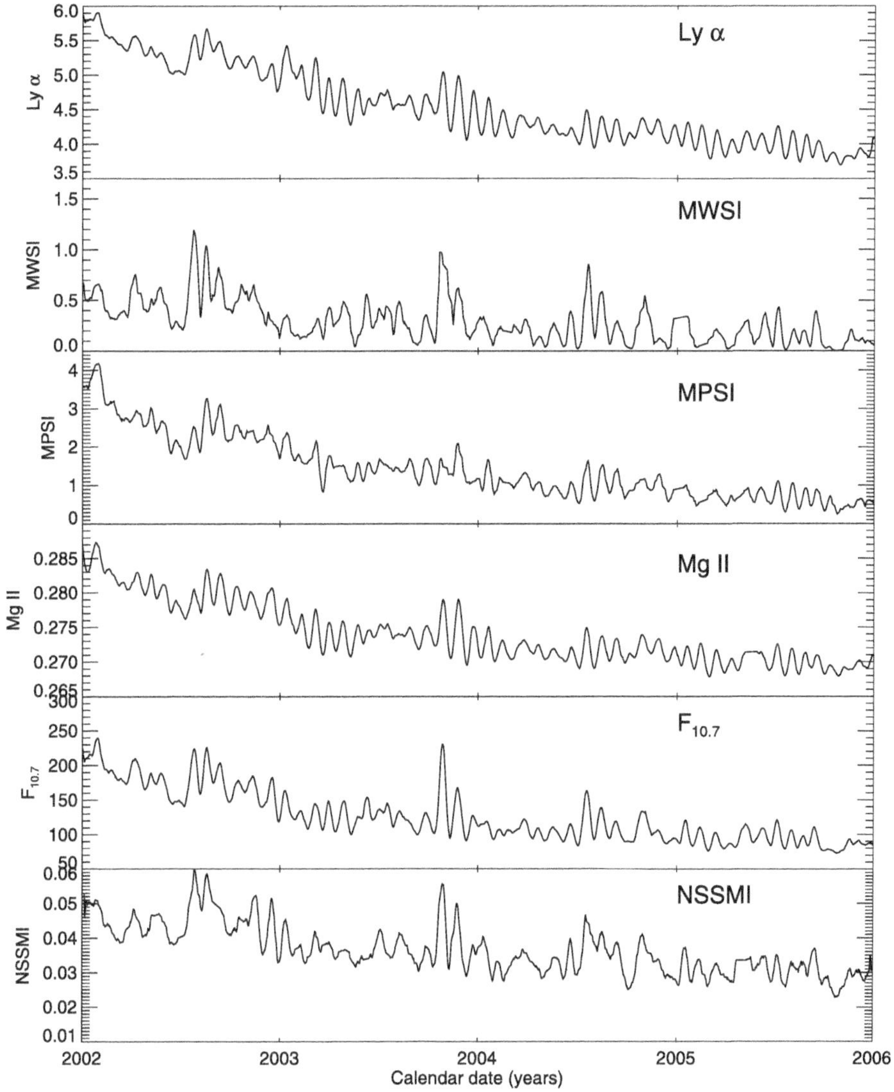

Figure 4 Ten-day smoothed averages of the simultaneous observations/calculations of the Lyα composite index, the Mount Wilson sunspot and plage indices, Mg II, $F_{10.7}$ and the near-side seismic magnetic index. The sequence spans from January 2002 to December 2005. The sign of the seismic magnetic index has been inverted to aid the comparison.

the near-side and the far-side values to compute the TSSMI. But the shift between the far-side and the near-side maps needs to be investigated, before the index is used as a proxy. A possible explanation comes from the empirical correction of the phase shift due to the superadiabatic layers at the near surface (Lindsey and Braun, 2000a). The different parts of the maps use waves which follow 1-skip, 2-skip or 3-skip paths, that is, they do not bounce, bounce one or twice before reaching the observation area; if the phase shift correction is not ideal, the different parts of the map will be affected differently.

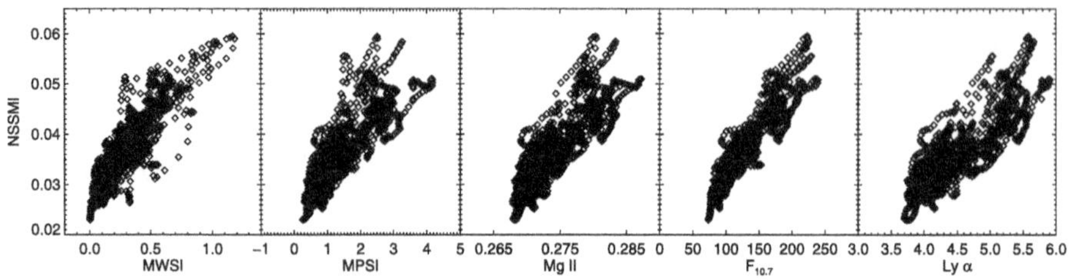

Figure 5 Scatter plots of the ten-day average NSSMI and each of the other front-side magnetic indices. A strong linear correlation can be observed in most of them, with the largest corresponding to that of $F_{10.7}$. The Pearson correlation coefficients are presented in Table 1.

Table 1 Pearson correlation coefficient between each of the standard front-side magnetic indices and the new seismic index.

Index	Correlation coeff. One-day values	Correlation coeff. Ten-day smoothed avg.
MWSI	0.86	0.87
MPSI	0.78	0.82
Mg II	0.79	0.84
$F_{10.7}$	0.89	0.92
Lyα	0.77	0.84

5. Discussion and Future Work

The comparison between the NSSMI and the standard near-side indices (see Figures 3, 4, and Table 1) shows the capability of seismic inferences to provide a new proxy for the solar magnetic activity. The helioseismology technique used to calculate far-side maps of the Sun, as well as the near-side maps calculated for this particular research, use waves that propagate following long ray paths. The sensitivity of such waves to small active regions is limited. This is due to the small perturbation that these active regions introduce in the wave propagation and the limited spatial resolution of the waves used in the analysis.

In general, we expect to find a larger correlation between the NSSMI and those indices that are sensitive to large, strong magnetic areas in the photosphere. The MPSI and the MWSI are calculated as the sum of the number of pixels with magnetic field intensity between 10 and 100 Gauss and larger than 100 Gauss and then divided by the total of number of pixels in the magnetogram (regardless of magnetic field strength), respectively. These are proxies of the strong photospheric magnetic fields, which are the main contributors to the phase shift of seismic inferences. Hence, we expect these indices to be closer to the NSSMI. The correlation with these two indices is high, however, we find a larger correlation with the $F_{10.7}$ index, which is the integrated emission from the solar disc at 2.8 GHz, *i.e.*, 10.7 cm wavelength, and serves as a proxy for the solar EUV emission. It mainly represents the contributions from sunspots and radio plages in the upper chromosphere in addition to the quiet-Sun background emission (Covington, 1969; Kundu, 1965).

A very strong correlation is also found with another chromospheric-related index, the Mg II. The Mg II core-to-wing ratio is derived by taking the ratio of the h and k lines of the solar Mg II feature at 280 nm to the background or wings at approximately 278 nm and 282 nm. The h and k lines are variable chromospheric emissions while the background

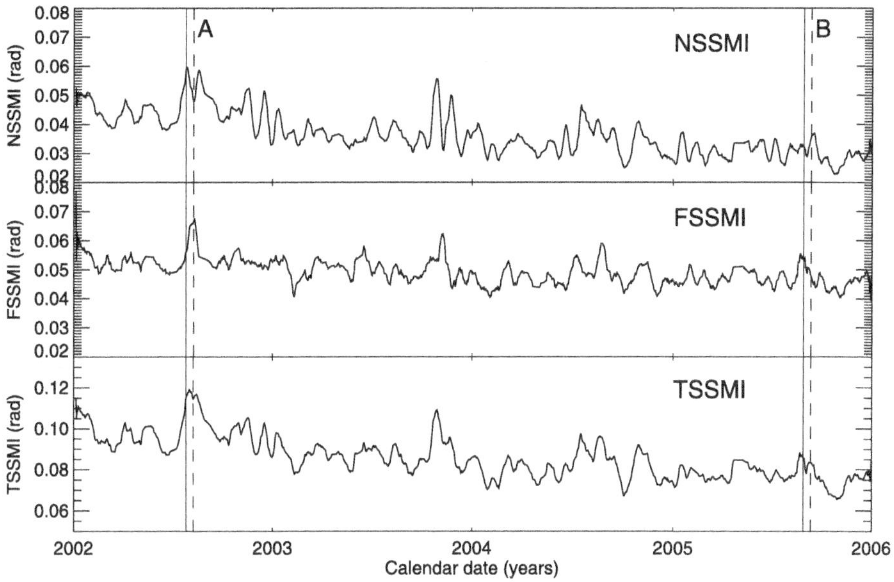

Figure 6 Values of the calculated near-side seismic magnetic index, the far-side seismic magnetic index and the combination of both into the total solar seismic magnetic index. The sequence spans from January 2002 to December 2005. A and B represents two scenarios where the index increases first in the near side and then in far side and *vice versa*, respectively.

emissions are more stable. This ratio seems to be a robust measure of chromospheric activity, mainly for solar UV and EUV emissions (Viereck *et al.*, 2001). At this point, we cannot explain these results, particularly given the fact that the quiet-Sun variation has been removed from the seismic maps; however, it provides the possibility of intercalibration between the near-side EUV indices and the seismic index. An in-depth comparison with longer series and filtering different areas of the seismic maps could help understand these results. The actual physical mechanism that produces the phase shift in the observed waves when they propagate through a strong magnetic field is still under investigation.

Finally, the correlation with the composite Lyα index is also reasonable given that Lyα emission serves as the best proxy for the irradiance originated from the transition region (Woods *et al.*, 2000; DeLand and Cebula, 2008).

In all cases, center-to-limb effects in the different standard proxies, combined with the location-dependent sensitivity of the seismic inferences, can contribute to lower the correlation. In fact, there may be an anticorrelation component with the emission-based proxies, since those incorporate a higher contribution of magnetic activity towards the limb, while the helioseismology inferences present stronger signatures close to disk center. New sets of artificial helioseismology data are currently available, which can be analyzed from different vantage points and will help to understand (and possibly correct for) the sensitivity function on the seismic maps.

The real potential of the seismic inferences is the ability to provide an index that comprises both the Earth-facing and the non-visible hemisphere solar magnetic activity (see Figure 6). The validation of the NSSMI *versus* the near-side indices means that the FSSMI, calculated with a very similar approach, provides useful information from an integrated perspective. The FSSMI contains information of the surface magnetic activity that will be facing Earth days in advance. In theory, solar models that use the integrated near-side indices as input to their calculations could use data provided by the far-side integrated index that would be facing Earth at a given time; so, if a far-side event is occurring, the Earth-facing part of the disk in, say six days, would be the relevant proxy for the six-day forecast. The part of the far-side disk that would be Earth-facing in ten-days would be the ten-day forecast and so on. It has been shown that short-term forecasting of UV irradiance can be improved when adding far-side seismic maps (Fontenla *et al.*, 2009). We expect that the integrated far-side seismic magnetic index will also help to complement the near-side indices for more accurate forecasting of the solar irradiance.

The TSSMI can be used for long-term studies of solar variability. The GONG project has been operating for almost 18 years, providing the capability of calculating the full-Sun seismic index for more than one solar cycle. So far, long-term studies of solar variability have been biased by the limitation of having only near-side observations. A comparison between long-term variation of the TSSMI and the standard near-side indices could provide insight as to have good the near-side indices have been at reproducing the full-Sun long-term variations. Any new knowledge in that respect will help to constrain the solar dynamo models. González Hernández, Scherrer, and Hill (2009) showed a variation of the wave propagation times through the solar interior that was highly correlated with the solar cycle, using integrated values of far-side seismic maps. We intend now to use the full-Sun approach to repeat the experiment, in an attempt to further understand the global *versus* local nature of the cycle dependence.

As stated before, the current seismic-holography method to calculate far-side and near-side maps using large ray paths has its limitations. So far, the technique is sensitive mainly to the contributions of medium-to-large size active regions. Two approaches are being investigated to improve the method, so it can account also for smaller active regions. The model used in helioseismic holography is being improved. Pérez Hernández and González Hernández (2010) developed a formalism that gives a more accurate representation of the wave field in the quiet Sun. Their preliminary results show a computational advantage in using these new Green's functions, as well as an increase of the signal to noise of $\sim 30~\%$. In addition, employing the 2×3-skip correlation seems to further increase the signal to noise in far-side seismic maps calculated using the time–distance technique (Zhao, 2007), we are currently exploring this possibility for seismic holography.

Future work includes the calibration of the full-Sun seismic magnetic index in terms of other solar indices/proxies like $F_{10.7}$, Mg II, and Lyα which, in turn, can be used as input to models such as the Jacchia–Bowman 2008 (JB2008; Bowman *et al.*, 2008) thermospheric density model used primarily for low Earth orbiting satellite drag calculations. No doubt other models that need full-Sun information will benefit from including different far-side data. The *Solar TErrestrial RElations Observatory* (STEREO) EUV images have been used to quantify the reliability of the seismic far-side maps (Liewer *et al.*, 2012), so integrated quantities from STEREO observations can also be used to validate the seismic index.

Acknowledgements The authors thank C. Lindsey for his extensive contribution to the NISP/GONG far-side pipeline and A. Malanushenko for the customization of part of the code. This work utilizes GONG data obtained by the NSO Integrated Synoptic Program (NISP), managed by the National Solar Observatory, which is operated by the Association of Universities for Research in Astronomy (AURA), Inc. under a co-operative agreement with the National Science Foundation. The data were acquired by instruments operated by the Big Bear Solar Observatory, High Altitude Observatory, Learmonth Solar Observatory, Udaipur Solar Observatory, Instituto de Astrofísica de Canarias, and Cerro Tololo Interamerican Observatory. SOHO is a project of international cooperation between ESA and NASA. This study includes data from the synoptic program at the 150-Foot Solar Tower of the Mt. Wilson Observatory. The Mt. Wilson 150-Foot Solar Tower is operated by UCLA, with funding from NASA, ONR, and NSF, under agreement with the Mt. Wilson Institute. We thank NOAA for the availability of the Mg II and $F_{10.7}$ data and LASP for the composite Lyα index. Part of the research has been supported by NASA grant NNH1AQ24I. Part of this work was supported by the Spanish National Research Plan under project AYA2010-17803.

References

Bertaux, J.L., Quémerais, E., Lallement, R., Kyrölä, E., Schmidt, W., Summanen, T., Goutail, J.P., Berthé, M., Costa, J., Holzer, T.: 1997, First results from SWAN Lyman-α solar wind mapper on SOHO. *Solar Phys.* **175**, 737. ADS:1997SoPh..175..737B, doi:10.1023/A:1004979605559.

Bowman, B.R., Tobiska, W.K., Marcos, F., Huang, C.: 2008, The thermospheric density model JB2008 using new EUV solar and geomagnetic indices. In: *37th COSPAR Scientific Assembly*, **37**, 367.

Braun, D.C., Lindsey, C.: 2001, Seismic imaging of the far hemisphere of the Sun. *Astrophys. J. Lett.* **560**, L189. ADS:2001ApJ...560L.189B, doi:10.1086/324323.

Braun, D.C., Duvall, J.T.L., Labonte, B.J., Jefferies, S.M., Harvey, J.W., Pomerantz, M.A.: 1992, Scattering of p-modes by a sunspot. *Astrophys. J.* **391**, 113. ADS:1992ApJ...391L.113B. doi:10.1086/186410.

Chou, D.-Y.: 2000, Acoustic imaging of solar active regions (Invited Review). *Solar Phys.* **192**, 241. ADS:2000SoPh..192..241C.

Covington, A.E.: 1969, Solar radio emission at 10.7 cm, 1947–1968. *J. Roy. Astron. Soc. Can.* **63**, 125. ADS:1969JRASC..63..125C.

DeLand, M.T., Cebula, R.P.: 2008, Creation of a composite solar ultraviolet irradiance data set. *J. Geophys. Res.* **113**, A11103. ADS:2008JGRA..11311103D, doi:10.1029/2008JA013401.

Dudok de Wit, T., Kretzschmar, M., Aboudarham, J., Amblard, P.-O., Auchère, F., Lilensten, J.: 2008, Which solar EUV indices are best for reconstructing the solar EUV irradiance? *Adv. Space Res.* **42**, 903. ADS:2008AdSpR..42..903D, doi:10.1016/j.asr.2007.04.019.

Fontenla, J.M., Quémerais, E., González Hernández, I., Lindsey, C., Haberreiter, M.: 2009, Solar irradiance forecast and far-side imaging. *Adv. Space Res.* **44**, 457. ADS:2009AdSpR..44..457F, doi:10.1016/j.asr.2009.04.010.

Fröhlich, C.: 1993, Relationship between solar activity and luminosity. *Adv. Space Res.* **13**, 429. ADS:1993AdSpR..13..429F, doi:10.1016/0273-1177(93)90515-D.

González Hernández, I., Scherrer, P., Hill, F.: 2009, A new way to infer variations of the seismic solar radius. *Astrophys. J. Lett.* **691**, L87. ADS:2009ApJ...691L..87G, doi:10.1088/0004-637X/691/2/L87.

Harvey, J.W., Hill, F., Hubbard, R.P., Kennedy, J.R., Leibacher, J.W., Pintar, J.A., Gilman, P.A., Noyes, R.W., Title, A.M., Toomre, J., Ulrich, R.K., Bhatnagar, A., Kennewell, J.A., Marquette, W., Patron, J., Saa, O., Yasukawa, E.: 1996, The Global Oscillation Network Group (GONG) project. *Science* **272**, 1284. ADS:1996Sci...272.1284H, doi:10.1126/science.272.5266.1284.

Ilonidis, S., Zhao, J., Kosovichev, A.: 2012, Detection of emerging sunspot regions in the solar interior. *Science* **333**, 993. ADS:2012Sci...336..296I, doi:10.1126/science.1215539.

Jain, K., Hasan, S.S.: 2004, Modulation in the solar irradiance due to surface magnetism during cycles 21, 22 and 23. *Astron. Astrophys.* **425**, 301. ADS:2004A&A...425..301J, doi:10.1051/0004-6361:20047102.

Kundu, M.R.: 1965, *Solar Radio Astronomy*, Interscience, New York.

Liewer, P.C., González Hernández, I., Hall, J.R., Thompson, W.T., Misrak, A.: 2012, Comparison of far-side STEREO observations of solar activity and active region predictions from GONG. *Solar Phys.* **281**, 3. ADS:2012SoPh..281....3L, doi:10.1007/s11207-012-9932-9.

Lindsey, C., Braun, D.C.: 2000a, Basic principles of solar acoustic holography (Invited Review). *Solar Phys.* **92**, 261. ADS:2000Sci...287.1799L, doi:10.1126/science.287.5459.1799.

Lindsey, C., Braun, D.C.: 2000b, Seismic images of the far side of the Sun. *Science* **287**, 1799. ADS:2000SoPh..192..261L.

Parker, D.G., Ulrich, R.K., Pap, J.M.: 1998, Modeling solar UV variations using Mount Wilson Observatory indices. *Solar Phys.* **177**, 229. ADS:1998SoPh..177..229P.

Pérez Hernández, F., González Hernández, I.: 2010, Green's functions for far-side seismic images: a polar-expansion approach. *Astrophys. J.* **711**, 853. ADS:2010ApJ...711..853P, doi:10.1088/0004-637X/711/2/853.

Scherrer, P.H., Bogart, R.S., Bush, R.I., Hoeksema, J.T., Kosovichev, A.G., Schou, J., Rosenberg, W., Springer, L., Tarbell, T.D., Title, A., Wolfson, C.J., Zayer, I., MDI Engineering Team: 1995, The solar oscillations investigation – Michelson Doppler Imager. *Solar Phys.* **162**, 129. ADS:1995SoPh..162..129S. doi:10.1007/BF00733429.

Tobiska, W.K.: 2002, Variability in the solar constant from irradiances shortward of Lyman-alpha. *Adv. Space Res.* **29**, 1969. ADS:2002AdSpR..29.1969T, doi:10.1016/S0273-1177(02)00243-0.

Viereck, R., Puga, L., McMullin, D., Judge, D., Weber, M., Tobiska, W.K.: 2001, The Mg II index: a proxy for solar EUV. *Geophys. Res. Lett.* **28**, 1343. ADS:2001GeoRL..28.1343V, doi:10.1029/2000GL012551.

Woods, T.N., Tobiska, W.K., Rottman, G.J., Worden, J.R.: 2000, Improved solar Lyman-alpha irradiance modeling from 1947 through 1999 based on UARS observations. *J. Geophys. Res.* **105**, 27. ADS:2000JGR...10527195W, doi:10.1029/2000JA000051.

Zhao, J.: 2007, Time–distance imaging of solar far-side active regions. *Astrophys. J. Lett.* **664**, L113. ADS:2007ApJ...664L.139Z, doi:10.1086/520837.

DOI 10.1007/978-1-4939-1182-0_7
Reprinted from *Solar Physics* Journal, DOI 10.1007/s11207-013-0431-4

Far- and Extreme-UV Solar Spectral Irradiance and Radiance from Simplified Atmospheric Physical Models

J.M. Fontenla · E. Landi · M. Snow · T. Woods

Received: 23 August 2013 / Accepted: 8 October 2013 / Published online: 15 November 2013

Abstract This article describes an update of the physical models that we use to reconstruct the FUV and EUV irradiance spectra and the radiance spectra of the features that at any given point in time may cover the solar disk depending on the state of solar activity. The present update introduces important modifications to the chromosphere–corona transition region of all models. Also, the update introduces improved and extended atomic data. By these changes, the agreement of the computed and observed spectra is largely improved in many EUV lines important for the modeling of the Earth's upper atmosphere. This article describes the improvements and shows detailed comparisons with EUV/FUV radiance and irradiance measurements. The solar spectral irradiance from these models at wavelengths longer than ≈ 200 nm is discussed in a separate article.

Keywords UV radiation · Transition region · Corona

1. Introduction

Spectra and images of the solar disk are used by Fontenla *et al.* (1999, 2009a, 2011, hereafter Articles 1, 2, and 3, respectively) to construct a set of models for the characteristic features of the solar atmosphere present on the disk throughout the solar cycle. These features are

Solar Origins of Space Weather and Space Climate
Guest Editors: I. González Hernández, R. Komm, and A. Pevtsov

J.M. Fontenla (✉)
NorthWest Research Associates Inc., Boulder, CO, USA
e-mail: jfontenla@nwra.com

E. Landi
Department of Atmospheric, Oceanic and Space Sciences, University of Michigan, Ann Arbor, MI, USA

M. Snow · T. Woods
Laboratory for Atmospheric and Space Physics, University of Colorado, Boulder, CO, USA

described by models of their physical atmospheric structures that produce radiance (*i.e.* emergent intensity) spectra that agree with the available observations. The complete radiance spectra (currently 0.1 nm to 300 micron) from each type of structure are then weighted using solar images and added to yield the total solar radiation in the direction of Earth, and thereby the solar spectral irradiance (SSI) at 1 AU. The spectra are computed at variable, extremely high, spectral resolution; they contain at least ten points covering every spectral line. The spectral resolution in narrow lines reaches $\Delta\lambda/\lambda \approx 2 \times 10^{-6}$, and in the continuum a lower resolution is used. In order to compare our computations with available observations, the computed spectra are degraded to match the resolution of the observations with various instruments. For the calculations, the Solar Radiation Physical Modeling (SRPM: Fontenla, Balasubramaniam, and Harder, 2007) set of tools is used.

This article compares calculations with key spectral features observed from space and shows that the present simple (one-dimensional, steady) models of mid-resolution areas of the solar surface, $\approx 2 \times 2$ arcsec^2, provide a fairly good representation of the SSI. Fine details of the spatial structure, *e.g.* loop structures, or of the spectral-line profiles often involve velocities that cannot be explained by these simple models. These details are important for understanding the internal physical structure of each individual feature but are not yet well understood. However, these details are of little importance for the applications to the Earth's upper-atmosphere modeling. The models presented here are a useful approximation to the averaging of the emission over those relatively small patches that are frequently observed in images of the full solar disk.

Our current scheme for determining the features present on the solar disk uses UV images from the *Atmospheric Imaging Assembly* (AIA) instrument onboard the *Solar Dynamics Observatory* (SDO) at several wavelengths. However, *Solar Heliospheric Observatory* (SOHO) images from the *Extreme ultraviolet Imaging Telescope* (EIT) instrument can easily be used as we did in the past, and *Solar TErrestrial RElationship Observatory* (STEREO) *Extreme Ultraviolet Imager* (EUVI) can also be used. For historical reconstructions, the Ca II K images can be used, although these do not directly provide information on the high-temperature components responsible for the UV and thus require some speculative assessment. We currently use both chromospheric and coronal images to discriminate between the features on the disk and be able to assign these features to atmospheric models. When the near-UV (NUV) and visible continuum are relevant, one should also use photospheric images to discriminate sunspot umbrae and penumbrae. Regarding sunspots, the distinction of, at least, these two features is very important because they have very different spectra. Also, the relative extent of umbra and penumbra are not well correlated and individual observations are necessary to determine their areas. However, in the present article we will not study sunspots because they are of little relevance for the range of wavelengths studied here.

Table 1 shows the indices of the models that we consider in the present as in Article 3 and partially in previous articles, the letter designations correspond to the solar features and a summary description of the feature is given. In this table sunspot models are omitted.

The behavior of the models temperature *vs.* height for layers below the chromosphere–corona transition region are not significantly changed from those given by Article 3, except to smooth out some irregularities. However, the addition of more species and levels produced some differences in the electron density and thereby in the density stratification that is computed, as in Article 3, using hydrostatic pressure equilibrium with the same additional acceleration term as in that article. This additional term is an upward acceleration that increases the height-scale for better matching the observations at the limb.

In the transition region and corona, the improvements in the atomic data along with an improved treatment of the energy balance introduce important changes in the temperature *vs.* height. The use of CHIANTI 7.1 (Landi *et al.*, 2013) electron-collision excitation

Table 1 Current models for the solar disk features.

Feature	Description	Photos.–chromos. model index	Corona model index
A	Dark quiet-Sun inter-network	1300	1310
B	Quiet-Sun inter-network	1301	1311
D	Quiet-Sun network lane	1302	1312
F	Enhanced network	1303	1313
H	Plage (that is not facula)	1304	1314
P	Facula (*i.e.* very bright plage)	1305	1315
Q	Hot Facula	1308	1318

and ionization rates largely improves the agreement with observations of many computed lines.

In the following, these changes are described and the resulting EUV/FUV spectra are compared with available observations of irradiance and radiance. UV spectral irradiance data are provided by the *Solar Stellar Irradiance Comparison Experiment* (SOLSTICE) instrument onboard the *Solar Radiation and Climate Experiment* (SORCE) satellite, and by the *EUV Variability Experiment* (EVE) instrument onboard the SDO satellite.

2. The Non-LTE Computations and Atomic Data

In the current calculation we have increased the number of levels for which full-NLTE is computed in detail and included more recent atomic data for many species. The improvements for the low ionization species, *e.g.* C, N, O, Si, Mg, at ionization stages II and III, now including collisional excitation data from CHIANTI 7.1, are more accurate than those used in Article 3 and earlier, which were derived according to the Seaton (1962) approximation. For several neutral species, we are now considering more levels. As a consequence of the increased number of levels, many lines that were not previously computed are now present in our spectrum.

For 198 higher ionization species, more levels and lines are now included in CHIANTI version 7.1 data. The complete data for the levels and transitions (with the only exception of proton collisional excitation) from this source have been used in the computation of the level populations using the effectively thin approximation. This approximation assumes that for most directions and important wavelengths the lines are optically thin, although for some paths (*e.g.* very close to the limb) this may not be the case. Also, this approximation assumes no optical pumping or interaction between transitions from different species, and is described in Article 3. Therefore, the effectively optically thin approximation in this article solves simultaneously the statistical equilibrium of all the level populations for each species at each height by including collisional rates and neglecting radiative rates other than spontaneous decay rates. The SRPM codes allow for including in the effectively thin approximation the radiative-transition rates driven by a prescribed illumination, *e.g.* external illumination, but these are not important for the layers and species considered here in which collisional transition and spontaneous-decay rates dominate. Also, the ionization of these species is computed according to the CHIANTI 7.1 data and procedure.

Table 2 shows the number of levels for the species computed here in full NLTE. These full NLTE calculations include low ionization stages in which it is estimated that the effectively

Table 2 Species currently computed with full NLTE radiative transfer.

Element	Abundance	Ion I levels	Ion II levels	Ion III levels
H	1.0	25	–	–
He	0.1	20	15	–
C	2.4×10^{-4}	45	27	*
N	0.9×10^{-4}	26	33	*
O	3.9×10^{-4}	23	31	*
Ne	6.92×10^{-5}	80	57	*
Na	1.48×10^{-6}	22	14	*
Mg	3.39×10^{-5}	26	14	54
Al	2.34×10^{-6}	18	14	32
Si	3.24×10^{-5}	35	14	60
S	6.92×10^{-6}	20	30	*
Ar	1.52×10^{-6}	48	57	*
K	1.20×10^{-7}	10	*	*
Ca	2.04×10^{-6}	22	24	34
Ti	7.94×10^{-8}	116	78	43
V	1.00×10^{-8}	120	41	40
Cr	4.36×10^{-7}	102	34	20
Mn	2.45×10^{-7}	85	40	28
Fe	2.82×10^{-5}	119	120	90
Co	8.32×10^{-8}	65	28	50
Ni	1.70×10^{-6}	61	28	40

* Indicates species that were computed using the effective optically thin approximation. Higher ionization of these and other elements was also computed using the CHIANTI 7.1 data.

thin approximation would not be accurate. Some species only become ionized to stages higher than II (*i.e.* charge more than one) within the lower transition region and therefore for them the effectively thin approximation is sufficient. This is the case of C III, N III, O III, and S III, which are now calculated in the effectively thin approximation but were computed in full-NLTE in previous articles.

Elemental ionization is calculated in full NLTE for the species as indicated in Table 2 with a valid number of levels. For these species, radiative interaction is expected to occur and the optically thin approach is not accurate. For the species where a valid number of levels in Table 2 is not specified, the effectively thin approximation using data in CHIANTI version 7.1 is used.

Collision rates from CHIANTI 7.1 were used, when available, for many species listed in Table 2 and all those not listed in the table and those not listed there. For the transitions not listed in CHIANTI the Seaton approximation was used to determine the collision excitation rates. In some cases, *e.g.* C I, N I, O I, and S I, some data were taken from those available in the literature as indicated in Article 3 and earlier articles. Collisional excitation by proton data are available for many species in CHIANTI 7.1 but after checking some cases it was found that its importance is secondary, at least in the tested cases.

Another improvement in the present calculations was the change of the photo-ionization cross-section for the first excited level of S I. This cross-section was previously assumed to be rather large from extrapolation of TOPBASE data (Seaton, 1987). However, we now estimate it to be smaller and adopted 6.0×10^{-18} cm^2 for its value at the head. The new value solves the issue with the computed edge at ≈ 134.55 nm, which was overestimated before.

There are still some issues with other FUV continua with edges at wavelengths longer than that of the Ly continuum. These issues are discussed in Section 5 below.

The calculations are carried out using the Solar Radiation Physical Modeling tools, now in version 2.2 (SRPMv2.2). These tools allow considering height-dependent abundance variations; however, our present calculations show no need to impose such variations. The present models use the same abundances throughout the photosphere/chromosphere/transition-region/low-corona and produce good results for the SSI reconstruction. Effects of uncertainties in ionization, *e.g.* due to velocities, or in atomic data or atmospheric models, can very well explain some particular lines that may not be well represented in our results. Of course abundance variations are to be expected above the layers that we consider because mixing may not be very effective there, *e.g.* within the solar wind depending on its acceleration mechanism. In this article we use the abundance values listed in Table 2, which are the same as in Article 3.

For the species with a valid number of levels in Table 2 the calculations were carried out assuming that the indicated levels contain sublevels of different quantum number J. The sublevels are assumed to be populated in LTE within each level. The transitions between levels are computed considering the fully resolved fine structure of the lines that compose these transitions.

The collisional data used to calculate C I non-LTE were taken from the recent calculations by Wang, Zatsarinny, and Bartschat (2013). The authors provided collision strengths for the lowest 22 levels in LS coupling, which correspond to 42 fine-structure levels (or sublevels as in the paragraph above). Given the much larger atomic model, the calculations by Wang, Zatsarinny, and Bartschat (2013) are likely more accurate than the earlier calculations using a smaller atomic model. Comparison with earlier calculations and with other calculations by the authors with more limited atomic models show differences. For the transitions involving levels above those computed by Wang *et al.*, we use collision rates following the Seaton formula. C I is a particularly complicated atom with many significant lines from high levels, and our calculations include 87 sublevels. However, these new C I data were not yet fully considered for calculating the autoionization and dielectronic recombination of C I, which is also importantly affected by the Lyman continuum and Ly α.

3. Structure of the Transition Region and Energy Balance

The updates of the atomic data mentioned in the previous section largely improve the comparison with observations of lines formed at temperatures $T < 10^5$ K, and lines formed at, and above, $T = 10^6$ K.

Transition-region lines formed at temperatures in the range $10^5 < T < 10^6$ K, *e.g.* of O IV and V and N IV and V in the range 40 – 80 nm, were overestimated by Article 3. The improvements in atomic data mentioned in Section 2 do not solve this issue, but additional considerations are needed. If the same energy-balance scheme of Article 3 were used, these transition-region lines would be overestimated by factors ranging from two to ten.

These lines show that the energy-balance transition-region calculation used in Article 3 produces too smooth a temperature increase with height. The transition region computed in Article 3 uses the Fontenla *et al.* (1990, 1991, 1993, hereafter FAL 1, 2, and 3, respectively) formulation with a unit filling factor and considers the Spitzer thermal conductivity of electrons, in addition to the energy transported by particle diffusion described in FAL which is negligible in the temperature range $T \geq 10^5$ but important at lower temperatures. This calculation yields too large an energy downflow, which the radiative losses from the observed spectrum cannot dissipate. The Spitzer (1962) electron heat conduction in the transition region at temperatures above $\approx 10^5$ K leads to very large emission compared with the observations.

Since the spectrum calculated assuming energy balance and using Spitzer conductivity is already larger that the observed in the temperature range $10^5 - 10^6$ K, but it matches the observed outside this range, any additional mechanical energy dissipation (while maintaining the Spitzer conductivity) would only produce even larger emission. Such larger emission would be required to maintain energy balance but would worsen the comparison of computations with observations.

Considering a filling-factor smaller than one leads to underestimation of the line emissions produced at temperatures below 10^5 K. Only an extremely sharp change in the filling factor at $T \approx 10^5$ K, from a value near unity in the part of the transition region below to a small value above, could reconcile the theory with the observations. We do not know of any observational evidence indicating such an abrupt change in filling factor within the transition region.

We found an explanation that is physically consistent with the observations and with energy balance in the transition region that consists in a reduction of the heat conduction from the value given by the Spitzer formula. Such a reduction was observed in many laboratory experiments, *e.g.* Brysk, Campbell, and Hammerling (1975), and was named "flux limiting". We considered the flux-limiting issues and found that these are relevant to the chromosphere–corona transition region for most of our models although of course the parameters of the plasma are not the same as in these laboratory experiments. Flux-limiting occurs in laboratory experiments when the fast electrons that supply the Spitzer formula electron conductivity cannot carry sufficient energy for energy balance, or when the Knudsen number describing the ratio of their free-path to the characteristic distance of temperature variation is not small enough. Note that these two reasons are just different ways of looking at the same issue of physical constraints on the applicability of the Spitzer formula. These conditions can easily occur in the solar atmosphere due to the low electron densities in the corona and the very steep temperature gradient in the chromosphere–corona transition region that the observations indicate.

Although improvements in the atomic data and in the ionization equilibrium may still be needed and elemental abundance variations may exist, only the reduction of the electron-conduction heat flux can solve the issue presented by *all* emission lines in the upper-transition-region and still remain consistent with those at lower temperatures. Such a reduction is needed to match the observed spectra and is also justified by basic physical considerations that are similar to those found in laboratory experiments.

In the quiet-Sun models the temperature gradient, and hence the heat flux, is smaller than in active regions but the density is also lower and therefore the mean free path of electrons becomes large. In the active-region models, the increased density shortens the mean free path but the temperature gradient is also larger. The physically consistent behavior resulting from consideration of flux limiting can only be described by a global calculation of the electron distribution function in the transition region (*e.g.* MacNeice, Fontenla, and Ljepojevic, 1991) but requires knowledge of the distribution function in the corona, which is likely non-Maxwellian. In principle, this distribution could be computed by taking into account the processes that supply the electron energy in the corona and sinks due to local dissipation and transport. Work on thermal conduction has been carried out in the context of the laboratory target preheating in laser-driven target implosions by Schurtz, Nicolai, and Busquet (2000) and references therein, and in the context of the solar wind by Bale *et al.* (2013) and references therein. An in-depth discussion of the validity of the Spitzer conductivity flux-limiting approaches has been given by Catto and Grinneback (2000).

Because a global calculation is outside the scope of the present article, and because we do not know the electron distribution function in the corona, which determines the downward transport through the transition region, we use here the following *ad-hoc* formula, which produces the physical asymptotic limits for very large and very small temperature gradients and is similar to that suggested by flux-limiting considerations:

$$K_{\mathrm{eff}} = K_{\mathrm{a}} + K_{\mathrm{i}} + K_{\mathrm{SH}}/\bigl(1 + \alpha F(T, n_{\mathrm{e}}) Z_T\bigr), \tag{1}$$

where K_{eff} is the effective energy transport coefficient that multiplied by the logarithmic gradient of the temperature [Z_T] gives the total heat flux. K_{a} and K_{i} give the heat-flux coefficient due to atoms conduction and diffusion carrying of ionization energy (see FAL articles), and K_{SH} is the electron conductivity given by the Spitzer (1962) formula, α is a constant coefficient, and $F(T, n_{\mathrm{e}})$ is the following function:

$$F(T, n_{\mathrm{e}}) = 2.7 \times 10^6 T^2 n_{\mathrm{e}}^{-1}; \tag{2}$$

all quantities are in cgs units. In this formula, the term $\alpha F Z_T$, where Z_T is the magnitude of the temperature gradient, parametrically describes the global issues. There are no compelling physical reasons for choosing the parametric form in Equation (2) over others, neither is there any good physical reason for picking one value of the parameter α over another value. We tried different values of α and concluded that a value of 7.0 produces the best agreement between the computed and observed line intensities. Thus, the parametric form and the value of the parameter α are defined empirically by matching the observations, but we note that the adopted function F is proportional to the Knudsen number.

There are further complications in the calculations arising from two issues. The first issue is that the flux-limiting mentioned above should only be applied to the electron conduction using the Spitzer formula, but it should not affect the H and He diffusion and therefore does not affect the ionization energy transport (*i.e.* the reactive energy transport; see FAL articles about these other transport processes). When flux limiting is applied to the electron conduction, the relative importance of this reactive transport becomes even larger. This complicates the calculations at the lower temperatures and we address it by an iterative procedure. Because of uncertainty in ionized He diffusion, here we just established the lower-boundary condition of our transition-region energy balance arbitrarily at $T \approx 6 \times 10^4$ K.

The second issue in the calculation is that a too large a flux-limiting function [F] can lead to a situation in which even an infinite (or extremely large) temperature gradient at one layer could be insufficient to provide for the energy dissipated below. We have used an *ad-hoc* approach to deal with this problem, which arises at high temperatures because our computed value for F increases steeply with increasing temperature and decreasing density. Of course, this issue is not a physical problem for energy transport in a general case, but only a problem for the flux-limiting approach. In reality, the heat flux could become highly nonlocal and only dependent on the global temperature differences but not on the temperature gradient at a particular location. In the present article we use an *ad-hoc* modification to the function F to avoid this problem.

Table 3a gives the values of logarithmic temperature gradient, free-streaming heat flux (see Wyndham *et al.*, 1982), Spitzer-formula heat flux, and the flux-limited heat flux in our calculation for each model at the layer of $T = 2 \times 10^5$ K. Most of the heat flux shown in that table produces emission by the lower part of the transition region and only a part of this flux reaches the region where the Ly α line center is formed. In our calculations for all models the flux-limited heat flux is about 0.18 of that given by the Spitzer formula, corresponding to $\alpha F Z_T \approx 4.4$. This is a result of the calculations and was not imposed.

Figure 1 shows the effect on the temperature structure of the lower part of the transition region of the model for feature B (network cell interior, see Table 1) using different values

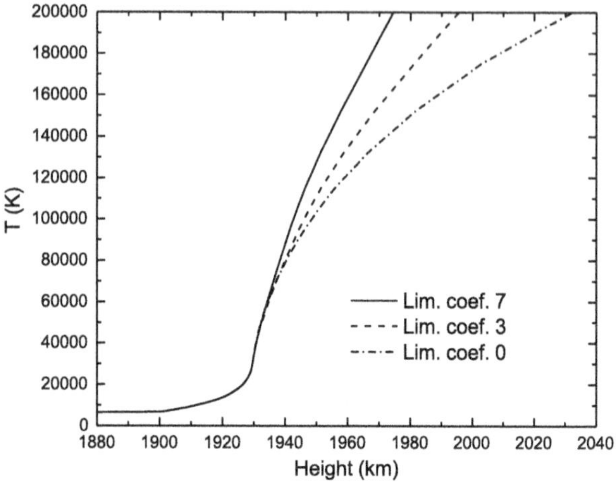

Figure 1 The lower part of the transition region in model 1301, see Table 1, calculated using different values of the flux-limiting coefficient [α]. The value 0 of the coefficient corresponds to the strict Spitzer formula for the electron conduction.

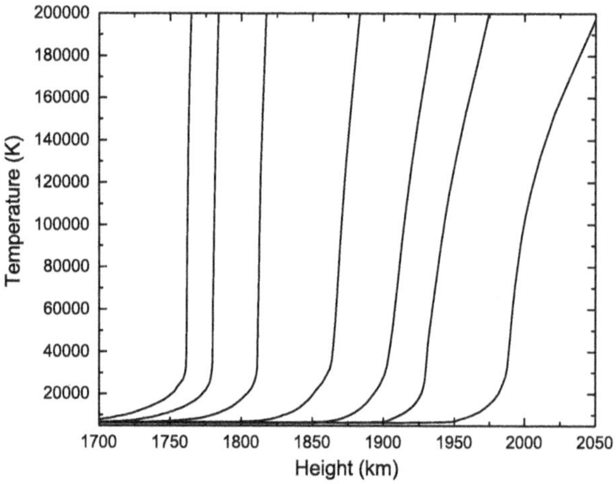

Figure 2 The lower part of the transition region for all models of the set indicated in Table 1, left to right models 1308, 1305, 1304, 1303, 1302, 1301, 1300.

of the flux-limiting parameter [α]. As the figure shows, increasing this parameter produces a steepening of the transition region, which does not affect the lower temperatures below say 8×10^4 K. The steepening effect also occurs in the upper part of the transition region (not shown in this figure) and significantly affects many lines throughout the spectrum bringing the calculated lines closer to the observed, *e.g.* the Ne VI 40.193 nm.

Note that, as the flux limiting becomes important, the total energy dissipated in the transition region by the electron heat-conduction decreases because the extension of this region decreases. The temperature gradient becomes stronger even when the heat flux decreases with respect to the non-flux-limited situation. All of these changes reduce the line emission when the flux-limiting parameter is increased. In this way flux limiting corrects the overestimation of the transition-region lines, bringing the calculations to better agreement with the observations than those of Article 3.

Figure 2 shows our final lower transition-region models that use limiting factor $\alpha = 7.0$. These are adopted in the remainder of this article for the various models. Tables 3a and 3b show the parameters in the current models for the point where the lower-layers models (photosphere/chromosphere/lower-transition-region) join the upper transition-region/low-

Table 3a Temperature gradient, heat flux at 2×10^5 K, and radiative flux above this level.

Model	grad ln T [$\mathrm{cm^{-1}}$]	F_{lim} [$\mathrm{erg\,cm^{-2}\,s^{-1}}$]	$-F_{\mathrm{s}}$ [$\mathrm{erg\,cm^{-2}\,s^{-1}}$]	$-F_{\mathrm{h}}$ [$\mathrm{erg\,cm^{-2}\,s^{-1}}$]	F_{r} [$\mathrm{erg\,cm^{-2}\,s^{-1}}$]
13x0	6.0×10^{-8}	1.0×10^{7}	2.2×10^{5}	4.0×10^{4}	8.4×10^{4}
13x1	1.1×10^{-7}	1.7×10^{7}	3.8×10^{5}	7.0×10^{4}	1.7×10^{5}
13x2	1.3×10^{-7}	2.6×10^{7}	4.7×10^{5}	9.0×10^{4}	3.1×10^{5}
13x3	2.7×10^{-7}	4.4×10^{7}	9.6×10^{5}	1.5×10^{5}	7.1×10^{5}
13x4	5.1×10^{-7}	8.7×10^{7}	1.8×10^{6}	3.4×10^{5}	2.2×10^{6}
13x5	8.6×10^{-7}	1.4×10^{8}	3.1×10^{6}	5.8×10^{5}	4.1×10^{6}
13x8	1.2×10^{-6}	2.0×10^{8}	4.3×10^{6}	8.0×10^{5}	6.0×10^{6}

Table 3b Other parameters at 2×10^5 K.

Model	p [$\mathrm{dyne\,cm^{-2}}$]	n_{e} [$\mathrm{cm^{-3}}$]	q_{r} [$\mathrm{erg\,cm^{-3}\,s^{-1}}$]	q_{m} [$\mathrm{erg\,cm^{-3}\,s^{-1}}$]	DEM [$\mathrm{cm^{-5}\,K^{-1}}$]
13x0	0.15	2.7×10^{9}	2.1×10^{-3}	1.8×10^{-4}	6.1×10^{20}
13x1	0.23	4.4×10^{9}	5.3×10^{-3}	5.2×10^{-4}	8.8×10^{20}
13x2	0.35	6.7×10^{9}	1.2×10^{-2}	1.2×10^{-3}	1.7×10^{21}
13x3	0.61	1.1×10^{10}	3.5×10^{-2}	3.5×10^{-3}	2.2×10^{21}
13x4	1.2	2.2×10^{10}	1.4×10^{-1}	1.4×10^{-2}	4.7×10^{21}
13x5	2.0	3.8×10^{10}	3.8×10^{-1}	3.7×10^{-2}	8.4×10^{21}
13x8	2.8	5.3×10^{10}	7.4×10^{-1}	7.4×10^{-2}	1.2×10^{22}

corona models, namely the layer with $T = 2 \times 10^5$ K. The parameters listed are derived from energy balance through the lower transition region and applied as a starting point for the energy balance of the upper transition region. These parameters are for the final set of models that use flux-limiting coefficient value $\alpha = 7.0$, which is the value that overall produces the best match between the calculations and observations of the spectra.

4. Structure of the Low Corona

The structure of the lower corona is similar to that in Article 3, however, the upper part of the transition region included in our models is now computed using the same flux-limiting formula and parameter α as for the lower part of the transition region. As before, we merge the temperature structure of the upper part of the transition region with an extended region at coronal temperatures that is shown in Figure 3.

In order to produce a smooth fit of the corona with the upper part of the transition region we use the following mechanical-heating formula:

$$q_{\mathrm{m}} = 10^{-22} n_{\mathrm{e}} n_{\mathrm{p}} \left(\frac{T}{10^6} \right), \tag{3}$$

where n_{e} and n_{p} are the electron and proton densities, respectively. We do not believe that this formula describes theoretically the coronal heating, but this formula has the approximate properties needed. At $T = 2 \times 10^5$ K the mechanical heating [q_{m}] is ≈ 0.1 of the radiative

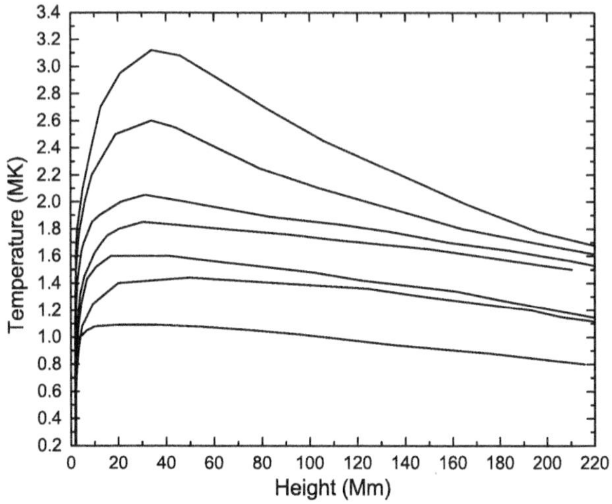

Figure 3 The structure of the coronal models adopted for various features described in Table 1. Top to bottom the model indices are 1318, 1315, 1314, 1313, 1312, 1311, and 1310.

losses [q_r], and only at $T \approx 6 \times 10^5$ K does the mechanical heating balances the radiative losses for all models. The same formula is used for all models, but the amount of mechanical heating is different because of the different pressure and maximum temperature in each of them. We adopt coronal layers above and around the point where $q_m \approx q_r$ independently of this formula and with a smooth temperature maximum and decay above it.

These temperature maximum and decay are determined in order to match the observations the best that we can do it currently. Unfortunately, there are not many observations of active-region spectra with high spectral resolution and reliable absolute calibration. As new observations become available we hope that the layers around and above the temperature maximum will be better determined. However, statistics of many observations are needed because many differences occur between various active regions. In addition these regions have very inhomogeneous temperature and density due to their loop fine structure, and therefore comprehensive spatial coverage and a statistical description are needed. Individual loops detailed structure is highly variable in space and time and have been modeled by other authors using very strong simplifications (*e.g.* Rosner, Tucker, and Vaiana, 1978; Serio *et al.*, 1981, also see review by Reale, 2010). Our models are focused on spectral-irradiance synthesis and do not attempt to model individual active-region loops; instead the present models are intended to describe the SSI relevant properties of an ensemble of such loops, over the term of several hours, which is relevant to SSI and its effects on Earth's upper atmosphere.

An issue remains with relatively cool structures extending through and sometimes over the low corona, namely the more or less stable prominences (with very complicated internal fine structure), and also the short-lived high-speed jets. We do not model these because both pose problems of their own resulting from complex interactions between the plasma, which is sometimes but not fully ionized, and magnetic fields. In the present work we neglect these structures in the context of overall solar irradiance, although they could have effects at some wavelengths due to absorption of underlying coronal emission.

Model 1310 corresponds to somewhat cooler regions of the lower corona often associated with coronal holes in X-ray images. However, this model does not represent all that can be called coronal hole. Prominences cool overlying material described above, which absorb radiation from the underlying low corona and therefore would produce a void in the X-ray emission. Such a void can be partially due to the absence of coronal material in the threads that generally form a prominence. However, underneath it and in between the threads, coronal material may exist whose emission is obscured by the less ionized material in the threads. Since in the present work we are not considering the cool material embedded in the corona, we do not address these cases in the present article.

Models 1311, 1312, and 1313 correspond to the magnetic structure of quiet Sun, and are the extensions of their counterparts in the low transition region and chromosphere supergranular structure described by models 1301, 1302, and 1303 (see Table 1). It is implied in the construction of the models that the coronal features are built as an upwards extension of the chromospheric models, but it is apparent that in images corresponding to high temperatures that the supergranular structure cannot be identified. This may be simply due to the closing of the magnetic fields in relatively low-altitude loops that form a more or less uniform background. Our use of the set of models for spectral-irradiance reconstruction assumes that at coronal layers the model 1311 for feature B is representative of the mix. Another alternative we have tried is to assume that the components are mixed in the same proportion as their chromospheric counterparts and form a background of quiet Sun that can be somewhat enhanced if the fraction of the more active features, *e.g.* 1313, increases. The issue of whether the so-called quiet-Sun background changes during the solar cycle is complex because, even far away from the chromospheric active regions, so-called interconnecting coronal loops (which connect relatively remote active regions) may cover much of the solar disk at very high-activity times. These interconnecting loops have a contribution to the EUV spectral irradiance that can hardly be assigned to quiet-Sun structures. Depending on their brightness in EUV images, in a pixel we may assign them to one of our active-region models (usually model 1314) but details on this image-processing issue are outside the scope of the present article.

Models 1314, 1315, and 1318 correspond to active regions. They are designed to approximately match the most common features of active regions outside of flares. Still, after the impulsive phase of flares, during flare decay a region on the disk may exhibit traits similar to some of these models. The three active-region models are characterized by different peak temperatures and also different pressures at the base (see Tables 3a and 3b).

Active regions are very heterogeneous and most likely all the components described by our models exist and are intermingled. However, for SSI modeling we use images to discriminate areas corresponding to one or another of the models because a mix would be very hard to assess from existing full-disk images. In the future, images with better spectral discrimination and calibration could improve this. Also, note that all of the models (when considering the entire chromospheric and coronal parts) contain a complete range of temperatures from the chromosphere up to a different maximum coronal temperature. This would correspond to a vertical continuity of the structure, different from observations showing curved loops; however, it can describe an ensemble of active-region compact loops. For high-altitude loops, the vertical structure is clear at different viewing angles and especially toward the limb. Depending on the observing angle these coronal loops may overlay regions that can or cannot be described as their loop footpoints. Because of this, we do not use the same identification of chromospheric for coronal features in the SSI reconstruction but instead compute the relative areas of the coronal components based on coronal images. The areas determined to be active regions from chromospheric or images of the lower part of the transition region (*e.g.* in He II 30.4 nm) can overlap or have an offset from the coronal active regions.

For our computation of the spectra from the coronal features, we use spherical coordinates over patches corresponding to the observed areas of the features. Note that for the lower layers (photosphere, chromosphere, and lower transition region) the radial extent is so small that a plane-parallel approach is sufficient and breaks down only very close to the limb where the contribution to the irradiance is very small. However, a much larger limb brightening would result if we had not considered spherical coordinates and optical thickness effects due to the long path. These issues were verified by computing cases near the limb and comparing the computed center-to-limb variation with that observed in images from SDO/AIA.

One issue remains to discuss concerning the areas just above the limb. At many XUV/EUV wavelengths these areas look bright in the observations, *e.g.* the well-known doubling of the intensity just above the limb. Detailed calculations using spherical coordinates show that for the quiet Sun their contribution to the irradiance is not very important. Our calculations match well the observed center-to-limb behavior near, at, and above the limb (except for the fainter regions well above the limb where instrumental effects are uncertain).

The consideration of the low corona just above the limb does not affect the irradiance for the quiet Sun. Active regions seen above the limb could contribute significantly when the disk is covered by quiet Sun. The result is just a temporal smoothing effect on the SSI rotational-modulation variations. When the active region comes into view its emission usually far surpasses that seen above the limb before, and when an active region disappears from the disk the change far surpasses that of its residual emission from above the limb. In any case, our methods could include the emission above the limb due to patches of active region by using three-dimensional radiative-transfer calculations if the horizontal extent of the active region is known or can be inferred from images.

Table 3a also shows the total radiative losses from the coronal models, *i.e.* the radiative flux [F_r] emitted by the upper-transition-region/low-corona of our models at $T > 2 \times 10^5$ K. These radiative fluxes are, for the coolest model, nearly twice the amount of downward energy flux by conduction, but become increasingly larger for the hotter models up to almost ten times the downward heat conduction. In terms of the emitted radiation this implies that the hot coronal lines emit more energy than the transition region for the cooler models and much more for the hotter models. The result in this table is different from what would occur if Spitzer conductivity were assumed, *i.e.* $\alpha = 0$, because in that case the hotter models would have upper-transition-region/low-corona radiation comparable with the downward energy flux while the cooler models would have much less upper-transition-region/low-corona radiation than downward energy flux. While the trends with increasing maximum coronal temperature are similar, the absolute values are quite different.

5. EUV/FUV Continua Issues

Here and in the following sections the SSI will be given in the usual units of $\mathrm{W\,m^{-2}\,nm^{-1}}$, which is equivalent to $10^2\ \mathrm{erg\,cm^{-2}\,s^{-1}\,\AA^{-1}}$.

The FUV continua computed from our models matches the observations at many wavelengths but not yet at all wavelengths. In the range 168 – 200 nm disagreement with observations remains. There are two other small ranges just short of 124 nm and of 50.4 nm where there is also some disagreement. The present section discusses these issues in detail for model 13x1 (resulting from combining chromospheric and coronal models 1301 and 1311), *i.e.* adding the spectrum from 1301 and 1311, but similar considerations apply to the other models of the set. Also, the same issue reflects on both the radiance and the solar spectral irradiance (SSI), and here we discuss the issue in both contexts.

Figure 4 shows a comparison of the computed and observed SSI for model 13x1. This figure shows three intervals of concern in the matching of the continuum. One of them is

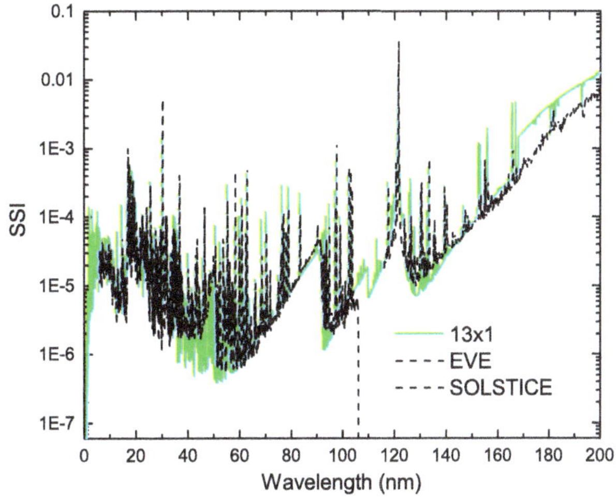

Figure 4 The overall SSI at 0.1 nm resolution computed for model 13x1 and the observations.

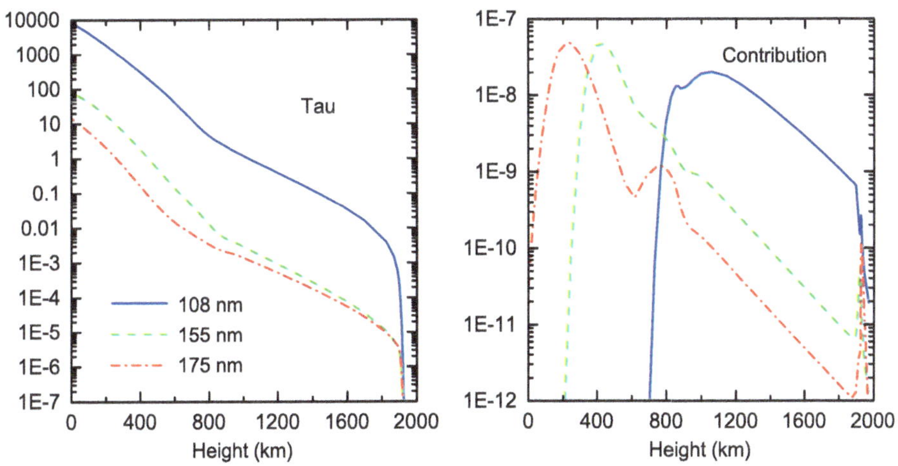

Figure 5 Optical depth (left) and "intensity contribution function" (right) at disk center as function of height for model 1301 at three selected continuum wavelengths.

wavelengths longer than ≈ 170 nm, the others are small ranges below 124 nm and below ≈ 50 nm. Details regarding the lines will be shown and discussed in Section 7; however, the continuum range just short of 124 nm is mixed up and interacts with the Ly α far wings.

Figure 5 shows the optical depth and the contribution functions for model 1301 at three wavelengths in the continuum. These wavelengths are selected to be representative of two of the issues mentioned above. For the purpose of this article the contribution function is redefined as the attenuated total emissivity (see Fontenla, Balasubramaniam, and Harder, 2007) but now divided by the emergent intensity. This change in the definition allows us to show this function at various wavelengths, with disparate emergent intensities, into the same figure and also ensures that for all curves the integral is unity. The emergent intensities at disk center for these wavelengths are: 457.3 for 175.0 nm, 297.1 for 155.5 nm, and 26.8 for 108.0 nm. Here and in the following all emergent intensities are given in the usual units of $\mathrm{erg\,cm^{-2}\,s^{-1}\,\AA^{-1}\,sr^{-1}}$, which is equivalent to $10^{-2}\ \mathrm{W\,m^{-2}\,nm^{-1}\,sr^{-1}}$, but see the detailed explanation about SSI and emitted intensity units in Fontenla *et al.* (1999) Appendix B.

A sharp edge occurs in the computed spectrum at the wavelength of the Si I continuum edge from the first excited state, namely 168.211 nm. This atomic level has very low energy and is the lowest of the singlet, and at shorter wavelengths from the edge the intensity is smaller than at longer wavelengths from the edge. The ground level of Si I is a triplet level with an ionization edge at 152.096 nm, and for this the intensity is a bit smaller at wavelengths longer than the edge than sorter from the edge, *i.e.* the opposite behavior to that of the 168.211 nm edge. Despite the correspondence to the ground level of Si I of the 152.0 nm edge, this jump is smaller than that of the 168.2 nm edge because the former is produced only a little bit above the temperature minimum. Instead, the 168.2 nm edge forms at a location of significant temperature gradient decreasing with increasing height. Another edge is shown in our calculation, but it is barely visible in Figure 6, at 198.596 nm, which corresponds to the second excited state of Si I, which also belongs to the singlet. In summary, our calculation shows the Si I edges from all of the lower states. In stark contrast, the spectra observed by SOHO/SUMER (emitted intensity at disk center) or by SOLSTICE (SSI) do not show these edges. Instead they show no obvious jump in intensity but only slight changes in slope at 152.0 nm. SUMER only covers up to 161.0 nm and cannot show the 168.2 nm edge, but SOLSTICE shows that at 168.0 nm, again, only changes in slope occur and important absorption lines only occur at wavelengths longer than ≈ 180 nm. The computed SSI spectra at wavelengths longer than the 168.2 nm edge show absorption lines and large continuum intensities that are absent in SOLSTICE spectra at wavelengths shorter than 200 nm.

The reason for the behavior in our computation is shown by Figure 5. At 155.5 nm and other wavelengths shorter than 168.2 nm, the continuum forms at higher altitude than that at 175.0 nm, and is formed in a region of the atmosphere with a small negative temperature gradient and at locations where the source function is affected by illumination from above and subject to NLTE effects. Instead, at 175.0 nm and other wavelengths longer than 168.2 nm, the continuum forms at the top of the photosphere or within the lower-chromosphere where the negative temperature gradient is large and non-LTE effects are small. Therefore, we conclude that, within the lower chromosphere, at or below the temperature minimum, some important opacity sources are not included in our calculation but are present in the Sun. We speculate that the gradual behavior and lack of observed jumps or absorption lines, as well as the relatively large density and low temperatures in the layers involved, point to molecular photodissociation opacity that is not yet identified. It is remarkable that even at the 0.1 nm resolution of SORCE/SOLSTICE only few and weak fluctuations due to lines are observed, and that a few groups of lines that our calculation show in absorption around 180 are shown in emission by SOLSTICE, see Figure 4. This behavior is inconsistent with what could be inferred from Avrett and Loeser (2008) Figures 1 and 6 that predict very large fluctuations in intensity *vs.* wavelength in the range 170 – 200 nm, which are not compared with observations in that article and seem to be incompatible with the SORCE/SOLSTICE 0.1 nm resolution data that we show in our Figure 4.

One of the problems for improving our understanding of this continuum is the lack of adequate computations of many molecular opacities in the FUV. Calculations exist for molecules in the interstellar medium, but these are not very useful for the solar atmosphere because of the substantial temperature and density of the low chromosphere affecting the populations of vibrational–rotational levels.

Let us now discuss the issues of the C I edge just short of 124.0 nm. The edge from the C I ground level occurs at 110.107 nm, a triplet level, and that from the first excited state at 124.027 nm, a singlet level. Both edges are clearly observable and our calculations overestimate the latter. Figure 5 shows that the continuum at 108.0 nm is formed almost entirely in the upper chromosphere as C II recombines into C I. For the edge at 124.0, *i.e.* within the Ly α red wing, some contribution to the emission occurs from the temperature-minimum region where C I is over-ionized by illumination from above. The issue with the overestimated C I edge arises from non-LTE effects and is more important here than in Article 3. As was explained in Section 2, we are now using a more recent calculation of the collisional excitation rates of C I, and these rates are generally larger than those used in Article 3. In particular the collision strength for the excitation from the ground level to the first excited is now about 100 times larger, at $T = 5000$ K, and a larger fraction of C I is in the first excited state throughout the chromosphere. This larger population of level 2 produces a somewhat larger ionization of C I because of the photo-ionization produced by Ly α pumping, and enhances the C I ionization through the upper chromosphere and at the temperature-minimum region. As a result of this, the emitted C I continua is enhanced by recombination into both levels 2 and 1. However, many of the photons emitted by recombination into level 1 are absorbed at larger heights leaving the strongest effect on the continua to level 2.

Our computed Ly α profile and corresponding non-LTE uses partial-frequency-redistribution and matches the observations at wavelengths longer than the 124 nm edge. The apparent enhancement of the Ly α line wings at wavelengths shorter than 124.0 nm results from the effects of the C I continuum. For C I non-LTE we use a discrete grid wavelength that accounts for the dependence of the C I ionization on Ly α, and we verified that improving the grid by doubling the number of points in the quadrature used does not affect importantly our results. Therefore, we attribute most of the increased recombination to the excited level of C I in our present models, at layers around the temperature minimum, to the increased collision rate of the inter-combination transition 1 – 2 discussed above. Considering the complicated nature of the C I atom and the interaction with Ly α, it is likely that further improvements in the collision rate, the C I ionization, or in the photo-ionization cross-sections from the C I levels could improve the agreement with the observations for the 124.0 nm continuum edge.

The third issue with the continuum is a minor one from the standpoint of SSI calculations. Article 3 indicated that the He I 50.4 nm continuum not shown in our calculations likely results from enhanced ionization of He I, *i.e.* increased presence of He II, at the top of the chromosphere due to illumination from the corona above. The observed slope of the He I continuum that can be discerned in between the coronal lines corresponds to a temperature of about 1.5×10^4 K, which corresponds to the top of the chromosphere. At this temperature there would not be much He II unless coronal illumination produces additional ionization of He I. However, from the irradiance standpoint, the large coronal emission lines are much more important than the He II recombination continua around 50.4 nm and therefore this continuum does not pose a very important issue.

In the present calculation we have approximately solved the previous minor issue of the gap between the merging Ly lines and the continuum edge by extending the continuum up to the wavelength of the center of the first Ly line which we do not explicitly compute. Our current H atom includes 25 levels and therefore the extension of the continuum is carried up to the Ly line center corresponding to level 26. This mimics the merging of the Ly lines into the continuum solving some issues about SSI in the small gap that Article 3 showed. Although overall this is a very small gap it is significant because of its effects on the telluric lines.

The present computation of the dominant Si I continua spanning the 124 – 168 nm range matches well the radiance and irradiance observations. Also, the computation of the H I continuum shorter than ≈ 91.3 nm matches well the observations of radiance and irradiance by SOHO/SUMER and SDO/EVE, respectively.

6. Comparison of Emitted Intensity (Radiance) with Observations

In this section we compare the computed spectra with radiance (*i.e.* emitted intensity) observations at solar-disk center. The comparisons use some quiet-Sun atlases derived from the following instruments: *Solar Ultraviolet Measurements of Emitted Radiation* (SUMER: Wilhelm *et al.*, 1995) onboard SOHO, *EUV Imaging Spectrometer* (EIS: Culhane *et al.*, 2007) onboard *Hinode*, and the *Coronal Diagnostic Spectrometer* (CDS: Harrison *et al.*, 1995) onboard SOHO. In all cases, we convolved the computed spectrum with the instrument resolution, which is known for the first two instruments but lesser understood for CDS. However, our profiles are computed for zero velocity and a "turbulent broadening velocity" that our models specify, and at the nominal vacuum wavelengths stated by NIST (Kramida *et al.*, 2013). Random velocities in the Sun would cause the profiles to be somewhat more broadened, and systematic velocities would shift them, but we do not attempt here to correct for that.

It is, of course, impossible to show here the spectra in all their detail. The graphs shown in detail in this article only include several key spectral lines, and more complete comparisons over the whole ranges of the atlases are posted at www.galactitech.net/SRPMrel2013/. The computed spectral data are also available there at the native, variable sampling (as mentioned in Article 3, and in the introduction of this article; the resolution is higher over the lines than in the continuum and the computations correspond to snapshots of infinitesimal bandwidth). These computed spectra can be convolved with any instrument resolution taking into account any given bandpass or instrumental profile. For all comparisons shown in this article we have used a filter with bandpass shape given by the cosine square truncated at the first zeroes on either side of the central wavelength. This type of bandpass is designated here as $\cos^2$, and the width of the bandpass is different when comparing our computations with different instruments. We consider this bandpass similar in some respects to a Gaussian but it does not require an arbitrary truncation, and it is also similar to the central part of a diffraction pattern. In the comparisons discussed here, only the full-width-at-half-maximum is given (hereafter FWHM), and we do not have reliable data on the various instruments' detailed profiles. We consider the $\cos^2$ bandpass as likely close to the true instrumental profile. When convolution with the $\cos^2$ bandpass is done, the sampling (*i.e.* output wavelength spacing) is $1/5$ of the FWHM nominally given. This sampling is often more closely spaced than the observational data.

In this section, we compare in detail important lines computed with the available quiet-Sun atlases from various instruments. Because the radiance observations are nominally spatially resolved data representative of quiet Sun, the criteria for selecting quiet-Sun areas becomes relevant and may not have been identical in all cases. In the following figures we indicate as 13x0, 13x1, 13x2, 13x3 the additions of the coronal and chromospheric parts of the corresponding models (of course including the complete transition region). For instance, model 13x1 $=$ 1301 $+$ 1311 indicates the sum of the spectrum from models 1301 and 1311. This is done assuming that at disk center the coronal part is optically thin and therefore the procedure adopted corresponds to appending the coronal model to the top of the chromospheric model forming a continuous vertical structure.

6.1. Comparison of Selected Lines with SOHO/SUMER

The SOHO/SUMER quiet-Sun atlas by Curdt *et al.* (2001) was constructed from many observations during solar minimum and therefore probably represents a weighted average between the various features that we represent by models 13x0, 13x1, and 13x2, and maybe

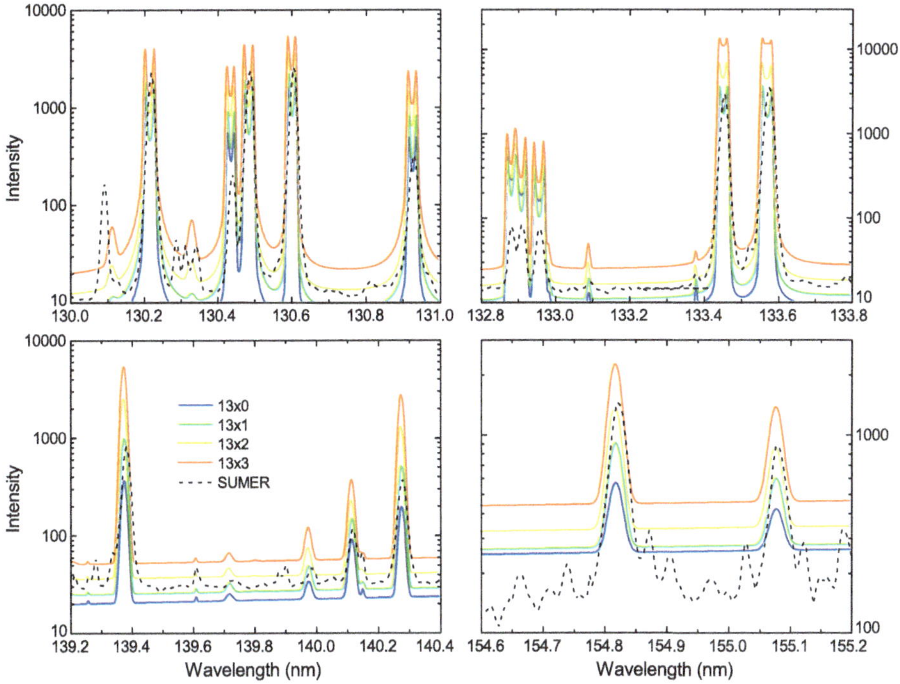

Figure 6 Top-left panel, the spectral range 130 – 131 nm showing the important lines of O I 130.217, 130.486, and 130.603 nm, and also Si I 130.437 and 130.828 nm. Top-right panel: the range 132.8 – 133.8 nm showing the important lines of C II 133.453, 133.566, and 133.571 nm, and weak C I lines from 132.8 to almost 133.0 nm. Bottom-left panel: the range 139.2 – 140.4 nm showing the important lines of Si IV 139.376, and 140.277 nm. Bottom-right panel: the range 154.6 – 155.2 nm showing the important lines of C IV 154.819 and 155.077 nm.

even a few percent of 13x3. The exact weights are not known to us although most likely the contribution from 13x1 is dominant and there is some sort of compensation between the brighter and fainter features. For this reason, in Figures 6 and 7 we compare the SUMER data with the three quiet-Sun components of our model set. For this comparison we convolve the computed spectra with a $\cos^2$ bandpass of 5.5 pm (*i.e.* 55 mÅ) FWHM.

Figure 6 (top panels) shows comparisons of the most important chromospheric lines of C and O that were observed by SUMER. A good match is obtained for the O I and C II strong lines shown in the upper panels, although the computed profiles display a weak reversal which is not shown by SUMER. It has been argued that macroscopic velocities could eliminate this reversal, but we do not want to argue this because SUMER sampling of line profiles is not very good and because observations by other instruments may display such reversals. A group of C I lines in the range 132.8 – 133.0 nm, in the top-right panel, is overestimated and this is consistent with the overestimate of the C I continuum that was discussed in Section 5. Therefore, the question becomes one of how accurate the new C I collision-strength data are for these lines. The bottom-right panel shows some differences in the continuum longer than 152.0 nm that were discussed in Section 5.

Figure 6 (bottom panels) shows important low transition-region lines of Si IV and C IV. The C IV lines are somewhat underestimated by the calculations shown here that use a flux-limiting coefficient of 7.0 and are better represented with a coefficient of 3.0; they are overestimated by assuming that the coefficient is equal to 0 (*i.e.* considering the full

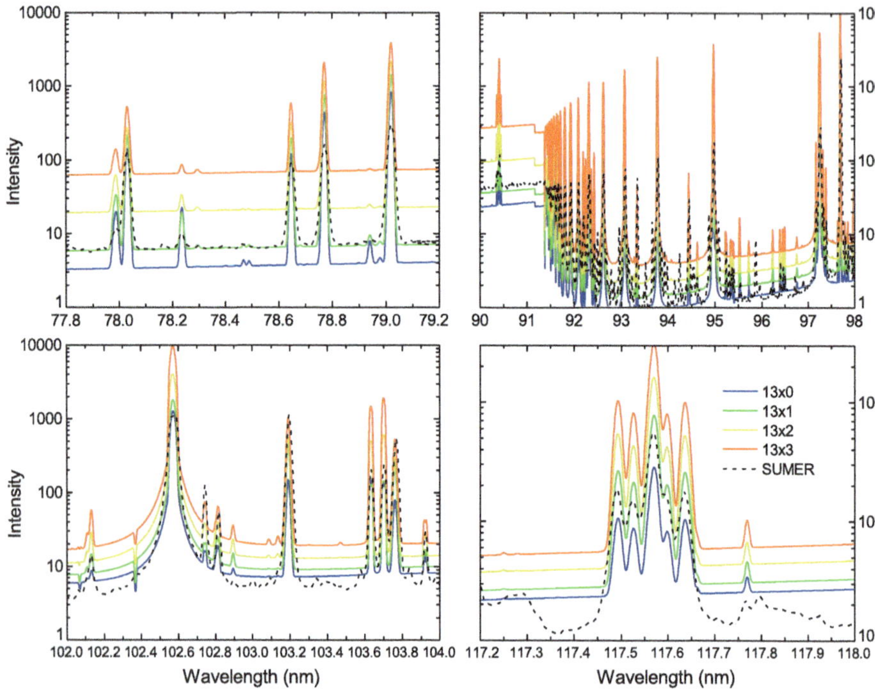

Figure 7 Top-left: the spectral range 77.8 – 79.2 nm showing Ne VIII 78.032, S V 78.647, and O IV 78.771 and 79.020 nm lines. Top-right: the range 90.0 – 98.0 nm showing the Ly continuum, Ly lines, and the prominent C III 97.702 nm line. Bottom-left: the 102 – 104 nm showing Ly β at 102.572 nm, O VI 103.191 and 103.761 nm, and several C II lines. Bottom-right: the range 117.2 – 118.0 nm showing a number of C III lines.

electron conduction by Spitzer's formula); see also Article 3. Looking at Figure 1, we see that the C IV lines are formed at the temperature where the flux limiting starts to produce its steepening effect on the transition-region temperature.

Figure 7 shows other key lines covered by the SUMER spectrum. The O IV lines in the top-left panel are slightly overestimated with respect to SUMER (but much less overestimated than they were in Article 3). An increase in the flux-limiting coefficient would further reduce these lines' intensities, but it would lead to underestimating other lines from similar temperature ranges. Thus, the issue of O IV is still not optimally solved, and we note that O IV forms within a narrow temperature range and its ionization equilibrium can be easily affected by velocities.

Figure 7 (top-right) shows the head of the Ly continuum, the high Ly lines, and many other lines. The head of the Ly continuum gap in Article 3 is now fixed by the extension of the Ly continuum described in Section 5 and is very well reproduced by our current calculations. However, the high Ly lines are somewhat overestimated by the present calculations and this may be due to an over-simplification in our H-diffusion calculations (taken from FAL 1). In that procedure the diffusion velocity of excited H atoms is made equal to that of H lower levels, and this may not be correct when the levels are close to the continuum but a complicated revision to that procedure may require that each level's transition rates be considered to determine each excited-level diffusion velocity. The large intensity optically thick lines Ly γ at 97.254 nm and C III 97.702 nm line are well reproduced by our calculations.

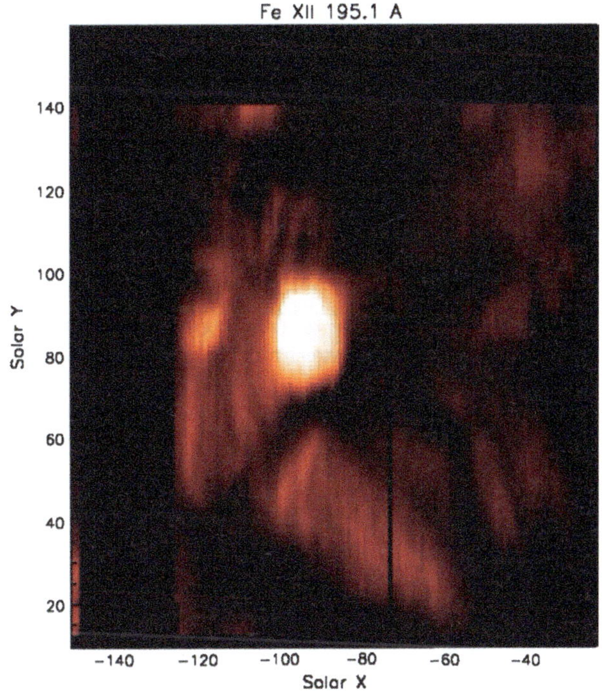

Figure 8 The region near disk center averaged to create the EIS spectral atlas. The image coordinates are in arcsec with respect to disk center and has been created using the log intensity of the Fe XII 19.512 nm line. Black stripes indicate missing data and have not been used in the calculation of the EIS average spectrum.

The optically thick Ly β line and other strong lines shown in the bottom-left panel are also well reproduced despite the C I continuum being overestimated as indicated in Section 5. This continuum may have affected the peak intensities of some weaker optically thin lines, whose peak intensities are underestimated. Also, the Ly β line wings may be affected by the C I continuum.

6.2. Comparison of Selected Lines with *Hinode*/EIS

The EIS spectrum was taken on 17 December 2008, and consisted of a $128'' \times 128''$ raster of the quiet Sun near disk center. The entire wavelength range covered by EIS was transmitted to the ground. The exposure time was 90 seconds at each slit position. The data have been reduced, calibrated, and cleaned from cosmic rays using the standard EIS software available in SolarSoft. Figure 8 shows the field of view including a bright point and some faint structures. The dark stripes in this figure correspond to missing data that have not been considered. The emission from all pixels with non-zero intensity was averaged together to increase the signal-to-noise.

For this comparison, we convolve the computed spectra with a $\cos^2$ bandpass of 5.5 pm (*i.e.* 55 mÅ) FWHM. From the present comparison it is apparent that the EIS data may be biased toward the lower temperature areas (model 13x0) than model 13x1 in our set of models. The 13x3 model perhaps corresponds to the bright point in this image.

Figure 9 shows two important lines that have also been targeted by imaging of the solar disk (*e.g.* by SDO/AIA, SOHO/EIT, and STEREO/EUVI). On the left panel, the Fe X line has relatively small changes between models 13x0, 13x1, and 13x2, *i.e.* the supergranular cell models of weak Ca II emission (see Article 3 and Table 1 in the present article). Therefore, the emission from all these models is close to the observed. However, on the

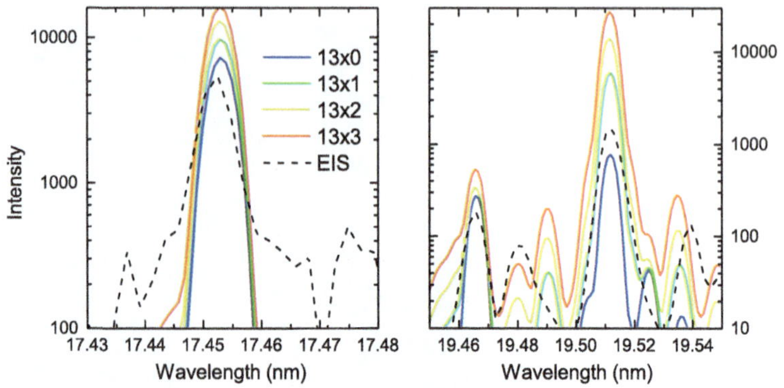

Figure 9 Left panel: the Fe X 17.453 nm line. Right panel: the Fe XI 19.512 nm and other lines including Fe VIII 19.466 nm. Contamination by overlapping diffraction orders is shown at the shorter wavelengths and low intensities are not relevant.

right panel, the Fe XII line shows much larger model sensitivity and the EIS observation is slightly higher than the model 13x0 (the weakest of the quiet-Sun models) and smaller than the model 13x1. We note that the maximum temperature in model 13x0 barely surpasses 1 MK, see Figure 3, while model 13x1 reaches 1.4 MK. Also in the right panel of Figure 9 is the Fe VIII 19.466 nm line, which shows little model sensitivity and is only slightly overestimated by our models. Other structure in the right panel is a line at 19.48 nm that is apparent in the EIS data but computed to be very weak in our models of quiet-Sun network from a blend of Fe IX, X, and XI lines with high excitation lower levels. In addition, we have a computed line at around 19.49 nm, which results from a blend of Fe XII 19.490 and Ni X 19.491 nm but is not clearly shown in EIS data.

Figure 10 shows many lines spanning from upper transition-region temperatures to the hot Fe XIII, and including the low transition-region He II 25.632 nm line. In the Fe XIII lines the contrast between our models 13x0 to 13x3 is rather large and model 13x0 is always closest to the EIS intensities. In the transition-region lines the contrast between 13x0, 13x1, and 13x2 is small, and the emission from these three models are all close to the EIS observations. For the He II 25.6 nm line, the EIS observation falls closer to 13x1 but 13x0 emission is not very different. In this figure we also note cases where lines are computed but not observed, and overall in the spectrum we also see EIS lines that do not show a computed equivalent line. We note that in some cases, especially for faint lines, the lines' wavelengths used were the theoretical ones in CHIANTI 7.1 data and may not be accurate, while in the cases where observed wavelengths are available we have used these.

6.3. Comparison with SOHO/CDS

The CDS instrument observed the solar spectrum in six spectral windows between 15.0 and 78.5 nm using two different spectrometers: the *Normal Incidence Spectrometer* (NIS), observing in the 30.7 – 37.9 nm and 51.3 – 63.3 nm ranges, and the *Grazing Incidence Spectrometer* (GIS), observing in the 15.1 – 22.1 nm, 25.6 – 34.1 nm, 39.3 – 49.2 nm, and 65.9 – 78.5 nm ranges. Due to the presence of ghost lines in the GIS spectrum, we only considered the NIS spectral ranges. The observation that we used to determine the quiet-Sun spectral atlas was taken on 6 October 1996, during solar minimum, at solar center, and was part of the standard NIS spectral-atlas program: NISAT. The NIS $2'' \times 240''$ slit was used at ten ad-

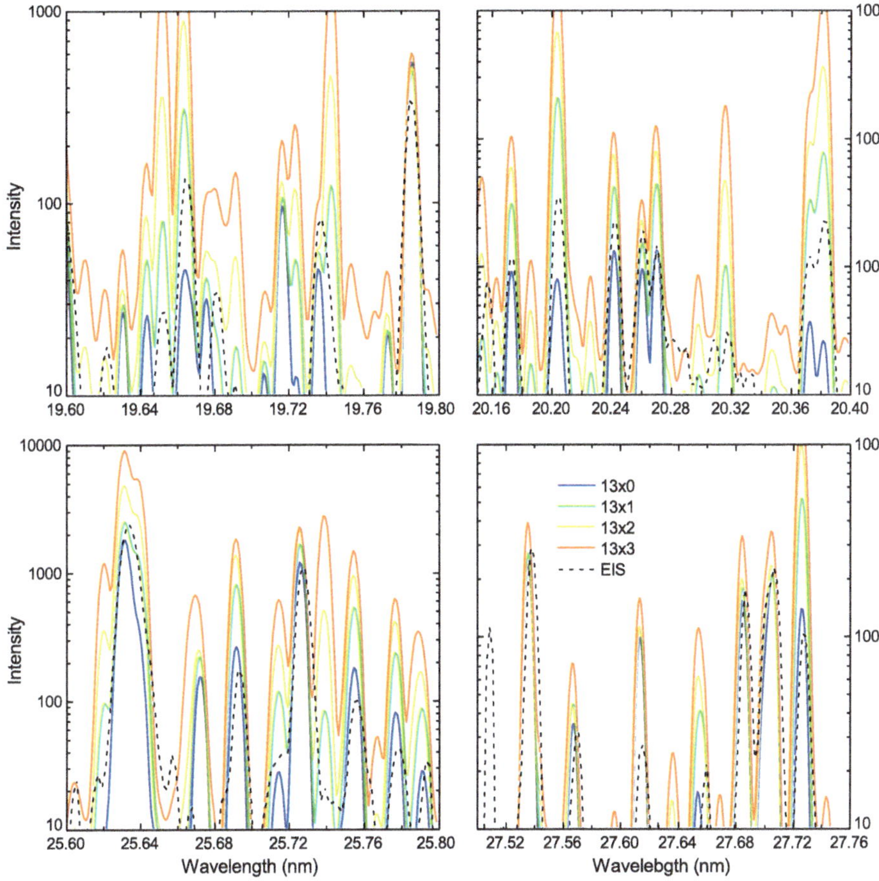

Figure 10 Top-left: the range 19.6 – 19.8 nm showing Fe XIII 19.652, Fe XIII 19.743, and Fe IX 19.786 nm lines. Top-right: the range 20.16 – 20.4 nm showing the Fe XII 20.174, Fe XIII 20.204, Fe XI 20.242, Fe XII 20.373, and Fe XIII 20.382 nm lines among others. Bottom-left: the range 25.6 – 25.8 nm showing He II 25.632, Fe XI 25.692, Fe X 25.726, Fe XI 25.755, and Fe XI 25.777 nm lines. Bottom-right: the range 27.5 – 27.8 nm showing Si VII 25.736, Mg VII 27.615, Mg V 27.658, Si VII/VIII 27.685, Mg VII/Si VIII 27.704, and Si X 27.726 nm lines.

jacent positions along the E–W direction, for a total field of view of $20'' \times 240''$. At each slit position, the solar spectrum was observed for 50 seconds. The entire field of view was averaged together in order to increase the signal-to-noise ratio. For this comparison we convolve the computed spectra with a $\cos^2$ bandpass of 55 pm (*i.e.* 550 mA) FWHM.

In this section we show only a couple of lines, but from these and other comparisons it seems apparent that the SOHO/CDS "quiet-Sun" data were selected, much like the SOHO/SUMER data, toward the feature we represent by 13x1. Therefore, Figure 11 only compares with the disk-center spectrum from this model.

Figure 11 shows that the O IV lines computed using model 13x1 are still somewhat larger than the observed by CDS, but the O V computed line matches the observed well. It is likely that the simple flux limiting that we used needs improvement probably in its temperature or density dependence. However, there may be other reasons for this discrepancy, *e.g.* O IV ionization treatment may need improvement.

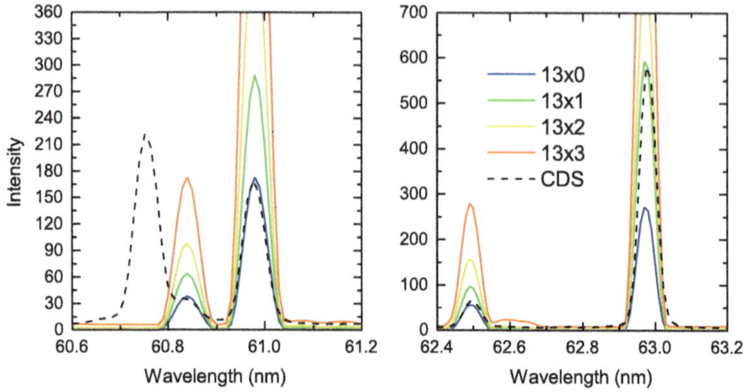

Figure 11 Left: the spectral range 60.8 – 61.2 nm showing O IV 60.840 and 60.983 nm lines. The spectral line shown at ≈ 60.75 nm is actually the He II 30.378 nm from the second-order diffraction. Right: the range 62.4 – 63.2 nm showing O V 62.973 nm line.

7. Comparison with Solar Spectral Irradiance Observations

The comparison of the complete SSI with model 13x1 was shown above in Figure 6, and the continua issues were discussed in Section 5. In the present section we discuss some details of the SSI that cannot be addressed by the graph in Figure 4 but need more detail on the lines.

For comparing the quiet-Sun EUV irradiance spectrum, we use the SDO/EVE rocket spectrum taken in April 2008 during a very quiet period, near the solar-activity minimum; however, on the day that the spectrum was taken a small active region was present on the disk and also coronal holes extended into low-latitude areas. The EVE instrument was calibrated before and after this rocket flight on beam line 2 (BL2) at the Synchrotron Ultraviolet Radiation Facility III (SURF-III) located in the National Institute of Standards and Technology (NIST) in Gaithersburg, Maryland, USA (see Chamberlin *et al.*, 2009 and also Woods *et al.*, 2009). We also use the SORCE/SOLSTICE spectrum taken in April 2005, which, although not at solar minimum, corresponded to fairly quiet conditions in terms of the chromospheric features observed on the disk. This instrument calibration was maintained in flight through comparisons with relatively bright stars that are believed to be constant (see Snow *et al.*, 2005, and McClintock, Snow, and Woods, 2005).

The highest resolution that was achieved by these high-quality and comprehensive SSI data was 0.1 nm. Although we do not know in detail the bandpass shape of the SSI instruments, we base our present comparison on a nominal resolution of 0.1 nm FWHM. Therefore we convolve our computed SSI for all our models with this profile using again a $\cos^2$ filter.

As was noted in Article 3, the Ca II K images show that the features corresponding to the chromospheric quiet Sun are more or less uniformly spread over the solar disk. The same is not quite true for the more recent images from SDO/AIA that show, in the 160 nm images, areas of depleted or increased network components that may relate to the growth and decay of active regions. However, this topic is beyond the present article and here we will show the irradiance that would be produced if the entire disk were covered by models 13x0, 13x1, 13x2, or 13x3. If the whole disk were covered by a uniform distribution of these features then the SSI can be calculated by a weighted sum of these components SSI. This corresponds to assuming the weights as independent of the disk position.

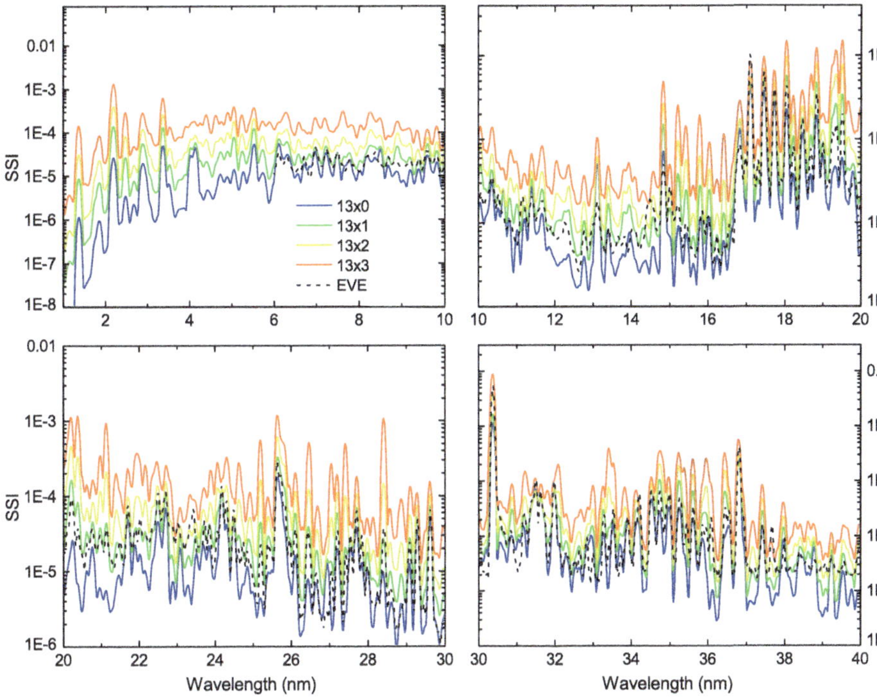

Figure 12 Comparisons of the SSI from the various components of the quiet Sun with the SSI observed by SDO/EVE in the range 1 – 40 nm.

The above assumption may not be as good for the coronal features because the presence of coronal holes is predominant over the polar high-latitude regions. However, this issue is beyond the scope of the present article and is related to assigning the weights of the various features as dependent on the position on the solar disk using images.

Figures 12 and 13 show the good agreement between the results that we obtain for an imaginary solar disk composed only of model 13x1 and the SDO/EVE measurements at a time when solar activity was near minimum. The tracings for the imaginary disk entirely composed of the other quiet-Sun component models show curves generally above and below the model 13x1 and probably compensate, but this is beyond the current article because the determination of the corresponding weights depends on the images of the solar disk.

Figure 14 last panel shows the issue with the continuum that was discussed in Section 5, and also two emission lines: an unresolved complex of C I lines around 165.8 nm and Al II 167.078 nm, whose computed irradiances are much larger than what is observed. We already mentioned the issue of C I in Section 5. The atomic data available for Al II are very old (from Aggarwal and Keenan, 1994) and may not be sufficiently accurate. These lines are examples of how critical for the present work is the availability of accurate collisional excitation rates.

8. Effects of Solar Activity on the SSI

In this section we show some comparisons with observed SSI in order to assess the quality of our models for estimating the SSI variability over the relatively short-term periods of rotational modulation, *i.e.* the variation of the SSI due to the non-uniformity of the distribution

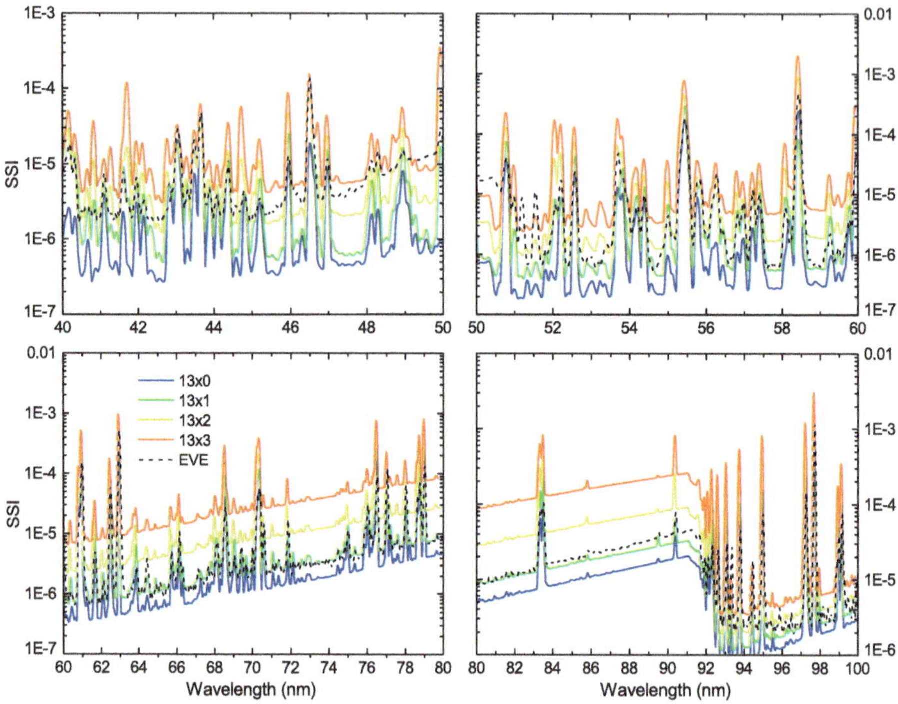

Figure 13 Comparisons of the SSI from the various components of the quiet-Sun with the SSI observed by SDO/EVE in the range 40 – 100 nm.

of active regions over the solar surface and their transiting of the solar disk as a consequence of the solar rotation.

It is impossible to separate the SSI variability from the images used to determine which features are present in each part of the solar disk. For our calculation it is only necessary to know which are the relative areas occupied by each feature at a given heliocentric angle [θ] or equivalently at each $\mu = \cos(\theta)$. These relative areas are determined, in our method, by a combination of images and μ-dependent thresholds. Some discussion on the image processing was given in Article 3, a detailed discussion of our method new sources for determining the relative areas, or weights, is beyond the scope of this article. So far we use only one image per day of each type and therefore intra-day variations on the solar surface or on the image quality may affect our measurements. Also, we note that we currently use two masks of features, one for the chromospheric/lower-transition-region and the other for the upper-transition-region/corona. The data that we present here are based on SDO/AIA images and essentially uses the 160 and 19.3 nm images for the chromosphere, and the 9.4 nm images for the corona. In this way, the coronal and the chromospheric SSI components are determined separately and the two components are added to form the total SSI, since we have verified that the coronal component is optically thin and effects on the SSI of absorption of the chromospheric component are negligible.

Figure 15 shows the comparison between the computed and observed SSI variation of the peak of a rotational modulation and its bottom, relative to the latter. The peak occurred on 23 October 2012 and for determining the bottom we averaged the two nearby minima on 9 October 2012 and 5 November 2012. Again we stress that the EVE data in this comparison

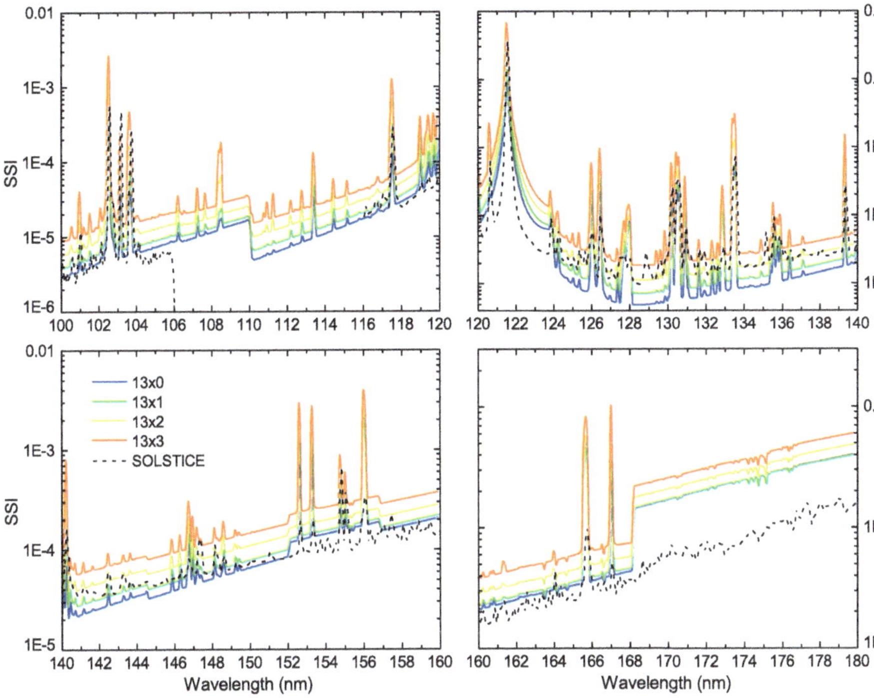

Figure 14 Comparisons of the SSI from the various components of the quiet Sun with the SSI observed by SORCE/SOLSTICE in the range 100 – 180 nm.

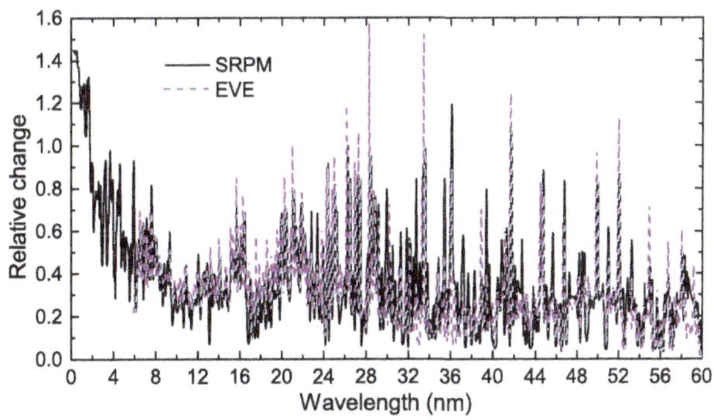

Figure 15 Comparison of the SSI relative variation between the top and bottom of a rotational modulation in October 2012.

corresponded to a daily average but the SRPM computed data corresponds to single-image snapshots from SDO/AIA using each filter. The figure shows fairly good agreement but the relative change is slightly underestimated in some lines by SRPM. A closer inspection showed that some of these differences occur because of the EVE data in those cases have a slightly higher spectral resolution than the 0.1 nm FWHM $\cos^2$ filter used to degrade the computed spectra. However, we believe that the variations in the lines near 28 and 34 nm were truly underestimated. These largest peaks in the relative variation correspond to the well-known Fe XV 28.3163 and Fe XVI 33.541 nm lines, which show very large enhancements in flares, and probably it is due to small flares that are included in the EVE daily average.

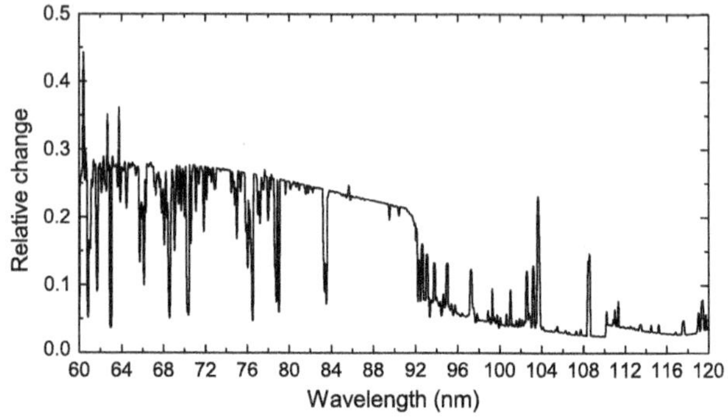

Figure 16 The SSI relative variation between the top and bottom of a rotational modulation in October 2012, at long wavelengths in the EUV.

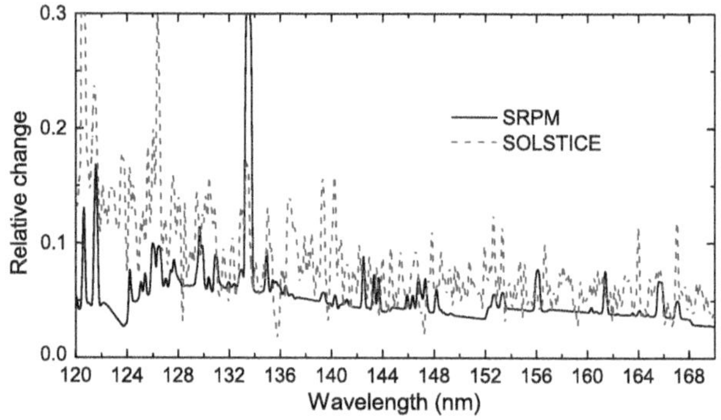

Figure 17 Comparison of the FUV SSI relative variation between the top and bottom of a rotational modulation, *i.e.* relative increase on 23 October with respect to 9 October 2012.

Also, note the large increase in the variability at short wavelengths and even at the smallest that we include in this graph, namely 0.22 nm, shorter than EVE can observe. This variability at the very short wavelengths is comparable to the largest variability EVE observed at the lines near 28 and 34 nm, but our calculations does not include flares. When flares are included, the variations at very short wavelengths are probably larger than those shown in Figure 15 and could have significant effects on photoelectrons produced in the Earth's upper atmosphere, see Peterson *et al.* (2012).

The EVE peak in the rotational modulation spectra seems to occur at a fraction of a tenth of a nm (*i.e.* fraction of an Å) shorter wavelengths in many lines; this shift maybe due to wavelength calibration uncertainty. Also, we note that our calculations give an even larger relative variation at wavelengths shorter than those shown in Figure 15. This figure could not include the shorter wavelengths because of the convolution imposed.

For comparisons at longer wavelengths, we need to rely on other data because the EVE version 3 that we have used does not yet include reliable degradation calibration at wavelengths longer than those shown here. An update of the EVE calibration for the year 2012 is forthcoming and will improve the longer wavelengths. Figure 16 shows the computed SSI variation in the range 60 – 120 nm, and it is interesting to note that the SSI in the Ly continuum shows an important variation due to rotational modulation in our calculations, and it also shows that its maximum amplitude does not occur at the head of this continuum.

Figure 17 shows the comparison of the computed rotational modulation with the SORCE/SOLSTICE data. For this comparison there were no observations on 5 November 2012, and the comparison is of the variation between the two days in October. Again here the SOLSTICE data correspond to a daily average. Although the original data from Level 2 processing is at 0.1 nm resolution we had to smooth the data at 0.2 nm resolution to reduce the noise. Of course we applied the same smoothing for the SRPM data; however, we note that effectively there is no noise in the computed data (only much lower numerical noise exists).

Figure 18 shows the masks used for the corona/upper-transition-region and the chromosphere/lower-transition-region in the computation of the total SSI. Note that although on the peak of the rotational modulation there are more hot active regions, at the minimum of the rotation plage are still present on the solar disk although in less total area than at the maximum of the rotational modulation.

9. Discussion

We have shown a set of physical models and comparisons between the computed and observed spectra that show that a semi-empirical set of models can account for the observed EUV SSI and its variation with solar activity. We consider the overall agreement good although in some details it can be improved by further work on atomic data, especially collision rates, minor improvements to the physical models, and perhaps a more complete physical description containing more components. However, it should be kept in mind that including more components would only improve the description of solar activity if there were a way to determine daily (or more frequently) the mix of components from solar images, and if their radiative interaction were either negligible or accounted for. Otherwise, a description with more components would not necessarily improve the results over those from the present set of models.

Moreover, because one of the important applications of the present research is in the forecasting of the EUV SSI, if more components were included it would also be necessary to forecast the expected evolution of the mix of components. This poses a question for the current set of features that Fontenla *et al.* (2009b) addressed using data from the farside of the Sun. Although these techniques have produced good results for Ly α, their forecasting power for other spectral regions needs to be further explored, and it would likely not be as precise unless EUV data from the farside is available because otherwise inference of the coronal features must rely on correlations with Ly α. For the present, STEREO/EUVI data provide additional input because the spacecraft observes the farside of the Sun, but in the long term the positions of the spacecraft or instrument failure will make their observations irrelevant in forecasting. We suggest that further consideration be given to keeping and improving ways for imaging the farside at all wavelengths possible and using a variety of techniques.

Also, more theoretical considerations of the physical mechanisms that produce coronal and chromospheric heating by magnetic fields would greatly enhance the forecasting power of even limited observations of the farside. If the evolution of magnetic-field structures could be forecast, then the connection with SSI requires the ability to infer from the magnetic field which coronal and chromospheric heating enhancements would result and which would be the spectra emitted by these magnetic features. This article shows that when the physical characteristics of the various features in the solar atmosphere are known, even in a highly simplified manner, it is possible to produce SSI spectra that match the observations fairly well.

The results presented here encourage the use of stellar models including chromospheres and coronae for the investigation of stars other than the Sun. Such studies yield important information that impacts the planets around such stars, *i.e.* "exoplanets", and can produce

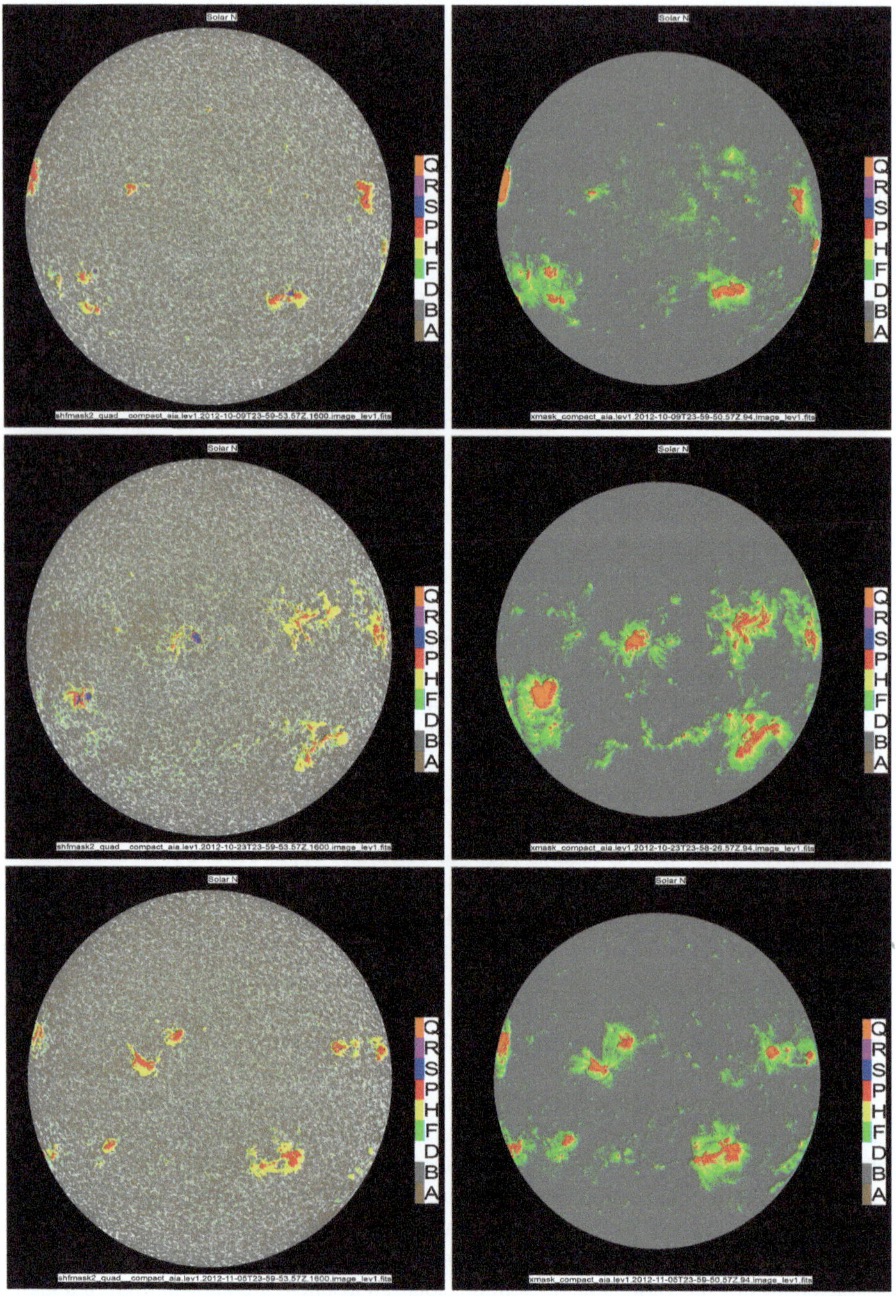

Figure 18 Image masks showing the features detected in the solar disk according to our classification in Table 1 and the SDO/AIA images. The left-side panels show the chromospheric masks, the right-side panels show the coronal masks. The corresponding dates are from top to bottom, 9 October, 23 October, and 5 November 2012.

important information about the EUV/FUV environment of such planets. For instance, Linsky *et al.* (2011) have shown that solar models with increasing heating rates fit stellar FUV continua (115.8 – 149.2 nm) for G stars with increasing heating rates.

Acknowledgements The work by JF was funded mainly by LWS–NASA grant NNX09AJ22G. Also, AF grant FA9453-13-M-0063 funded in a small part the work by JF, EL, and TW. We acknowledge the NIST Database that was funded (in part) by the Office of Fusion Energy Sciences of the US Department of Energy, by the National Aeronautics and Space Administration, by NIST's Standard Reference Data Program (SRDP), and by NIST's Systems Integration for Manufacturing Applications (SIMA) Program. We also acknowledge the CHIANTI team for providing atomic data that made this work possible. CHIANTI is a collaborative project involving George Mason University, the University of Michigan (USA) and the University of Cambridge (UK). *Hinode* is a Japanese mission developed and launched by ISAS/JAXA, with NAOJ as domestic partner and NASA and STFC (UK) as international partners. It is operated by these agencies in cooperation with ESA and NSC (Norway). Data are provided by courtesy of NASA SDO/AIA and EVE science teams.

References

Avrett, E.H., Loeser, R.: 2008, *Astrophys. J. Suppl.* **175**, 229.
Aggarwal, K.M., Keenan, F.P.: 1994, *J. Phys. B* **27**, 5321.
Bale, S.D., Pulupa, M., Salem, C., Chen, C.H.K., Quataert, E.: 2013, *Astrophys. J. Lett.* **769**, L2.
Brysk, H., Campbell, P.M., Hammerling, P.: 1975, *Plasma Phys.* **17**, 473.
Catto, P.J., Grinneback, M.: 2000, *Phys. Lett. A* **277**, 323.
Chamberlin, P.C., Woods, T.N., Crotser, D.A., Eparvier, F.G., Hock, R.A., Woodraska, D.L.: 2009, *Geophys. Res. Lett.* **36**, L05102. doi:10.1029/2008GL037145.
Culhane, J.L., Harra, L.K., James, A.M., Al-Janabi, K., Bradley, L.J., Chaudry, R.A., *et al.*: 2007, *Solar Phys.* **243**, 19. doi:10.1007/s01007-007-0293-1.
Curdt, W., Brekke, P., Feldman, U., Wilhelm, K., Dwivedi, B.N., Schühle, U., Lemaire, P.: 2001, *Astron. Astrophys.* **375**, 591.
Fontenla, J.M., Avrett, E.H., Loeser, R.: 1990, *Astrophys. J.* **355**, 700.
Fontenla, J.M., Avrett, E.H., Loeser, R.: 1991, *Astrophys. J.* **377**, 712.
Fontenla, J.M., Avrett, E.H., Loeser, R.: 1993, *Astrophys. J.* **406**, 319.
Fontenla, J.M., Balasubramaniam, K.S., Harder, J.: 2007, *Astrophys. J.* **667**, 1243.
Fontenla, J.M., White, O.R., Fox, P.A., Avrett, E.H., Kurucz, R.L.: 1999, *Astrophys. J.* **518**, 480.
Fontenla, J.M., Curdt, W., Haberreiter, M., Harder, J., Tian, H.: 2009a, *Astrophys. J.* **707**, 482.
Fontenla, J.M., Quémerais, E., González Hernández, I., Lindsey, C., Haberreiter, M.: 2009b, *Adv. Space Res.* **44**, 457.
Fontenla, J.M., Harder, J., Livingston, W., Snow, M., Woods, T.: 2011, *J. Geophys. Res., Atmos.* **116**, D20108. doi:10.1029/2011JD016032.
Harrison, R.A., Sawyer, E.C., Carter, M.K., Cruise, A.M., Cutler, R.M., Fludra, A., *et al.*: 1995, *Solar Phys.* **162**, 233. ADS:1995SoPh..162..233H, doi:10.1007/BF00733431.
Kramida, A., Ralchenko, Yu., Reader, J. (NIST ASD Team): 2013, *NIST Atomic Spectra Database* (ver. 5.0), http://physics.nist.gov/asd. Cited April 2013. National Institute of Standards and Technology.
Landi, E., Young, P.R., Dere, K.P., Del Zanna, G., Mason, H.E.: 2013, *Astrophys. J.* **768**, 94.
Linsky, J.L., Bushinsky, R., Ayres, T., Fontenla, J., France, K.: 2011, *Astrophys. J.* **745**, 25.
MacNeice, P., Fontenla, J.M., Ljepojevic, N.N.: 1991, *Astrophys. J.* **369**, 544.
McClintock, W.E., Snow, M., Woods, T.N.: 2005, *Solar Phys.* **230**, 259. ADS:2005SoPh..230..259M, doi:10.1007/s11207-005-1585-5.
Peterson, W.K., Woods, T.N., Fontenla, J.M., Richards, P.G., Chamberlin, P.C., Solomon, S.C., Tobiska, W.K., Warren, H.P.: 2012, *J. Geophys. Res., Atmos.* **117**, A05320. doi:10.1029/2011JA017382.
Rosner, R., Tucker, W.H., Vaiana, G.S.: 1978, *Astrophys. J.* **220**, 643.
Reale, F.: 2010, *Living Rev. Solar Phys.* **7**, 5. doi:10.12942/lrsp-2010-5.
Schurtz, G.P., Nicolai, D., Busquet, M.: 2000, *Phys. Plasmas* **7**, 4328.
Seaton, M.J.: 1962, *Proc. Phys. Soc.* **79**, 1105.
Seaton, M.J.: 1987, *J. Phys. B, At. Mol. Phys.* **20**, 6363S. doi:10.1088/0022-3700/20/23/026.

Serio, S., Peres, G., Vaiana, G.S., Golub, L., Rosner, R.: 1981, *Astrophys. J.* **243**, 288.

Snow, M., McClintock, W.E., Rottman, G.J., Woods, T.N.: 2005, *Solar Phys.* **230**, 295. ADS:2005SoPh..230..295S, doi:10.1007/s11207-005-8763-3.

Spitzer, L. Jr.: 1962, *Physics of Fully Ionized Plasmas*, Interscience, New York.

Wang, Y., Zatsarinny, O., Bartschat, K.: 2013, *Phys. Rev. A* **87**, 012704.

Wilhelm, K., Curdt, W., Marsch, E., Schühle, U., Lemaire, P., Gabriel, A., *et al.*: 1995, *Solar Phys.* **162**, 189. ADS:1995SoPh..162..189W, doi:10.1007/BF00733430.

Woods, T.N., Chamberlin, P.C., Harder, J.W., Hock, R.A., Snow, M., Eparvier, F.G., Fontenla, J., McClintock, W.E., Richard, E.C.: 2009, *Geophys. Res. Lett.* **36**, L01101. doi:10.1029/2008GL036373.

Wyndham, E.S., Kilkenny, J.D., Chuaqui, H.H., Dymoke-Bradshaw, A.K.L.: 1982, *J. Phys. D, Appl. Phys.* **15**, 1683.

DOI 10.1007/978-1-4939-1182-0_8
Reprinted from *Solar Physics* Journal, DOI 10.1007/s11207-012-0184-5

Survey and Merging of Sunspot Catalogs

Laure Lefevre · Frédéric Clette

Received: 12 July 2012 / Accepted: 29 October 2012 / Published online: 22 November 2012

Abstract In view of the construction of new sunspot-based activity indices and proxies, we conducted a comprehensive survey of all existing catalogs providing detailed parameters of photospheric features over long time intervals. Although there are a fair number of such catalogs, a global evaluation showed that they suffer from multiple limitations: finite or fragmented time coverage, limited temporal overlap between catalogs, and, more importantly, a mismatch in contents and conventions. Starting from the existing material, we demonstrate how the information from parallel catalogs can be merged to form a much more comprehensive record of sunspots and sunspot groups. To do this, we use the uniquely detailed Debrecen Photoheliographic Data (DPD), which is already a composite of several ground-based observatories and of SOHO data, and the USAF/Mount Wilson catalog from the Solar Observing Optical Network (SOON). We also outline our cross-identification method, which was needed to match the non-overlapping solar active-region nomenclature. This proved to be the most critical and subtle step when working with multiple catalogs. This effort, focused here first on the last two solar cycles, should lead to a better central database that collects all available sunspot group parameters to address future solar-cycle studies beyond the traditional sunspot-index time series [R_{i}].

Keywords Catalogs · Surveys · Sun: photosphere · Sunspots · Methods: data analysis · Statistical

1. Introduction

So far, the main sunspot time series available for studying solar activity on long timescales has been the International Sunspot Index [R_{i}], currently derived from a worldwide network

Solar Origins of Space Weather and Space Climate
Guest Editors: I. González Hernández, R. Komm, and A. Pevtsov

L. Lefevre (✉) · F. Clette
Royal Observatory of Belgium, 3 Rue Circulaire, 1180 Brussels, Belgium
e-mail: laure.lefevre@oma.be

F. Clette
e-mail: frederic.clette@oma.be

of more than 80 visual observers. Based on the number of groups [N_G] and spots [N_S] present on the solar disk through Wolf's formula $R_Z = 10 \times N_G + N_S$ (Wolf, 1850), it forms the longest solar-activity record at our disposal. Other parallel sunspot series have been produced, such as the Boulder Sunspot Number (NOAA/USAF), the American Sunspot Number (AAVSO), the Group Sunspot Number (Rg: Hoyt and Schatten, 1998), or sunspot areas (Hathaway, 2010). Other modern solar indices also show a good correlation with the sunspot index, such as the 10.7-cm radio flux ($F_{10.7}$: Tapping and Detracey, 1990, started in 1947), irradiance measured from space (spectral or the total solar irradiance: TSI) or the photospheric magnetic field (observed systematically since the early 1970s). However, except for the Group Sunspot Number, those series are much shorter than the R_i series, spanning only a few recent solar cycles. Therefore, as multi-secular information on past cycles is needed to understand and predict the evolution of solar activity of the Sun–Earth coupling (climate) and to constrain physical models of the solar cycle, only sunspot information can fulfill the new research needs.

A prime limitation in the sunspot series and all unidimensional index time series is that the instantaneous index value actually mixes the contributions of multiple groups and spots captured at different stages of their individual evolution. Still, various recent research efforts, such as the development of advanced solar-dynamo models, need additional information about the location and internal properties and dynamics of active regions. It turns out that we have only recently acquired the technical means to retrieve and efficiently use this extended, but also much more voluminous, information. This extra sunspot information can be extracted from many past solar images and drawings. Some of these drawings, as well as images in white light, Hα, or even Ca II K can often be found directly on observatory websites: the Kanzelhöhe (Otruba, 2006: http://cesar.kso.ac.at/) and Kandilli (http://www.koeri.boun.edu.tr/astronomy/) databases, for example, are very extended and go back to respectively 1944 and 1946. However, most data acquired before the second half of the twentieth century have not been digitized yet and are largely unexploited. Thus, the only information directly at our disposal is condensed in a limited number of historical catalogs that were mostly built many decades ago while the images and drawings themselves were produced, leading to multiple shortcomings and gaps that will be described in this article.

The purpose of the work presented here is to extend the exploration of the mid- and long-term solar activity from the well-known R_i time series towards the complete parameter space characterizing sunspots and sunspot groups by validating and merging the information of available catalogs. In Section 2, we first present a survey of existing solar photospheric data. Sections 2.2 and 2.3 describe in more detail our comparative analysis of the USAF and DPD catalogs. Section 2.3 focuses in particular on the very detailed measurements that make the DPD catalog unique. Section 3 then describes the matching process between the two parallel catalogs, its algorithm, and its outcome, while Section 4 discusses possible applications and future improvements of this combined catalog.

2. Catalog Survey and Selection

As noted in the introduction, given the limited amount of digitized data available before the mid-twentieth century, we chose to focus first on the information readily available and usable in the form of digital catalogs of sunspot parameters. As a starting point, we considered a base of 18 catalogs that can be found on the National Geophysical Data Center (NGDC) website (www.ngdc.noaa.gov/stp/spaceweather.html).

2.1. Survey of Existing Sunspot Catalogs

The columns in Table 1 correspond to the different catalogs, while the rows list the parameters contained in each catalog. Distinct catalogs that are presented in the same format with the same contents are grouped in the same column. The list of parameters displayed here is non-exhaustive: parameters that are unique to one catalog or appear only rarely were left aside. We also highlight the temporal coverage of each catalog, which sets the temporal window for each sunspot parameter.

This panorama shows the parameters that are included in almost all catalogs and are thus considered as primary sunspot information, namely: the time of observation (with varying formats and conventions), the size of sunspot groups and their positions, as well as the number of spots in the group. While identification numbers and a morphological classification of sunspot groups are also often included, other physical sunspot properties such as sunspot area or magnetic classification and other conventional parameters such as the image quality are missing in most catalogs. Even when they appear, sunspot parameters and measurements are not normalized and standardized. Different schemes, methods, and definitions were used; some of them are even unique to one catalog, making catalogs incompatible even when they share similar sunspot parameters.

Moreover, we can distinguish two main kinds of catalogs: "group" catalogs providing information on the group properties and evolution (Rome, USAF, RGO) and more comprehensive "sunspot" catalogs providing the description of single sunspots (Kodaikanal, DPD). Our survey showed that those two kinds of catalogs contain distinct, nonoverlapping information. In either case, the information is incomplete and gives a truncated view of sunspot activity: either a global view of group properties and histories, with only a very crude and schematic description of the internal group dynamics, or a close-up view of actual sunspots but without any direct notion of their clustering and collective behavior.

Therefore, in order to correct this inadequacy, we considered that it was definitely worthwhile to build a composite catalog, by combining catalogs belonging to both categories. As a testbed case, we decided to use the detailed information from the Debrecen Photoheliographic Data (DPD: Mezo and Baranyi, 2005; Gyori *et al.*, 2005, 2011) and the US Air Force multi-station catalog (USAF) data to build a new, more complete sunspot catalog. Those two catalogs were chosen for their comprehensiveness and complementarity and because they provide a good match in temporal coverage, with an intersection spanning the period 1986 to 2010, *i.e.* a bit more than two solar cycles.

2.2. The USAF Catalog

The US Air Force maintains a worldwide network of ground-based solar observatories known as the *Solar Observing Optical Network* (SOON) to carry out 24-hour/day synoptic and solar patrol observations. SOON has been in operation since 1977, and the resulting catalog, accessible on the NGDC website (ftp.ngdc.noaa.gov/STP/SOLAR_DATA/SUNSPOT_REGIONS/USAF_MWL), contains exploitable data from 1981 to 2010. SOON produces daily sunspot drawings on an 18-cm diameter projected image. Figure 1 shows the relative contribution of each SOON station to the total number of observations over the 1986 – 2010 interval considered in the merging operation. You can see that from 1987 to 1991, there were seven stations observing at a similar level. On the other hand, the Culgoora station (Australia) ceased observing after April 1992, thus showing a diminished distribution for this particular year. Currently, only three SOON telescopes are still in operation, those at Holloman Air Force Base, Learmonth in Australia, and San Vito in Italy.

Table 1 Survey of sunspot catalogs.

Parameters Temporal coverage	RGO 1874 1982	Kodaikanal 1906 1987	Mt. Wilson 1927 2011	ROB 1940 2011	Rome 1958 2000	Taipei 1964 1994	USSR 1968 1991	SGD 1969 1982	Catania 1978 1999	DPD 1980 2011	Yunnan 1981 1992	Coverage 1874 2011
Date	y, m, d, dec	PT	UT	UT	UT	UT	y, m, d, dec	y, m, d	UT	UT	UT	1874 2011
Groups area	U, U + P	U	U + P	Ø	U, U + P	U, c	U + P, c	U + P, c	U + P	U, U + P	U + P	1874 2011
[µsd, µsh]	p, c	p, c	c	Ø	p, c	U + P, p, c		Ø	p, c	p, c	p, c	
Spots area [µsd, µsh]	Ø	spots	Ø	Ø	Ø	Ø	Ø	Ø	Ø	spots	largest spot	1906 2011
Nb spots	Ø	•	1981 2011	•	•	•	•	•	•	•	Ø	1906 2011
Latitude	•	•	•	•	•	•	•	•	•	•	•	1874 2011
Longitude	•	•	Ø	•	•	•	•	•	•	•	•	1874 2011
LCM	•	Ø	•	•	Ø	Ø	•	•	•	•	•	1874 2011
ID nb	local	Ø	Mt. Wilson	Ø	local	local	local	MtWilson	Ø	•	local	1874 2011
	< 1976	Ø	NOAA	Ø	•	•	•	•	Ø	NOAA	•	1981 2011
Classification	local	Ø	modified	modified	McIntosh	Zurich	Ø	Zurich	Zurich	Ø	McIntosh	1940 2011
	< 1976	Ø	McIntosh	McIntosh	•	•	Ø	Brunner	Ø	Ø	Ø	
Magnetic class	•	Ø	•	Ø	Ø	Ø	Ø	•	Ø	Ø	Ø	1874 2011
Dist. from disk center	•	Ø	Ø	Ø	•	•	•	Ø	Ø	•	•	1874 2011
Position angle	•	Ø	Ø	•	•	•	Ø	Ø	•	•	Ø	1874 2011
Image	Ø	•	•	•	•	Inverted	Ø	•	•	Ø	Ø	
Quality						1981, 80						1906 2011

This table summarizes the parameters available in a majority of the catalogs. Missing information is represented by slashed circles, available information by large dots. Some secondary parameters are not represented here because they occur too rarely. Mount Wilson data include Kandilii, Tashkent, USAF, Voroshilov, USNO, and Boulder data. They all follow the Mount Wilson format. µsd and µsh stand for millionth of the solar disk and hemisphere, respectively. U and U + P stand for umbral and whole-spot areas, respectively, while "p" and "c" stand for *projected* (as seen on the solar disk) and *corrected* (as appearing on the Sun). "Local" means the ID or the Classification of the group is valid only for this observatory/dataset. RGO: Royal Greenwhich Observatory, ROB: Royal Observatory of Belgium, SGD: Solar Geophysical Data, USSR: Russian Books, DPD: Debrecen Photoheliographic Data.

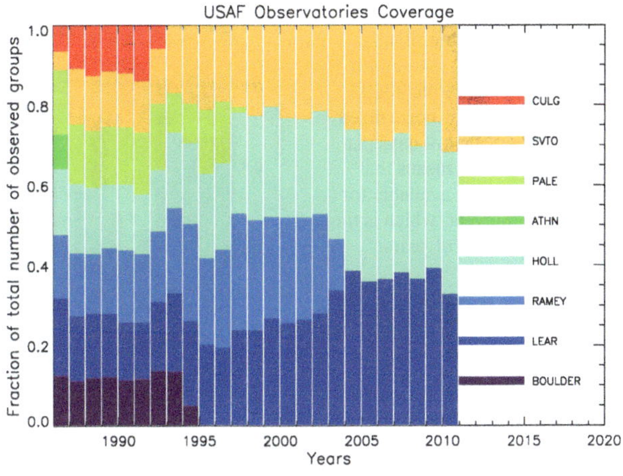

Figure 1 Relative contribution of each USAF/SOON station to the resulting sunspot catalog in terms of the total number of observations per year during the 1986 – 2010 period. The ordinate axis is normalized to the total number of groups included in the catalog for each year. Stations are Boulder (US), Learmonth (Australia), Ramey (US), Holloman (US), Athens (Greece), Palehua (US), San Vito (Italy), and Culgoora (Australia). The global percentage of observed days over this time period is 86.3 %.

Groups are uniquely numbered by the National Oceanic and Atmospheric Administration (NOAA) during their existence over one solar rotation (http://www.swpc.noaa.gov/ftpdir/latest/SRS.txt): we refer to this number as the *NOAA group number*. The report of each SOON station is included separately in the daily coded USSPS message sent daily to all regional warning centers of the International Space Environment Service (ISES). The main catalog entries are: *NOAA group number*, size of group [μsh], number of spots, group longitudinal extent [GLE, degrees], heliographic latitude, longitude from central meridian [LCM], as well as the modified Zurich/McIntosh type (McIntosh, 1990) and the Mount Wilson magnetic classification (Bray and Loughhead, 1964). Areas and positions are measured by hand, with disk overlays, which limits the accuracy of measurements (the minimum group size is 10 μsh).

Two peculiarities in the catalog structure must be taken into account regarding the group identification. Firstly, the catalog often contains several entries per day for one particular group, each one coming from the raw observing report of one of the SOON stations. As observing times often differ by several hours, different McIntosh types are sometimes attributed to the same group for the same date. In addition, if a group identified by the SOON observatories was not given an official NOAA number (different period of observations or seeing conditions between NOAA and SOON), the catalog lists it as a new group, with a suffix appended to the NOAA number of the closest NOAA-identified group.

2.3. The DPD Catalog

The DPD catalog should eventually cover the period 1970 to the present. The construction and digitization work was recently performed in the framework of the SOTERIA project (Lapenta and SOTERIA Team, 2007) and is still in progress. As of 2012, the preliminary catalog is complete from 1986 to 2011. Most of the data in this catalog are based on photographic and CCD images from the Gyulia and Debrecen Observatories (Figure 2). Solar images from several other observatories, including SOHO/MDI images (Scherrer *et al.*, 1995), are used to fill in the gaps in the observations from the primary stations.

The DPD catalog lists all groups and spots for each day, with an excellent temporal coverage: 98.9 % of all days in the studied period. It provides extensive details for each spot and group of spots: time of observation, NOAA number of the corresponding group,

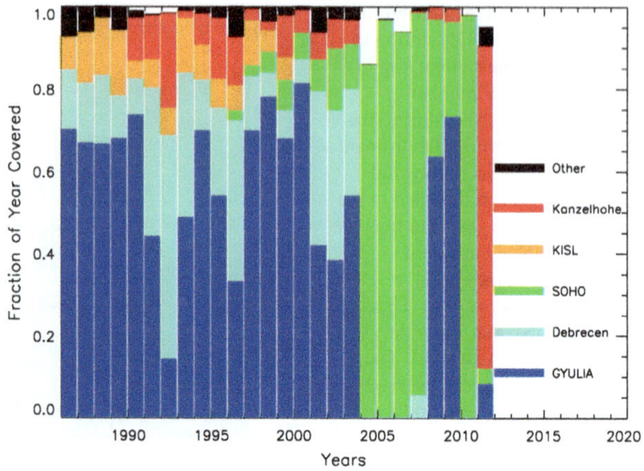

Figure 2 Relative temporal coverage of each contributing station in the DPD dataset, for each year from 1986 to 2010. The ordinate axis is normalized to the total number of days per year. A value of 1 means that observations were available for all days of the year, which is the case in a majority of years.

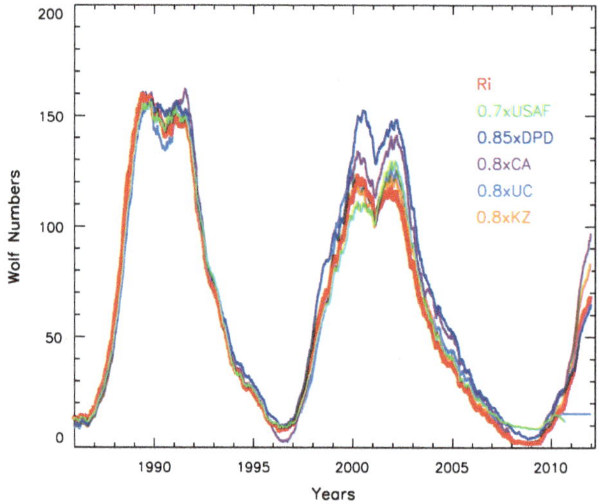

Figure 3 Comparison of the International Sunspot Number [R_i] (red) with Wolf numbers [$10 \times N_G + N_S$] computed from the number of groups [N_G] and spots [N_S] listed in the DPD (blue) and USAF (green) catalogs, as well as individual stations, namely: Catania (purple), Uccle (light blue), and Kanzelhöhe (orange). All numbers have been scaled to match R_i over Solar Cycle 22, thus providing a scaling *relative* to Cycle 22. The corresponding scaling ratios are indicated in the legend.

projected and *corrected* areas of umbrae [U] and whole spot [U + P], heliographic latitude [B] and longitude [L], LCM, position angle [B_0], and distance from the disk center. The accuracy on spot positions is about 0.1 heliographic degrees, while the accuracy on whole-spot areas is about 10 % (Gyori *et al.*, 2005). The *projected* area is the observed area as it appears in the solar disk, while the *corrected* area is the effective area on the Sun itself. The DPD catalog and documentation on its format can be found at http://fenyi.solarobs.unideb.hu/DPD/index.html.

In order to assess the consistency of this dataset, we compute the Wolf number, as defined above, using the sunspot and sunspot group counts based on the DPD sunspot lists [SN_{DPD}] and the USAF group data. In Figure 3, we compare both of them to the International Sunspot Index [R_i] as well as other individual stations from the worldwide R_i-index network. The close match of the global cycle shape and the ratio between the maxima of Solar Cycle [SC] 22 and Cycle 23 [$r_{22/23} = \frac{\langle \mathrm{Wolf}(1989-1991)\rangle}{\langle \mathrm{Wolf}(2000-2002)\rangle}$] are our prime indicators. While the USAF-based series shows slight deviations in shape, its $r_{22/23}$ closely matches the reference R_i ratio and the ratios from individual stations ($r_{22/23}(R_i) = 1.4 \pm 0.1$, $r_{22/23}(\mathrm{USAF}) = 1.3 \pm 0.1$, $r_{22/23}(\mathrm{Catania}) = 1.2 \pm 0.1$, $r_{22/23}(\mathrm{Uccle}) = 1.3 \pm 0.1$, and $r_{22/23}(\mathrm{Kanzelhöhe}) = 1.4 \pm 0.1$,

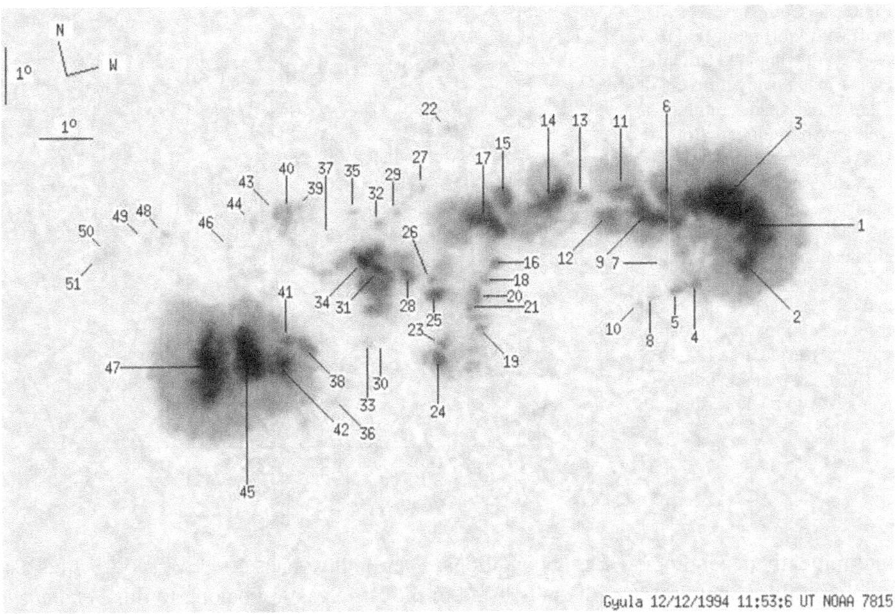

Figure 4 Large and complex sunspot group 7815 on 12 December 1994, classified as McIntosh type Ekc in the USAF catalog. Individual features are tagged with their entry number in the DPD catalog entries. Features 10, 30, 37, 49, and 51 are listed as *transient* features (umbra [U] = 0 and whole-spot area [U + P] = 0). Features 7, 8, 22, 27, 33, 36, 39, 44, 46, 48, and 50 are cataloged as isolated *penumbrae without umbrae* [U = 0 and U + P ≠ 0].

Figure 3). By contrast, we observe a double discrepancy between SN_{DPD} and all other Wolf numbers. First, there is a difference in cycle shape (a significant shoulder on the rising phase of SC23, flat maximum without Gnevyshev gap in SC22). Secondly, the Cycle 22/23 ratio is abnormally low ($r_{22/23}(\mathrm{DPD}) = 1.1 \pm 0.1$, Figure 3).

The source of this difference resides in the details of the DPD catalog, which lists about 15 % more sunspot groups than, *e.g.* the USAF catalog. In order to build the DPD catalog, a dedicated software was developed by the Debrecen team to detect the contours delineating the umbra–penumbra and the photosphere–penumbra boundaries in photographic or CCD images of sunspot groups (Gyori, 1998; Gyori *et al.*, 2011). Those measurements are then converted to umbral and penumbral areas and heliographic coordinates of spot centers. By reducing human intervention, this method reduces the processing time and costs but, foremost, improves the homogeneity of the output values. However, due to the interplay between seeing effects and the intensity thresholding used, the splitting between the umbral area [U] and penumbral area [P] given in the DPD catalog is not trivial. Baranyi *et al.* (2001) show that DPD umbral areas are systematically underestimated compared to umbral areas from the Rome observatory, while whole-group areas show a very good agreement.

Figure 4 shows an example of a very complex group recorded in the Debrecen image collection (http://fenyi.sci.klte.hu/DPD/1994/19941212/19941212_7815.html). Next to the main spots, the image contains several weakly contrasted features that are cataloged either as *transient* structures, *i.e.* features that are not observed in all the frames taken for a particular day, or as isolated *penumbrae without umbrae*. The latter should not be considered as spots in a strict observational sense: part of those marginal objects may be real tiny spots without penumbrae or pores, but at the limit of the spatial resolution and detectable only

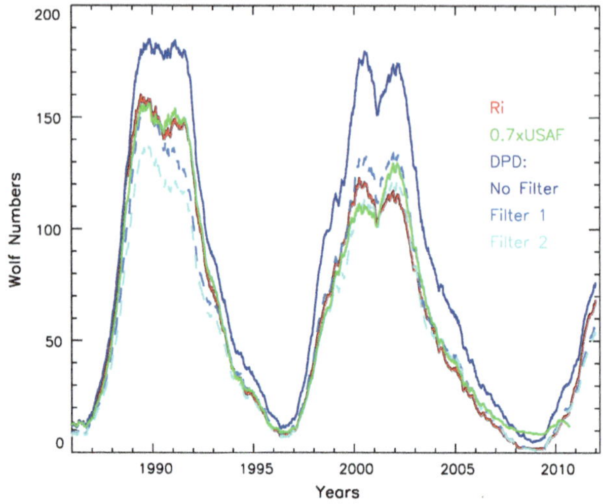

Figure 5 Comparison of R_i (red) with Wolf numbers [$10 \times N_G + N_S$] computed from the DPD catalog using different data filters (three shades of blue), as well as Wolf numbers from USAF data (green). The dark blue corresponds to the whole unfiltered catalog. The light-blue dashed line is obtained after elimination of transient features and *penumbrae without umbrae* and the cyan dashed line results from the elimination of features with a corrected area U + P ≤ 1 µsh, *i.e.* we keep only features with a whole-spot area ≥ 2 µsh.

intermittently depending on seeing conditions. Their inclusion as effective spots is thus disputable. Still, about 50 % of all spots listed in the DPD catalog belong to this *penumbrae without umbrae* category. So, despite their small dimensions, they play a significant role in the overall catalog statistics. Moreover, a neighboring category, spots without penumbra, is strongly under-represented: such spots, which usually form groups of McIntosh type A and B, account for less than 0.1 % of all listed spots. This imbalance suggests that a substantial part of the smallest spots ended up either in the *penumbrae without umbrae* category, when unresolved, or among spots with rudimentary penumbra, *i.e.* having a small though non-null penumbra to umbra ratio.

Our investigation thus indicates that the DPD catalog is highly comprehensive, to the point of including pores or features that do not usually qualify as sunspots in long-term visual observations. To correct for this bias, we filtered the data according to different criteria, involving a combination of both the total spot size and the umbra/penumbra ratio. The main results are shown in Figure 5.

When first excluding transient features and *penumbrae without umbrae*, the Cycle 22/23 ratio rises from less than 1.1 to 1.2, which is equivalent to the Catania ratio, although it remains slightly too low compared to ≈ 1.4 for R_i. The Pearson correlation coefficient also increases from 94 to 95 %. We also removed spots with a *corrected area* lower than 2 to 5 µsh. We found that when the lower size limit is increased from 2 to 5 µsh, the correlation between SN_{DPD} and R_i starts to decrease again. Likewise, the ratio between the maxima of Cycle 22 and 23 varies around 1.1, while the number of features taken into account is drastically reduced. We thus conclude that the optimum filtering consists in only rejecting *transient* features and *penumbrae without umbrae*, *i.e.* features with a dubious identification as sunspots. However, as those marginal features may be relevant to other topics of solar research, our output catalog will be provided in both filtered and unfiltered versions.

3. Merging Catalogs

After this sunspot-level selection process, which is possible only with the DPD catalog, we take advantage of the temporal overlap between both catalogs to merge the contents, *e.g.* by

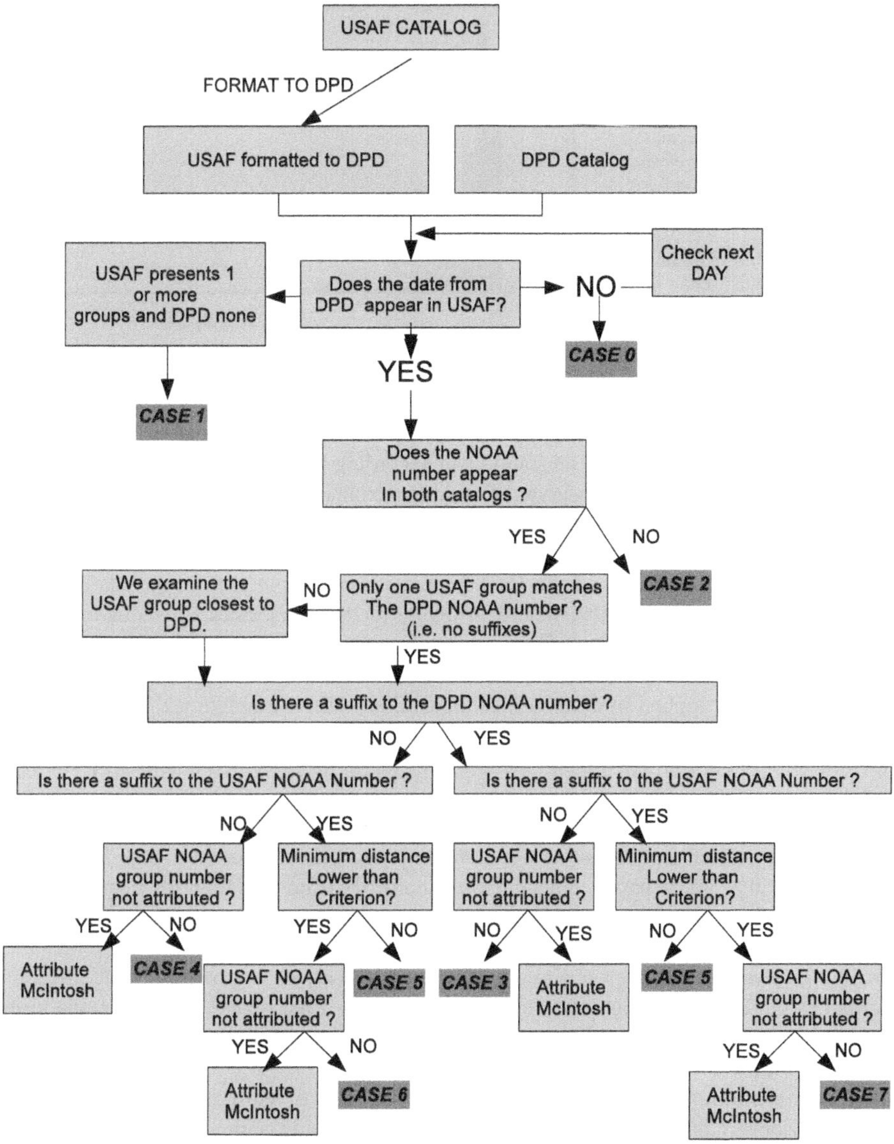

Figure 6 Flowchart of the automated group matching algorithm between DPD and USAF catalogs. As the DPD catalog has the best temporal coverage, it is taken as a reference; *i.e.* each group from the DPD catalog is associated with its counterpart in the USAF catalog, if available and correct. Cases 0 through 7 correspond to the eight situations where human intervention is required to complete the cross identification: they are developed in Table 2. The distance criterion is explained in Figure 7.

adding essential parameters such as the USAF modified McIntosh morphological classification (McIntosh, 1990) to the DPD sunspot group catalog. This requires the development of a global method for matching the sunspot groups recorded in both catalogs. An overall flowchart of this matching algorithm is given in Figure 6.

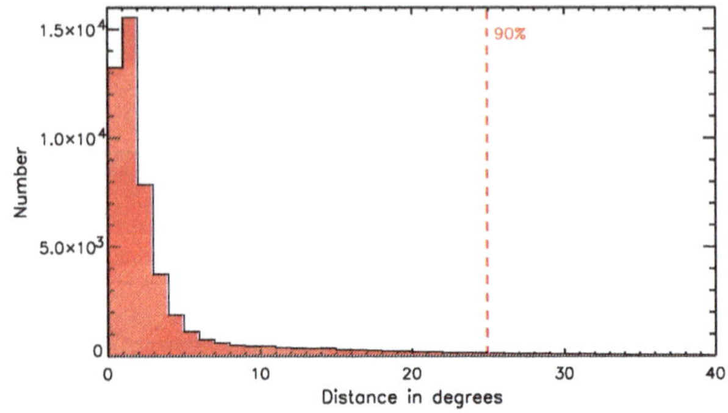

Figure 7 Histogram of the Euclidean distance between the positions (heliographic longitudes and latitudes) listed for the same group in the USAF and DPD catalogs (in degrees). From this distribution, we can derive a distance criterion for group matching. A rather conservative limit is 25° (dashed red line); 90 % of groups have a distance below this value.

As the USAF catalog includes observations from several stations on each date, we first weight the observations from each station according to the difference in location and time between that station and the DPD station and according to its image quality (see Table 1), keeping only one station per observation. Taking advantage of this daily redundancy in the USAF data enables a closer match with the DPD information. Then for each day, we associate each group listed in the DPD dataset with a group from the USAF catalog, using their NOAA group number and the Euclidean distance computed from their heliographic latitude and longitude. The statistical distribution of those distances allows us to derive a maximum distance criterion (see Figure 7; only groups separated by less than 25° are considered as valid matches). In this process, the main difficulty comes from the different group splitting methods applied in those two catalogs: this leads to ambiguities in the respective numbering schemes when suffixes are added to the official NOAA group number. The group identification algorithm must then extend the search to all possible nomenclatures in both catalogs.

Once this matching of groups is completed, it then becomes possible to add other complementary parameters such as the Mount Wilson magnetic class or the USAF longitudinal extent, as well as group parameters computed indirectly from the DPD single-spot data.

Overall, about 80 % of the DPD sunspot groups have an immediate correspondence in the USAF group catalog (Table 2). The remaining 20 %, where the matching is problematic, mainly involves "orphan" groups in the DPD catalog. These are groups without any counterpart in the USAF catalog (11.3 %), mostly because of the higher level of details in the DPD catalog. The remaining mismatches also largely occur when the group positions in the USAF and DPD catalogs are too far apart (3.7 %).

As an illustration, Figure 8 shows two contrasted cases of sunspot-group matching. The left panel shows a group from the DPD catalog that is absent from the USAF catalog: NOAA 6234 is actually listed as a *penumbra without umbra*, which can explain why it is not found in the USAF data. However, the right panel shows at least four *penumbrae without umbrae* groups (NOAA 6390, 6399, 6408, and 6411) that appear in the USAF catalog, and should thus have been classified as real groups (with at least one spot) in the DPD catalog. These differences could help to further improve the current filtering based only on the DPD information. In those marginal cases, the presence or absence of a sunspot group in the USAF catalog can be exploited to remove the ambiguity for many *penumbrae without umbrae* entries in the DPD catalog. However, this advanced process is complex and cannot solve all cases, because, with the USAF data, it can only rest on group information and not the properties of single sunspots.

The global success rate of this automatic matching process reaches almost 97 % (when excluding groups filtered from the DPD catalog). Therefore, human intervention proved nec-

Table 2 Results of the matching between the DPD and USAF datasets.

Case	Number	Percent	Description
Total Groups	54 973		45 656 effective groups after filtering described in Section 2.3.
Matching cases	44 113	80.2 %	96.6 % of effective groups
Non-matching	**10860**		Total of all cases below, except 1 (**boldface**)
CASE 0	**523**	0.9 %	No USAF observation for DPD
CASE 1	141	0.3 %	No DPD observation for USAF
CASE 2	**6241**	11.3 %	USAF gives no corresponding NOAAs
CASE 3	**1968**	3.6 %	USAF = "" and DPD = "a" + Already used
CASE 4	**19**		Only one match but already used
CASE 5	**2011**	3.7 %	DIST > criterion
CASE 6	**0**		DIST ≤ criterion but already used
CASE 7	**98**	0.2 %	USAF = "A" and DPD = "a" + Already used
10860: Penumbrae	7299		Penumbrae without umbrae
Transient	2018		Transient features
Diverse	1543		Unknown: Investigation needed (3.4 %)

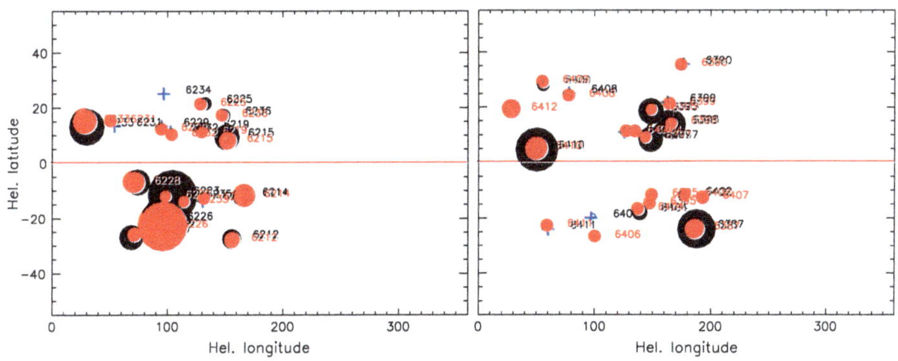

Figure 8 Synoptic maps of the Sun built from the catalogs' data for 25 August 1990 (left panel) and 10 December 1990 (right panel) during the matching process. USAF group entries (red disks) are overplotted on top of DPD group entries (black disks). The disk size is proportional to the group size. Blue crosses mark DPD groups belonging to the problematic *penumbrae without umbrae* category. NOAA group numbers are printed on the right side of each group (group 6234 on the left panel corresponds to the blue cross).

essary only for a few large complex active regions in order to complete the merging process. Table 3 shows a section from the resulting catalog in its unfiltered version: the format is largely inspired by the DPD catalog, but with many additional fields at the end of the group section ("g") and with the number of groups appended to the date line ("d"). In the filtered version of the catalog, the weighted area and positions of the groups are automatically adjusted. Filtered and unfiltered versions of the catalog, as well as a summary catalog containing only the group information, can be found at http://sidc.oma.be/merged-sunspot-catalog/.

To further assess the quality of this merging, we compare the group longitudinal extents [GLE, in degrees] extracted from the DPD data to the ones from corresponding groups in the USAF catalog. In particular, in the new catalog, we can now choose all the groups that have a

Table 3 Sample section extracted from the unfiltered merged catalog combining the information from the DPD and USAF datasets: data for 3 March 1986 at 07:33:00 (UT).

d	UT	KISL		98	639	78	499	2446492.81458			−22.05	−7.24	2									
g	UT	NOAA		Up	U + Pp	Uc	U + Pc	lat	long	LCM	PA	DCS	NS	Zpc	E0	E1	E2	DA	E	W	E3	M
		4717		98	628	78	493	−0.59	35.93	−50.12	84.74	0.7727	7	DKC	4.93	5.80	1.96	38.9	75	3	6	B
s	UT	4717	1	49	171	38	133	−0.20	36.45	−49.60	84.14	0.7678										
s	UT	4717	2	46	302	37	242	−0.47	34.89	−51.16	84.80	0.7844										
s	UT	4717	3	0	3	0	3	1.11	34.54	−51.51	82.89	0.7909										
s	UT	4717	4	0	14	0	10	0.08	39.36	−46.69	83.11	0.7352										
s	UT	4717	5	0	41	0	31	−0.86	38.66	−47.39	84.53	0.7416										
s	UT	4717	6	3	84	3	64	−1.64	37.08	−48.97	85.88	0.7583										
s	UT	4717	7	0	13	0	10	−2.12	35.47	−50.58	86.80	0.7755										

“d” stands for date, “g” for group, and “s” for spot. In section g, the parameters are as follows. Up and U + Pp: umbral and whole-spot projected areas in μsh. Uc and U + Pc: same for corrected areas. LCM: longitude from the central meridian. PA: polar angle. DCS: distance from the center of the Sun [in solar radii]. NS: number of spots. Zpc: modified McIntosh morphological type. E0: longitudinal extent from the DPD data. E1: diagonal of the rectangle containing the DPD spots. E2: DPD dipole extent. E3: USAF longitudinal extent. E and W: east- and west-side sunspot areas of dipolar groups in μsh. M: Mount Wilson magnetic class.

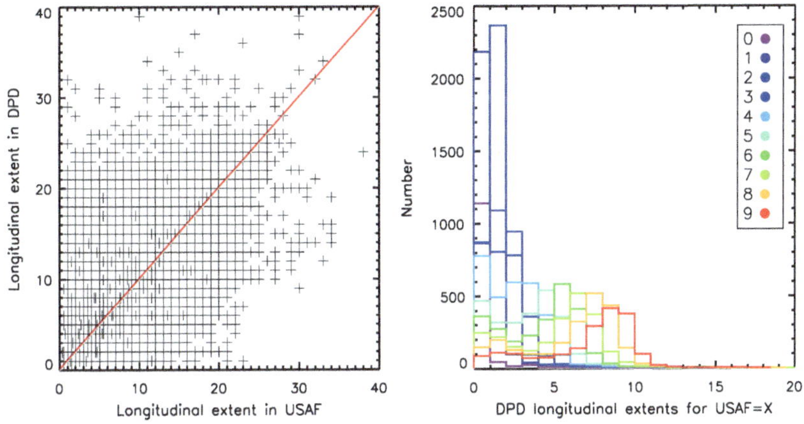

Figure 9 Left panel: Scatterplot of the longitudinal extent [GLE, in degrees] of corresponding groups in the DPD catalog (vertical axis) and the USAF catalog (horizontal axis). The red diagonal marks the unit ratio ($x = y$). Right panel: Distributions of DPD-GLE for successive values of the corresponding GLE in the USAF catalog, in integer steps between zero and nine degrees. Penumbrae without umbrae and transient groups are not included. The zero bin in the USAF data corresponds to an absence of measurement.

fixed USAF extent and plot the distribution of DPD extents within this collection of groups. Figure 9 shows the direct comparison (left panel) and the distributions of DPD extents for subsets of USAF groups having a fixed integer extent between zero and nine degrees (right panel). Globally, USAF-GLE and DPD-GLE are in good agreement: the Pearson correlation is more than 70 % (significance > 99.99 % for more than 35 000 sample values). This is no overall scaling bias, as each DPD-GLE distribution is centered on the same GLE value in the USAF catalog (right panel). The only exception appears for the smallest groups, for which DPD-GLE are slightly larger than USAF-GLE (*i.e.* more outlying spots fall outside of the USAF ones). The overall dispersion can be attributed to differences in the time when the group was measured, *i.e.* outlying spots appearing and disappearing, as well as the method of measurement, which is much more precise for the DPD catalog. Once again, this is evidence that the smallest features are taken into account more in the DPD catalog than in other catalogs, and the measurements by the USAF network are made at a lower resolution.

4. Applications and Discussion

Our study and the resulting catalog, which combines the largest number of sunspot parameters available so far, provide a good reference for the construction of future, modern catalogs based on the wealth of information available in recent photographic and CCD images. Ensuring that such catalogs always include the same information will help to maintain common standards and to make the combined use of sunspot catalogs much easier and efficient – a major improvement compared to the current situation.

The methods that were developed to assemble those two rather recent catalogs can also be transposed to the recovery and merging of the much more sparse information in historical collections of visual and photographic sunspot observations. Thus, they provide useful tools to reconstruct and extend past fragmentary and randomly formatted sunspot lists.

Beyond this general quest for standardization, the merged catalog also leads to immediate practical applications. It can now form the base for the construction of advanced long-term solar activity indices (flare index, CME rate, Ca II K plage index, Mg II core-to-wing ratio) and solar flux proxies ($F_{10.7}$ radio flux, total and spectral solar irradiance, solar open magnetic flux, solar wind), in response to a growing scientific demand for direct information about the secular evolution of solar activity. This information is needed for constraining the latest physical models of solar dynamo and also for analyzing the long-term Sun – Earth relations, including climate change.

In this context, it turns out that the Sun has been undergoing a Grand Maximum throughout all modern measurements (Solanki *et al.*, 2004; Vonmoos, Beer, and Muscheler, 2006; Steinhilber, Abreu, and Beer, 2008), while the last cycle, Solar Cycle 23, marked a return to activity levels that were encountered before the mid-twentieth century (Lefèvre and Clette, 2011) and thus most modern solar measurements. Therefore, in spite of the current abundance of solar data, we actually lack direct information about the Sun in regimes lower than the high regime that prevailed over the last 60 years. The only direct solar information at our disposal comes from sunspot observations. However, so far, long-term proxies designed for secular scales have been almost exclusively based on the sunspot-number time series. This strongly limits the reliability and diversity of proxy reconstructions, as the sunspot index mixes in an instantaneous snapshot the properties of all groups visible at a specific time. The independent evolution of each group is thus inaccessible. Moreover, the sunspot index series only includes temporal information, missing the spatial patterns of active regions tracing each solar cycle (longitudinal and latitudinal distributions, magnetic dipoles).

Therefore, the much wider array of sunspot parameters included in our merged catalog opens much wider possibilities to improve and refine current indices and proxies. As the time span of this dataset is limited to the last 35 years, we can use it primarily to validate the relation between various target solar indices and fluxes and the array of sunspot parameters available in our catalog. Then, once established and validated using actual modern measurements, those relations open the way to the reconstruction of new long-term proxy series resting on all past historical sunspot observations, as those proxies make use exclusively of sunspot information which was also available back in the distant past when no other solar information was available. Just as an initial illustration of this potential, we explored a few cross-relations.

As methods we used principal component analysis (PCA: Chatfield and Collins, 1990), or singular value decomposition (SVD: Golub and Van Loan, 2000), as they can help in understanding the linear relations between solar parameters through multidimensional scaling (Chatfield and Collins, 1990). The notion of mutual information (Kraskov, Stögbauer, and Grassberger, 2004) can also help measure the degree of similarity between any kind of parameters for non-linear relations. The two-dimensional (2D) maps shown in Figure 10 were obtained by applying SVD to different sunspot and sunspot groups parameters from our catalog, as well as standard indices like the International Sunspot Number [R_i] and the flux at 10.7-cm [$F_{10.7}$]. We also included the Wolf number extracted from the DPD data with and without filtering (SN and SN0, respectively). All of the described variables were first normalized with respect to time and then subjected to SVD/PCA. In this mapping, all included parameters are projected on two axes which maximize the variance among them. The remaining fraction of the total variance is small and contained in the other non-mapped dimensions. The axes themselves and their units do not match any physical dimension, but the mapping is such that the relative distance between the different parameters is inversely proportional to the correlation between their corresponding data series. Another interesting property of SVD/PCA is the fact that linear combinations of different quantities will be connected by a straight line.

The left panel of Figure 10 demonstrates this very useful property for linear combinations $SN0 = 10 \times N_{G0} + N_{S0}$ or $SN = 10 \times N_G + N_S$. Note how the filtering that rejects features

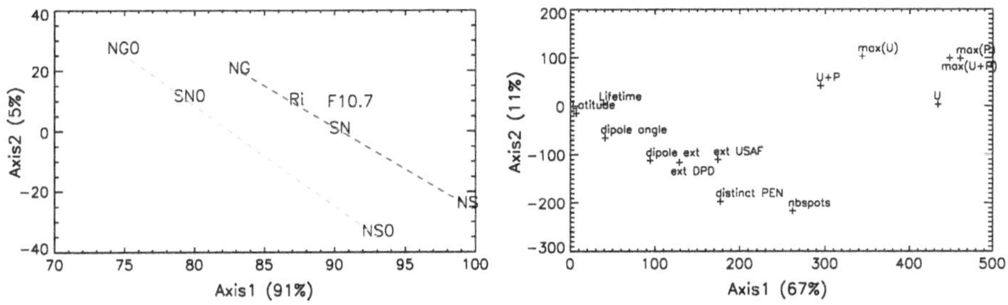

Figure 10 Principal component analysis (PCA) of sunspot parameters. Left panel: PCA for eight different indices of solar activity: R_i, the 10.7-cm radio flux ($F_{10.7}$) and the DPD data (unfiltered "0" suffix, filtered: no suffix), number of groups (N_G, N_{G0}), number of spots (N_S, N_{S0}), and Wolf number (*SN* and *SN*0). Right panel: PCA of thirteen different sunspot group parameters. Lifetime, latitude, and dipole angle are self-explanatory. Dipole ext. is the dipole extent of the group, *i.e.* the distance between the two magnetic poles (main spots). ext. USAF is the longitudinal extent from the USAF catalog, while ext. DPD is the same for the DPD catalog. distinct PEN is the number of distinct penumbrae inside the group, and nbspots is the number of spots inside the group. U + P and U are the total and umbral area of the group, while max(U + P), max(U), and max(P) are maximum total, umbral, and penumbral areas. The output of these PCAs is reduced to the two dimensions over which the similarities and dissimilarities between parameters (*i.e.* the variances) are largest. In the resulting 2D maps, the variance along each axis is indicated in the axis label (in parentheses).

with dubious sunspot identification in the DPD catalog seems to drastically increase the correlation of the DPD Wolf numbers with R_i and also $F_{10.7}$. The right panel of Figure 10 shows the mutual relations among the sunspot-group parameters themselves. In this map, sunspot parameters are clearly concentrated around two clusters: parameters located in the upper right are rather dispersed and thus appear as rather independent, while parameters within the large band in the lower left are largely interchangeable, as they are distributed along a common line. This well-defined clustering already indicates that different subsets of sunspot properties are containing distinct information and are probably associated with distinct underlying processes. This brings a first clear confirmation that the multi-parametric sunspot description provided by the catalog indeed carries additional information, by distinguishing aspects that are mixed in the classical sunspot index series. The in-depth interpretation of this well-defined clustering in different classes of sunspot properties goes beyond the scope of this article and will be the topic of further investigations.

Our catalog also offers for the first time the possibility to investigate how groups of different sizes and morphologies differ in terms of the various individual properties of spots and groups. For instance, we can extract the distribution of spot sizes inside groups of different McIntosh types, A to H (Figure 11). Here, all distributions have a simple profile, peaking around five and then decreasing exponentially, except for types C and H, which also feature an excess for large sizes above 50 (Figure 11, left panel). The difference is even clearer in the right panel of Figure 11, which represents only the size of the largest spot in each group: C and H groups both present a bimodal distribution not seen in other groups. In both cases, this is a clear indication that those two classes are actually mixing two distinct populations of sunspot groups, with or without a major spot in the same size range as large D, E, or F groups. This conclusion matches well an equivalent result obtained for sunspot lifetimes by Lefèvre and Clette (2011). There, C-class groups also featured a unique broad and flat distribution combining the simple unimodal distributions of small A – B groups and large D – E – F groups. This suggests that the current sunspot classifications do not represent the actual populations of sunspot groups. It also shows that the comprehensive information provided by this catalog could help define improved classification schemes for active regions.

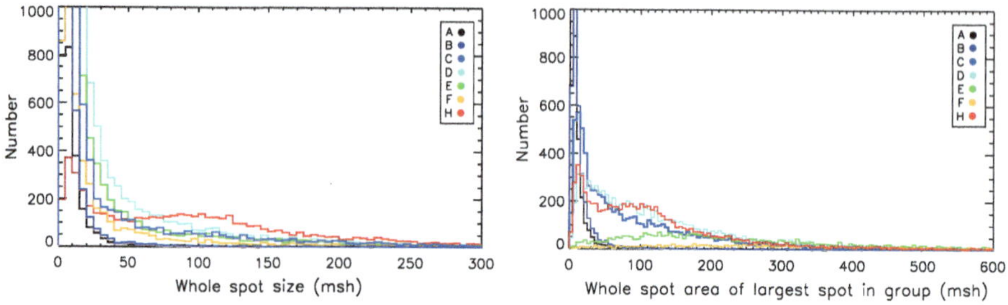

Figure 11 Left panel: Distribution of the whole-spot size [corrected U + P, in μsh] inside groups of McIntosh classes A to H for period 1986 – 2010. Right panel: Distribution of the size of the largest spot [corrected U + P, in μsh] inside groups of McIntosh classes A to H for the same period. Bin size is 5 μsh in both plots.

The completion of this merged catalog marks a primary step. In the near future, the first objective will be to extend the current catalog backwards in time, by using complementary catalogs providing a partial temporal overlap with the current catalog. Table 1 shows that the Rome and Kodaikanal datasets will help cover approximately two more cycles. The Royal Greenwich Observatory (RGO) stands out as the next logical backwards extension before 1976, but the merging will be more challenging, mainly because the morphological classification is a local numerical scale that cannot be directly converted in the more recent McIntosh classification. In addition, another useful sunspot group catalog is currently being completed by our team, and is based on the drawings from the Uccle Solar Equatorial Table: USET (ROB column in Table 1). Covering a long temporal interval of 70 years (1940 – 2012), it straddles the temporal intervals of the current catalog and the RGO catalog and provides a good homogeneity, as it was built recently in a single run. As it largely follows the same standards and includes both the McIntosh and Zurich classifications, the group-matching method developed here can be applied directly and will allow one to associate the NOAA number to the USET entries, a label that is currently missing and will enable a wider compatibility of the USET catalog with other solar catalogs. Conversely, the USET catalog can then be used to expand our current catalog and also to validate it, in particular at the transitions between the successive base catalogs embedded in the overall merged catalog.

When the effort started here has reached its full maturity, in the form of a coherent catalog of detailed sunspot parameters going back to the nineteenth century, solar physicists will have at their disposal a unique tool to probe and understand the past long-term evolution of the solar cycle and solar output, which in turn may improve our ability to predict the future evolution of solar activity.

Acknowledgements This work was funded by the European Community's Seventh Framework Program (FP7/2007-2013) under grant agreement No. 218816 (www.soteria-space.eu). We also wish to thank all the staff at the Konkoly Heliophysical Observatory for their invaluable help. We also acknowledge the support from the Belgian Solar-Terrestrial Center of Excellence (STCE) funded through the Belgian Science Policy Office (BELSPO).

References

Baranyi, T., Gyori, L., Ludmány, A., Coffey, H.E.: 2001, Comparison of sunspot area data bases. *Mon. Not. Roy. Astron. Soc.* **323**, 223 – 230. doi:10.1046/j.1365-8711.2001.04195.x.
Bray, R.J., Loughhead, R.E.: 1964, *Sunspots*, *International Astrophysics Series*, Chapman and Hall, London.
Chatfield, C., Collins, A.J.: 1990, *Introduction to Multivariate Analysis*, Chapman and Hall, London.

Golub, G.H., Van Loan, C.F.: 2000, *Matrix Computations*, Johns Hopkins Press, Baltimore.
Gyori, L.: 1998, Automation of area measurement of sunspots. *Solar Phys.* **180**, 109 – 130. ADS: 1998SoPh..180..109G, doi:10.1023/A:1005081621268.
Gyori, L., Baranyi, T., Ludmány, A.: 2011, Photospheric data programs at the Debrecen observatory. *IAU Proc.* **273**, 403.
Gyori, L., Baranyi, T., Muraközy, J., Ludmány, A.: 2005, Recent advances in the Debrecen sunspot catalogues. *Mem. Soc. Astron. Ital.* **76**, 981.
Hathaway, D.H.: 2010, The solar cycle. *Living Rev. Solar Phys.* **7**(1). http://www.livingreviews.org/lrsp-2010-1.
Hoyt, D.V., Schatten, K.H.: 1998, Group sunspot numbers: a new solar activity reconstruction. *Solar Phys.* **181**, 491 – 512. ADS:1998SoPh..181..491H, doi:10.1023/A:1005056326158.
Kraskov, A., Stögbauer, H., Grassberger, P.: 2004, Estimating mutual information. *Phys. Rev. E* **69**(6), 066138. doi:10.1103/PhysRevE.69.066138.
Lapenta, G., SOTERIA Team: 2007, SOTERIA: SOlar-TERrestrial Investigations and Archives. AGU Fall Meeting Abstracts, A338.
Lefèvre, L., Clette, F.: 2011, A global small sunspot deficit at the base of the index anomalies of solar cycle 23. *Astron. Astrophys.* **536**, L11. doi:10.1051/0004-6361/201118034.
McIntosh, P.S.: 1990, The classification of sunspot groups. *Solar Phys.* **125**, 251 – 267. doi:10.1007/BF00158405.
Mezo, G., Baranyi, T.: 2005, HTML presentation of the Debrecen photoheliographic data sunspot catalogue. *Mem. Soc. Astron. Ital.* **76**, 1004.
Otruba, W.: 2006, Solar monitoring program at Kanzelhöhe observatory. *Sun Geosph.* **1**(2), 020000.
Scherrer, P.H., Bogart, R.S., Bush, R.I., Hoeksema, J.T., Kosovichev, A.G., Schou, J., Rosenberg, W., Springer, L., Tarbell, T.D., Title, A., Wolfson, C.J., Zayer, I., MDI Engineering Team: 1995, The solar oscillations investigation – Michelson Doppler Imager. *Solar Phys.* **162**, 129 – 188. doi:10.1007/BF00733429.
Solanki, S.K., Usoskin, I.G., Kromer, B., Schüssler, M., Beer, J.: 2004, Unusual activity of the Sun during recent decades compared to the previous 11 000 years. *Nature* **431**, 1084 – 1087. doi:10.1038/nature02995.
Steinhilber, F., Abreu, J.A., Beer, J.: 2008, Solar modulation during the Holocene. *Astrophys. Space Sci. Trans.* **4**, 1 – 6. doi:10.5194/astra-4-1-2008.
Tapping, K.F., Detracey, B.: 1990, The origin of the 10.7 CM flux. *Solar Phys.* **127**, 321 – 332. doi:10.1007/BF00152171.
Vonmoos, M., Beer, J., Muscheler, R.: 2006, Large variations in Holocene solar activity: constraints from ^{10}Be in the Greenland Ice Core Project ice core. *J. Geophys. Res.* **111**, 10105. doi:10.1029/2005JA011500.
Wolf, R.: 1850, Mittheilungen über die Sonnenflecken III. *Astron. Mitt. Eidenöss. Sternwarte Zurich* **1**, 27 – 50.

DOI 10.1007/978-1-4939-1182-0_9
Reprinted from *Solar Physics* Journal, DOI 10.1007/s11207-013-0416-3

Sunspot Group Development in High Temporal Resolution

J. Muraközy · T. Baranyi · A. Ludmány

Received: 31 August 2012 / Accepted: 10 September 2013 / Published online: 23 October 2013

Abstract The *Solar and Heliospheric Observatory/Michelson Doppler Imager* – Debrecen Data (SDD) sunspot catalogue provides an opportunity to study the details and development of sunspot groups on a large statistical sample. In particular, the SDD data allow the differential study of the leading and following parts with a temporal resolution of 1.5 hours. In this study, we analyse the equilibrium distance of sunspot groups as well as the evolution of this distance over the lifetime of the groups and the shifts in longitude associated with these groups. We also study the asymmetry between the compactness of the leading and following parts, as well as the time profiles for the development of the area of sunspot groups. A logarithmic relationship has been found between the total area and the distance of leading–following parts of active regions (ARs) at the time of their maximum area. In the developing phase, the leading part moves forward; this is more noticeable in larger ARs. The leading part has a higher growth rate than the trailing part in most cases in the developing phase. The growth rates of the sunspot groups depend linearly on their maximum total umbral area. There is an asymmetry in compactness: the number of spots tends to be smaller, while their mean area is larger in the leading part at the maximum phase.

Keywords Sunspots · Solar activity

Solar Origins of Space Weather and Space Climate
Guest Editors: I. González Hernández, R. Komm, and A. Pevtsov

J. Muraközy (✉) · T. Baranyi · A. Ludmány
Heliophysical Observatory, Research Centre for Astronomy and Earth Sciences, Hungarian Academy of Sciences, P.O. Box 30, 4010 Debrecen, Hungary
e-mail: murakozy.judit@csfk.mta.hu

T. Baranyi
e-mail: baranyi.tunde@csfk.mta.hu

A. Ludmány
e-mail: ludmany.andras@csfk.mta.hu

1. Introduction

According to the generally accepted conception, sunspot groups, or in more general terms solar active regions, emerge from large global toroidal magnetic fields generated at the bottom of the convective zone. The ideas about the causes of emergence are diverse, but it is also widely accepted that the process of emergence is driven by buoyancy and influenced by the strength, twist, and curvature of the flux tubes and the ambient velocity fields. The rising magnetic-flux ropes are mostly imagined to be Ω-shaped formations; their top arches protrude from the convective zone, and the intersections of the flux ropes with the photosphere are observed as sunspots. The directly observable surface properties of sunspot groups primarily provide pieces of information about this complex process, *e.g.* positions, sizes, developments, time profiles of rising and decaying phases, tilt angles, fragmentations, leading–following asymmetries, morphology, internal motions, cycle dependencies, and relations to the velocity fields.

The typical time profile of the sunspot-group development has been known for a long time; its growing phase is usually shorter than its decaying phase. The two phases are governed by two different mechanisms. They were first examined both empirically and theoretically by Cowling (1946). He made calculations based on simple electromagnetic assumptions and obtained an expected decay time of about 300 years. The expected rise time was comparable to this result. These results indicated that these processes cannot be described by simple assumptions based on the conductivity of plasma. Further studies assumed different kinds of motion fields; a detailed summary of these mechanisms is given by Fan (2009).

Considerable effort has been devoted to finding the most realistic theoretical description of the interaction between the magnetic and velocity fields resulting in those phenomena and processes which are directly observable at the surface. Fan, Fisher, and DeLuca (1993) found possible theoretical reasons for the empirical finding that the subgroup of leading polarity tends to be more compact than the trailing part: they assume that the Coriolis force drives the flow in the rising flux from the leading part to the following one, and this was confirmed by Abbett, Fisher, and Fan (2001). Fan, Fisher, and McClymont (1994) also confirmed that the magnetic field of the leading leg in the emerging loop is stronger than in the trailing leg. Moreno-Insertis, Caligari, and Schüssler (1994) and Caligari, Moreno-Insertis, and Schüssler (1995) found that the unstable flux tube ascends with a geometrical asymmetry: the leading leg is more inclined to the vertical direction than the trailing leg. Caligari, Schüssler, and Moreno-Insertis (1998) compared the consequences of two different initial conditions of buoyant ascent, the mechanical equilibrium *vs.* temperature balance, and found that the resulting leading–following asymmetry is different in the two cases. Later three-dimensional work (Abbett, Fisher, and Fan, 2000, 2001) found further stabilising effects on the rising tubes from the initial magnetic field and its twist and curvature, as well as rotation and convection.

The theoretical works provide several features that may be observable at the surface, as is summarised by Fisher *et al.* (2000). The present investigation focuses primarily on the developing phases of the sunspot groups until their largest extension.

2. Statistical Study of Sunspot Group Details

2.1. The Observational Material

The data of the *Solar and Heliospheric Observatory/Michelson Doppler Imager* – Debrecen Data (SDD) sunspot catalogue were used (Győri, Baranyi, and Ludmány, 2011). This sunspot catalogue is more detailed than the Greenwich Photoheliographic Results (GPR)

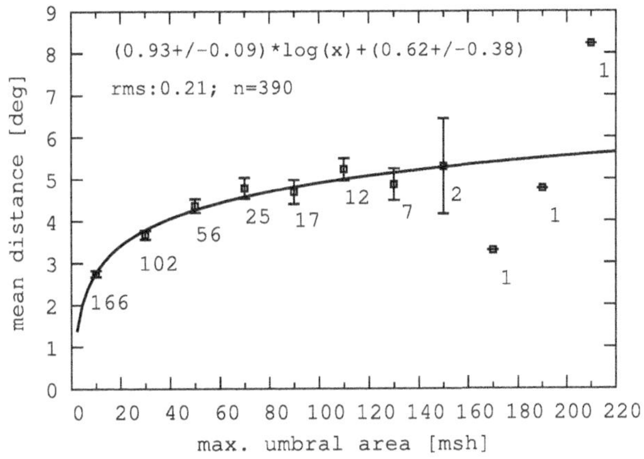

Figure 1 Distances measured in heliographic degrees between leading and following polarity regions depending on the total umbral areas of sunspot groups measured in millionths of a solar hemisphere [MSH]. The means are plotted in the middle of the bins with the numbers of cases. The bins are 20 MSH wide.

and its continuation, the Debrecen Photoheliographic Data (DPD). These traditional catalogues provide sunspot data on a daily basis and do not contain magnetic data. They are indispensable for long-term studies of the solar activity, but the investigation of internal details in sunspot groups requires higher temporal resolution and also magnetic data. The SDD meets these requirements. The data of position, area, and magnetic field for all observable sunspots and sunspot groups are given in 1 – 1.5 hour intervals. The catalogue covers the entire time interval of the SOHO/MDI mission: 1996 – 2010.

In the present work, unless otherwise stated, the selection criteria of sunspot groups are as follows: only the positions between central meridian distances (CMD) of $\pm 60^\circ$ are considered, the group had to be observable on the solar disc for at least six days within this CMD range, it should have reached its maximum in this longitudinal range by requiring that at least two days after maximum were observed, and the total area on the day after maximum is 10 % less than the maximum area. These criteria resulted in a sample of 390 sunspot groups.

2.2. Distance of Leading–Following Subgroups

Following the emergence of the sunspots at the photosphere, the distance between the leading and following polarity parts grows in parallel with the growth of the total spot area (Gilman and Howard, 1986). The leading–following distance might be a parameter of the achieved state of maximum area at the time of the largest size.

Figure 1 shows the relationship between the total umbral area of sunspot groups and the distance between their leading and following parts at the time of maximum area. The distance is computed between the "centre of mass" of both leading and following parts taking the umbral areas as masses. The mean distances with their errors have been computed in selected bins; their width was 20 millionths of a solar hemisphere [MSH]. The number of cases is indicated at each bin. Figure 1 shows that there is a clear logarithmic relationship between the maximum area reached and the distance between leading and following spots; the function is indicated in Figure 1. The bigger the group in its most developed state, the stronger the stretching effect. This may indicate the role of magnetic tension in forming the longitudinal extension of the group.

2.3. Longitudinal Shifts

After their emergence, the sunspot groups move in a longitudinal direction. The shifts from the first appearance until the maximum state have been computed and plotted in Figure 2.

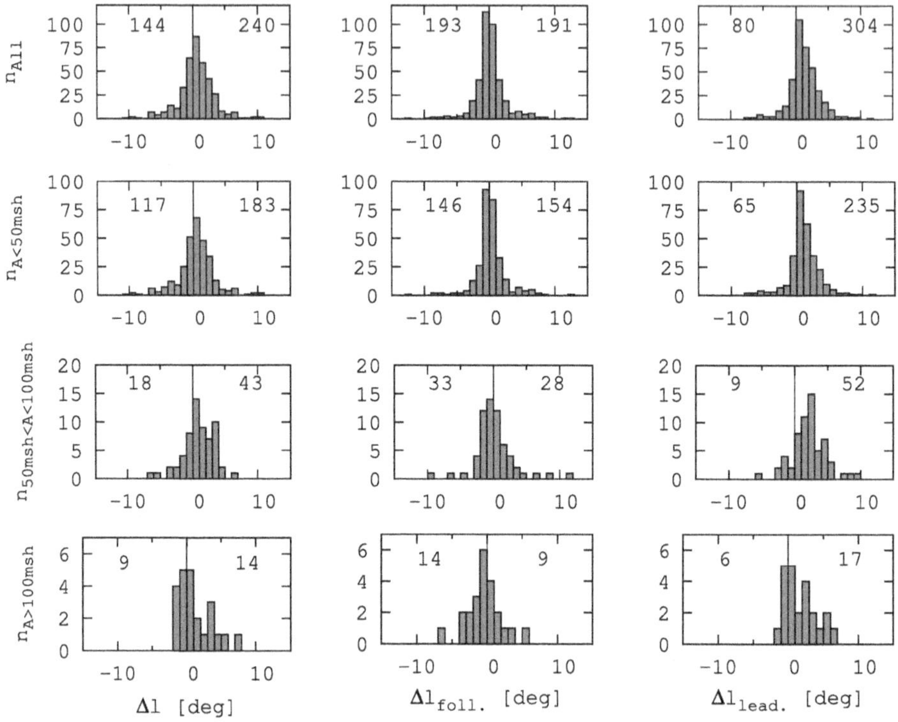

Figure 2 Distributions of the longitudinal shifts between the first appearance and maximum measured in heliographic degrees. The rows correspond to the sizes of the groups measured in MSH; the upper row shows the distributions of all cases, the other three rows show the distributions of three groups of sizes separately. The columns distinguish between the parts of active regions, left column: entire groups, middle and right columns: following and leading parts of the sunspot groups. The numbers in the positive and negative halves show the cases of forward and backward shifts.

The left column of the figure shows the longitudinal shifts for the total groups and groups of different sizes: $A < 50$, $50 < A < 100$, and $100 < A$ MSH. The central and right columns show the same data for the following and leading parts of the groups. The numbers in the positive and negative domains indicate the cases of forward and backward shifts.

The majority of the groups tend to move forward, but the backward and forward shifts depend on the size of the group. The topmost panels of Figure 2 show all cases, here the forward motion is predominant in the leading parts (80 % of all cases) and less significant in the following parts (50 %). Instead of the numbers of positive and negative cases, the means and medians of the shifts of the subgroups are more informative; see Table 1 and in graphical representation Figure 3. In the smallest groups ($A < 50$ MSH) the forward motion of the leading parts until the maximum is small; its mean value is about one heliographic degree, and the mean shift of the following parts is almost negligible. Thus, the smallest groups remain very close to the position of first appearance; their mean shift is about 0.4 degree. The following part in the middle group also does not move, and in the largest group slightly recedes. The forward motion of the entire group is most clearly expressed in the two largest groups. The means of these forward shifts are 0.909 ± 0.302 (for sizes between $50 < A < 100$ MSH) and 1.145 ± 0.482 (for sizes of $A > 100$ MSH). One can conclude that the diverging motion of the two parts is mainly produced by the forward motion of the leading part.

Table 1 Averages (upper table) and medians (lower table) of longitudinal shifts in heliographic degrees for sunspot groups of different umbral areas [A] indicated in MSH.

Sample	Entire groups	Following parts	Leading parts
All	0.516	0.104	1.251
$A < 50$	0.388	0.150	1.024
$50 < A < 100$	0.909	0.087	2.149
$100 < A$	1.145	−0.455	1.836
All	0.480	−0.020	1.060
$A < 50$	0.405	0.000	0.915
$50 < A < 100$	0.870	−0.290	2.110
$100 < A$	0.150	−0.850	1.470

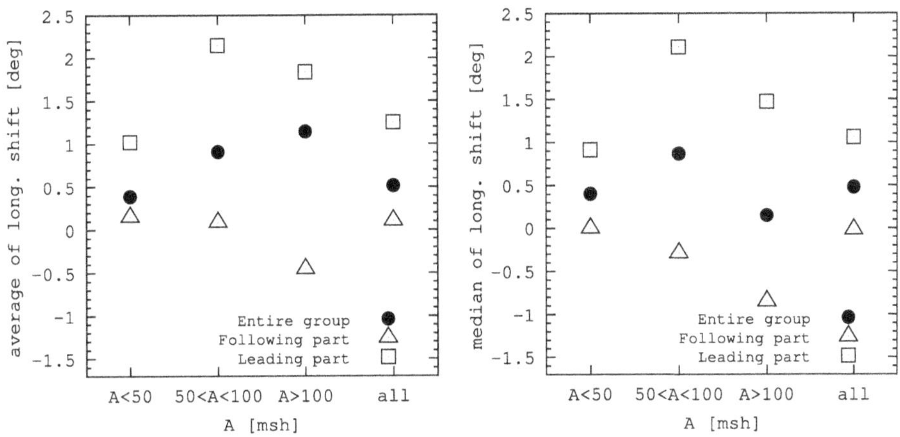

Figure 3 Left panel: average of longitudinal shifts in the three size ranges and the entire sample for the leading and following parts and the entire group. Right panel: the same as in the left panel for the medians of longitudinal shifts.

2.4. Growth Rate

The growing phases of the leading and following parts also show differences in time. The growth rates were determined in a fairly straightforward way: the maximum value of the total area was divided by the time interval between the first appearance and maximum area of the group. The results are depicted in Figure 4, for the entire group (left panel) and the following and leading parts (middle and right panels). The horizontal axes show the total umbral areas of the entire groups and the following and leading parts, respectively. The rise times refer to the maxima of the relevant subgroups. It can be seen that larger groups grow faster and that the leading growth rate is higher than the following one. The most important property is that the growth rate depends on the maximum area linearly. The present method is a simple procedure for the estimation of the growth rate; we will return to this relationship by using the temporal profiles derived in Section 2.6.

2.5. Asymmetric Compactness

The levels of compactness in the leading and following parts of the sunspot groups are usually different. The asymmetry indices [AI] of sunspot groups were computed for the spot

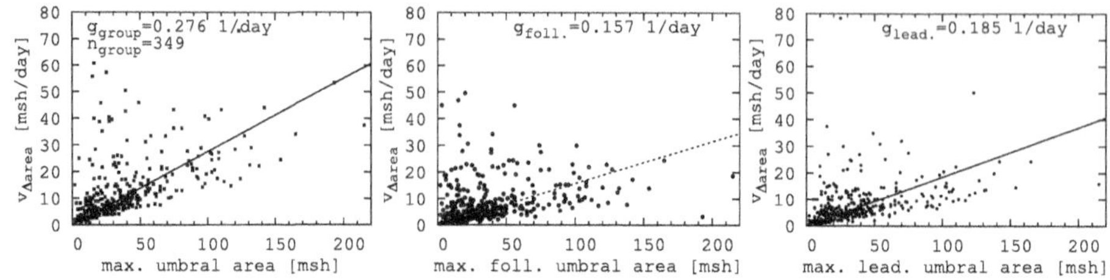

Figure 4 Growth rates [g] of sunspot groups depending on their maximum umbral area measured in MSH; g means the daily growth expressed as a fraction of the total umbral area. Left panel: dependence of the growth rates on the maximum umbral area of the entire groups, middle and right panels: the same relationships in the following and leading parts.

numbers [SN] in the leading [SN_L] and following [SN_F] parts with the formula

$$AI_{SN} = \frac{SN_L - SN_F}{SN_L + SN_F} \quad (1)$$

and the same formula was used for the computation of the asymmetry index [AI_{Ar}] of average sunspot areas [Ar] in the leading and following parts. Both parameters were considered at the time of the maximum of the sunspot group development, and only the umbrae were considered. Figure 5 shows the relation between the asymmetry indices AI_{SN} and AI_{Ar}. Figure 6 shows the histograms of both asymmetry indices for the umbrae.

The trend of the diagrams shows that most cases belong to the upper left quarter; *i.e.* in a typical distribution the leading part contains fewer spots than the following part, but the average area is larger in the leading part. The distribution of the points in the diagram shows a linear relationship, and the fitted regression line is as follows:

$$AI_{Ar} = (-1.02 \pm 0.06)AI_{SN} + (0.11 \pm 0.01) \quad (2)$$

The linear relationship between the two asymmetry indices has a simple meaning: if the leading or following part of the sunspot group contains more spots than the other part, then the mean area of these spots is smaller. However, more importantly, the regression line intersects the vertical axis at $AI_{Ar} = 0.11$, which means that even if the number of spots is the same in the two parts, the area asymmetry index is positive. This offset means that the mean sunspot area is typically 25 % larger in the leading part than in the following part. This can be considered as a mean measure of the asymmetric clustering of the high-density magnetic flux.

The number of nonzero cases is indicated in all quarters. This shows that besides the predominant upper left quarter the cases in the other quarters also cannot be neglected. The dots in the lower right quarter contribute to the linear relationship. 29 % of all cases are in the domains of the other diagonal; however, these cases cannot be considered as distinct configuration types, as they are simply members of the scatter around the regression line in Figure 5. In other terms, the compactness asymmetry is better represented by the offset of Equation (2) than by the numbers of cases in the domains of Figure 5.

2.6. Time Profiles of Sunspot Group Development

Development and decay are important characteristics of active-region dynamics. They are governed by different physical processes. The emergence is driven by buoyancy, while the

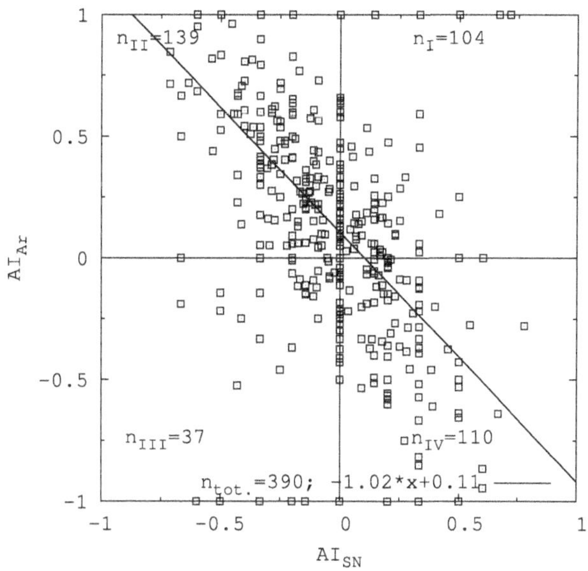

Figure 5 Relation between leading/following asymmetry indices of numbers [AI_{SN}] and mean areas [AI_{Ar}] of umbrae in sunspot groups.

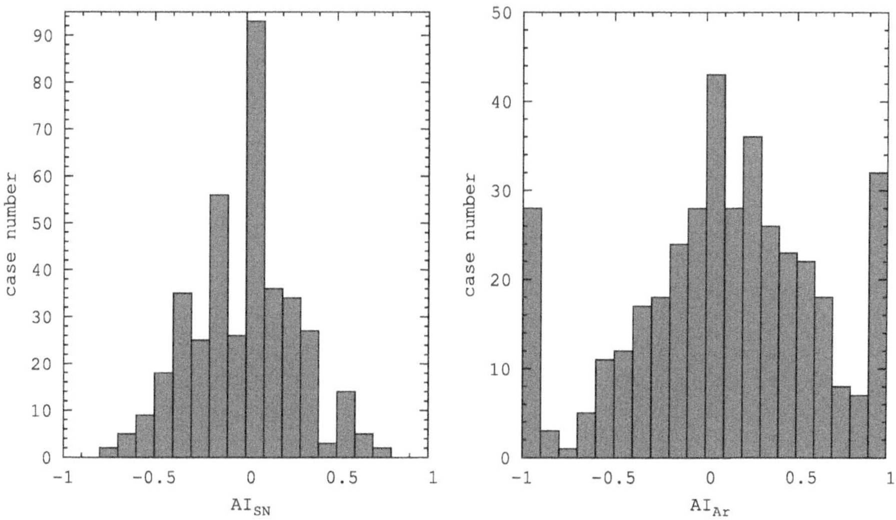

Figure 6 Histograms of the leading–following asymmetry indices. Left panel: numbers of umbrae [AI_{SN}], right panel: mean areas of umbrae [AI_{Ar}].

decay results from the impact of turbulent erosion (Petrovay, Martínez Pillet, and van Driel-Gesztelyi, 1999). These processes can also be mixed during the development of the active region. Hathaway and Choudhary (2008) could only follow the development curve of the total area of a sunspot group with the one-day resolution allowed by the GPR. The SDD enables us to investigate the heading and trailing parts separately in 1.5-hour resolution.

A list has been compiled for those sunspot groups that were observable from their first appearance to their decaying phase. This criterion is more strict than those formulated in Section 2.1, and therefore this sample is more restricted than that analysed above; it contains

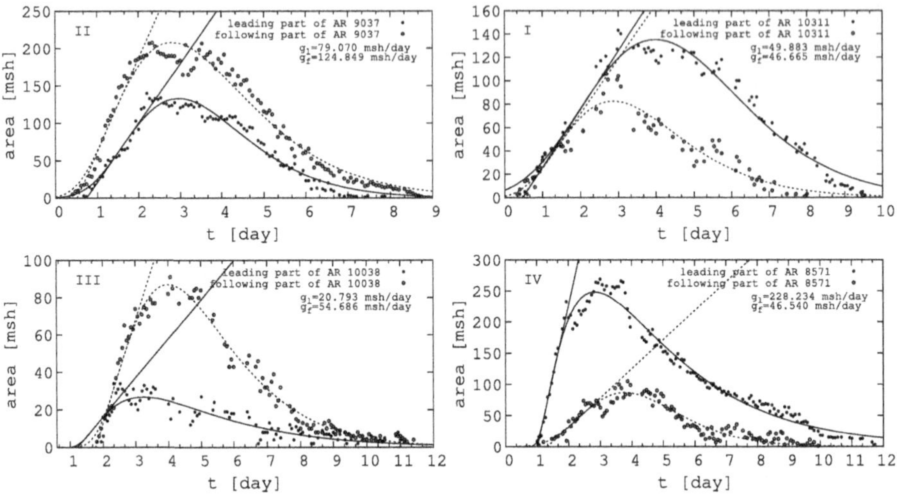

Figure 7 Four cases for the relative time profiles of the leading and following parts in sunspot groups with the slopes of the fitted function in Equation (3) at the inflection points.

223 groups. The other difference is that in this study the umbra plus penumbra [U + P] areas are considered, because their variation is more smooth than that of the umbrae. The following asymmetric Gaussian function has been fitted to the time series of their total area data:

$$f(t) = H \exp\left(-\frac{(t - t_{\mathrm{M}})^2}{D(1 + A(t - t_{\mathrm{M}}))}\right) \tag{3}$$

where H and t_{M} are the height and time of the maximum, and D and A determine the width and asymmetry, respectively. This formula is a somewhat modified version of our previously applied function (Muraközy and Ludmány, 2012) and another formula applied by Du (2011). Its advantage over the two-component bell function (*i.e.* two half-Gaussians for the ascending and descending phases) is that the heights and times of the maxima as well as the ascending and descending slopes can be obtained by appropriate fitting to the data, and these parameters can be compared directly for the leading and following parts. For two-component Gaussians, the maximum should be determined separately.

The time profiles of leading and following subgroups were treated separately. The following properties were examined: ratio of leading/following maxima [$H_{\mathrm{L}}/H_{\mathrm{F}}$], difference between the times of leading and following maxima [$t_{\mathrm{L}} - t_{\mathrm{F}}$], the growth rates of leading and following parts, and the areal dependence of all of these data. In this case the growth rate was defined as the slope of the function in Equation (3) at its inflection point. Concerning the relationships of leading–following maxima, the following cases were distinguished. The leading maximum can be: i) higher and later, ii) smaller and later, iii) smaller and earlier, iv) higher and earlier than the following maximum. Figure 7 shows one example for each case with the fitted Equation (3) functions and the lines indicating the slopes at the inflection points.

A further type of leading–following asymmetry can be studied by comparing the differences between the heights and times of maxima. Figure 8 shows the comparison of the Equation (1) asymmetry indices applied to the maximum areas and times of maxima.

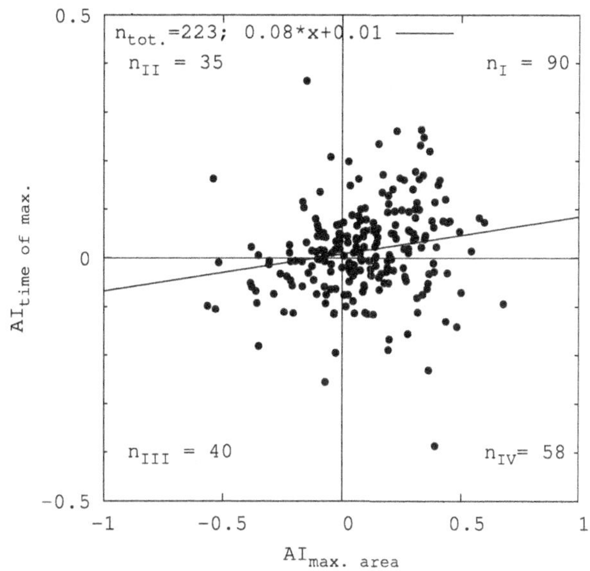

Figure 8 Comparison of leading–following asymmetry indices computed for the maximum areas and times of maxima. The number of cases is indicated in each segment.

Figure 8 shows that the most typical relative position of leading and following maxima is the later and higher maximum of the leading part (90 groups out of the sample of 223 groups), but none of the other three cases (35, 40, and 58 groups) can be neglected.

We also examine statistically the parameters of the curves separately. Figure 9 shows the areal dependence of the growth rates in the leading and following parts. In this case the growth rate is defined as the greatest steepness of the fitted Equation (3) curve, but the relationship between the fitted lines of Figure 9 is similar to that of Figure 4, which is based on a comparatively simplified method and a larger sample. The leading and following growth rates are 0.185 *vs.* 0.157 day^{-1} when using the method of Section 2.4, whereas they are 0.25 *vs.* 0.22 day^{-1} with the present time-profile analysis. Of course, this difference is due to the use of different methods; the recent method should give higher values because it considers the steepest moment of the development.

In Figure 10, the area dependences and distributions are summarised for the relationships between the leading–following parameters of the fitted Equation (3) curves. The area means the total $\mathrm{U}+\mathrm{P}$ area of the group at maximum. None of the examined relationships depends on the area, but the right-hand panels show the histograms of the cases. The most unambiguous leading predominance is exhibited by the ratio of maximum areas; in two-thirds of the cases the maximum area of the leading part is larger. The other two histograms are more intriguing. The leading/following ratio of growth rates is mostly larger than one, but the peak is just below one (upper panel). The difference between the times of maxima is mostly positive (the maximum of the leading part is later), but the peak of the histogram is just below zero.

3. Discussion

Earlier investigations of sunspot-group development used the classic photospheric data, the Mount Wilson observations (Howard, 1992a), and their comparison with the Greenwich catalogue data (Lustig and Wöhl, 1994). These works presented growth and decay rates of the

Figure 9 Comparison of the growth rates in the leading and following parts of sunspot groups measured in MSH day^{-1}. The data are obtained as the steepest slopes of the curves from Equation (3) fitted to the temporal profiles of the area data in the leading and following parts.

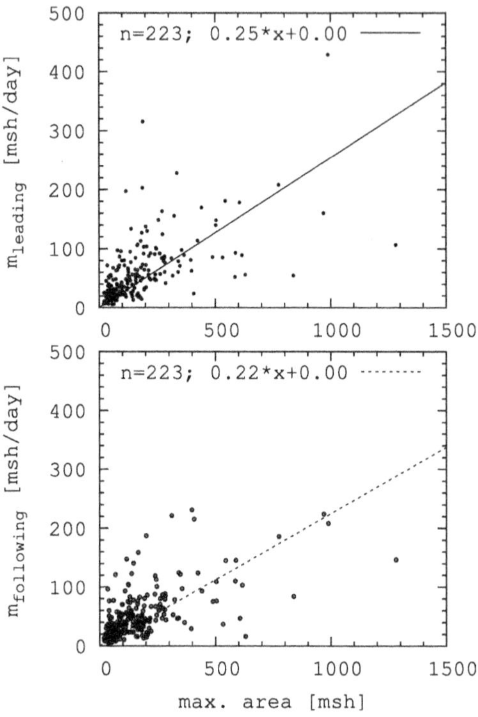

sunspot groups, but with the restriction of the daily resolution and missing magnetic data, the results were presented for large statistical samples. The SDD catalogue offered several specific advantages allowing the present work to go into deeper detail of the processes. The first advantage is the temporal resolution of 1.5 hours without nocturnal interruption, which allows us to follow the developments and motions with the precision that is necessary here to determine the reliable temporal profiles of the developments and the times of maxima. The second advantage is the magnetic information of the spots, which makes the separation of leading and following parts much more reliable. Earlier works (*e.g.* Howard, 1992b) had to separate the groups to spots at longitudes westward and eastward from the position of the area-weighted centroid of the group. This method may result in false separations in some cases. Another benefit of the polarity data is the reliable separation of entire sunspot groups. In preparing the SDD, considerable effort has been devoted to distinguish between two active regions emerging close to each other. This is not possible in the classic sunspot catalogues, and the unresolved cases distort the statistics. The third advantage is the availability of the data for both the sunspots and sunspot groups, which was indispensable for the present asymmetry studies.

Section 2.2 presents results for the distances between the leading and following parts at the time of the maximum area of each sunspot group. According to Figure 1, the growth of the mean leading–following distance is proportional to the logarithm of the maximum total umbral area of the group. This relationship may indicate an impact of the magnetic tension in stretching the sunspot group. The clarification of the role of the magnetic tension will require more detailed statistics of umbral areas along with the relevant flux-density data. This study is out of the scope of the present article and will be the topic of a subsequent work. Precedents of this study are sporadic in the literature, and they treat the distance of

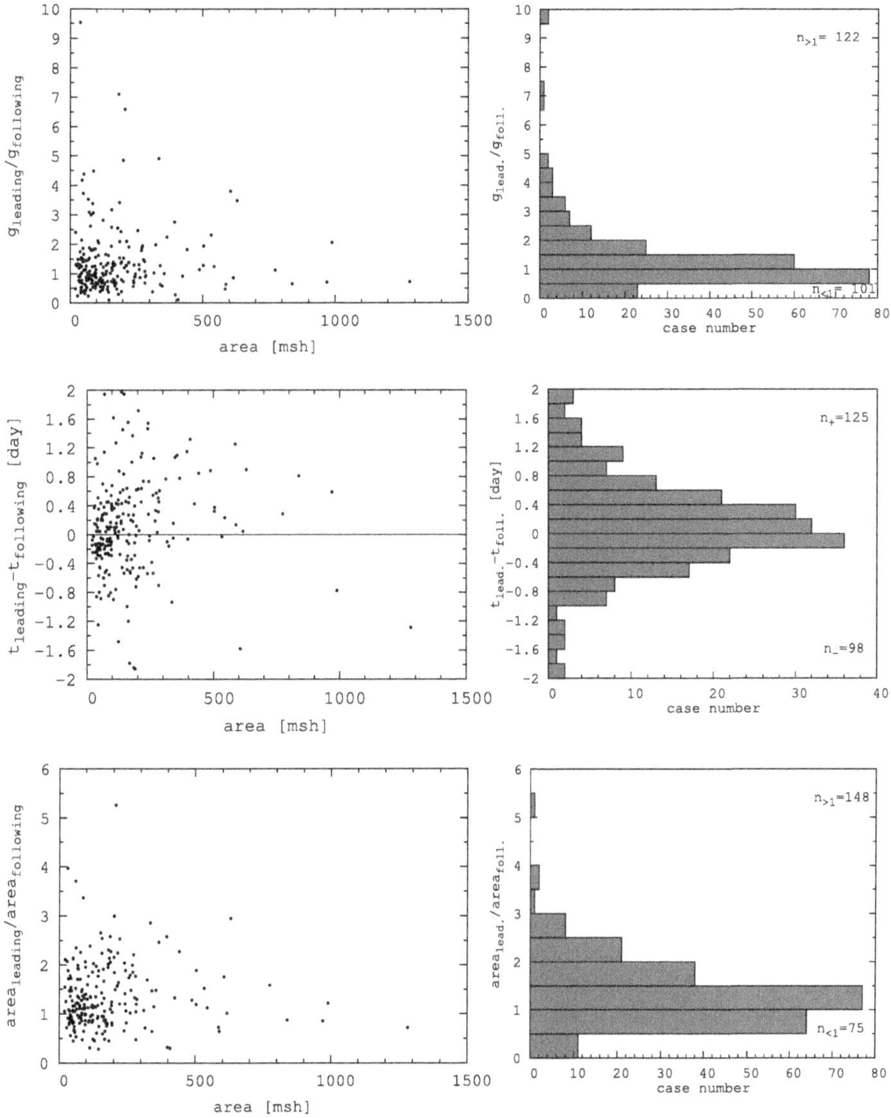

Figure 10 Leading–following relationships of three parameters of the curves fitted using Equation (3): the growth rates, the times, and the values of maxima. Area dependences (left panels) and histograms of cases (right panels) are shown for the following relationships: ratios of leading–following growth rates (upper row), differences of leading–following times of maxima (middle row), ratios of leading–following maxima (bottom row).

the leading–following parts in the context of sunspot group tilts without analytic formulation of its relationship to the maximum total size of the group. The plot presented by Howard (1992b) (his Figure 8) weakly resembles our Figure 1, but in that study the separation of leading and following spots dispensed with the polarity data and the work was not restricted to the maximum state. Tian *et al.* (1999) presented a diagram about the relationship of the polarity separation and the total magnetic flux, regardless of the phase of development, but with emphasis on the tilts.

The results of Section 2.3 show a leading–following asymmetry in the development. It is worth comparing the data of Figure 2 and Table 1 with the theoretical results of Moreno-Insertis, Caligari, and Schüssler (1994) and Caligari, Moreno-Insertis, and Schüssler (1995), *i.e.* that the leading field lines are more inclined to the radial direction than the following ones. The measurements of van Driel-Gesztelyi and Petrovay (1990) and Cauzzi and van Driel-Gesztelyi (1998) on asymmetric magnetic fields with respect to the neutral line can also be explained by this property. The present data show that during the growth phase the following part mostly remains at the position of appearance (nearly vertically according to the results mentioned), whereas the leading part shifts forward – this is the source of the group lengthening. During this process, the inclination difference may decrease because Cauzzi and van Driel-Gesztelyi (1998) found decreasing asymmetry with the aging of the active regions. The larger the group, the larger the forward shift of the leading part. The groups of smallest size (below 50 MSH) seem to remain at the starting position; *i.e.* they proceed with the ambient plasma. This may be related to the theoretical finding of Schüssler and Rempel (2005) about the dynamical disconnection of the emerged flux from the roots; this process may be more effective in the case of smaller groups.

Up to now this behaviour – the growing polarity separation during the sunspot group development – has received little attention. Numerous theoretical works have been devoted to the magnetic-flux emergence – their detailed overview is given by Fan (2009) – but in the photosphere they mostly focus on the tilt angles. However, the empirical works were restricted by the lack of detailed datasets. The earliest works were case studies. Expansion velocities were reported by Nagy and Ludmány (1980) for one active region and by Chou and Wang (1987) for 24 bipolar regions but not restricted to the time interval prior to the maximum. Howard (1989) presented polarity separation data for 7629 active regions regardless of their area and phase of development and without direct magnetic data. Strous *et al.* (1996) presented mean diverging velocities of spots in a single active region but not restricted to the time interval between the birth and maximum. Schüssler and Wöhl (1997) investigated 3793 active regions over 108 years and found that secondary groups tend to emerge westward from the earlier emerged primary ones in the case of large groups. Like our results in Section 2.3, this can be interpreted by the result of Moreno-Insertis, Caligari, and Schüssler (1994) that a larger amount of emerging flux is more asymmetric; *i.e.* it is more inclined to the radial direction in the western leg than in the eastern one. Nevertheless, the phenomenon of secondary groups is different from the shifts presented in Section 2.3, which apparently contains the first detailed data about the displacements of the opposite polarity regions in developing sunspot groups.

Section 2.4 presents growth rates of sunspot groups separately for the leading and following parts. Figure 4 shows the linear relationship between the maximum total area and the daily mean growth rate. For the entire group this is 27.6 % of the actual total area independent of the size. This linearity may contribute to the description of buoyant motion. The buoyant force is estimated to be $B_0^2/2\mu H_p$ (Fan, Fisher, and DeLuca, 1993), where B_0 is the initial value of the magnetic field and H_p is the external pressure scale height. This means that a stronger magnetic field can be expected to result in a higher emerging velocity. Both the magnetic field and the emerging velocity are represented here by proxy measures, by the total umbral area, and the growth rate, but their linear proportionality implies that at the surface the emerging velocity is the same for all active-region magnetic fields independent of their sizes (the strengths of the magnetic fields) or their presumed emerging velocities within the convective zone. This can happen in such a way that even if the velocities are different during the emergence they might apparently be equalised before reaching the surface, perhaps by the different or varying drag force.

Section 2.5 presents the asymmetry of compactness between the leading and following parts of sunspot groups. Figure 5 shows that in the most typical cases the leading part contains fewer spots than the following part but that their mean area is larger than that of the spots in the following part. This is consistent with the results of Fan, Fisher, and DeLuca (1993) that a significant asymmetry is produced in the rising magnetic flux rope by the Coriolis force; the leading part is more compact and the following part becomes fragmented. Figure 5 also shows that in the case of equal spot number in the two parts, the area asymmetry is $\mathrm{AI}_{\mathrm{Ar}} = 0.11$ for umbrae. This means that on average a 25 % larger area comprises high-density magnetic flux in the leading part than in the following part. The rest of the magnetic flux is dispersed in the ambient facular clusters. This may be compared to the result of Yamamoto (2012), who analysed the area asymmetries of the opposite polarity parts in active regions on the basis of magnetograms. His asymmetry parameter differs from our Equation (1): $A = \log(S_{\mathrm{F}}/S_{\mathrm{L}})$, where S_{F} and S_{L} denote the areas of following and leading polarities. He found that the average area–asymmetry ratio was distributed between -0.2 and 0.4; the peak of the distribution was at 0.03. The conversion of this $A = 0.03$ value to the AI index defined by Equation (1) gives -0.037. By converting to percentages, one obtains that in the most typical cases the area of following polarity region is 7 % higher than that of the leading polarity. A direct comparison of this asymmetry data with ours cannot give reliable assessments about the amounts of magnetic fluxes within and out of the spots, because that work analysed the active regions at the centre of the solar disc, whereas we considered them at their maximum state.

Section 2.6 describes the statistics of the fitting parameters in Equation (3) fitted to the leading and following time profiles of the selected 223 sunspot groups, a sample of 446 curves. A comparison of Figures 4 and 9 gives qualitatively similar results for the growth rates. The main point is the linear dependence on the total area as discussed above. Further statistical properties of the fitting parameters are that the maximum area of the leading part is higher in two-thirds of the cases, and the leading part reaches its maximum later than the following part in 56 % of the cases.

4. Summary

The obtained results can be summarised as follows.

i) The distance between the leading and following parts of the sunspot groups increases with increasing total area [A] measured in the most developed state of the group. This dependence is described by a logarithmic function, and it may mean that magnetic tension plays a role in the longitudinal extent of the sunspot group.
ii) The dependence of the growth rate on the maximum umbral area is linear for the whole group as well as the leading and following parts; this was obtained by both the simplified method (Section 2.4) and the time-profile analysis (Section 2.6). This linearity means that close to the surface the emerging speed is independent of the amount of emerging magnetic flux.
iii) The longitudinal shift of the whole group during the growth phase shows dependence on its total area, but the following part mostly remains close to the starting location and the leading part shifts forward. The mean value of the shift of the leading part is about $\Delta L \approx 2^{\circ}$ until the maximum in the groups of maximum size exceeds 50 MSH.

iv) The following asymmetries have been found between the leading and following parts. In the state of maximum area the compactness is different: the leading part contains fewer and larger spots than the following part. The mean area of the leading spots is about 25 % larger than that of the following spots. In two-thirds of all cases, the maximum area of the leading part is higher than that of the following part. The development is also different in the two subgroups; in most cases the areal dependence of the growth rate is stronger in the leading part and the time of maximum is later in the leading part.

Acknowledgements The research leading to these results has received funding from the European Space Agency project, ESA PECS 98081. The results have been presented at the NSO Workshop 26, Sunspot, New Mexico, and JM wishes to acknowledge the support provided by the organisers. The authors are grateful to the unknown referee for the substantial improvement of the manuscript.

References

Abbett, W.P., Fisher, G.H., Fan, Y.: 2000, *Astrophys. J.* **540**, 548. ADS:2000ApJ...540..548A, doi:10.1086/309316.

Abbett, W.P., Fisher, G.H., Fan, Y.: 2001, *Astrophys. J.* **546**, 1194. ADS:2001ApJ...546.1194A, doi:10.1086/318320.

Caligari, P., Moreno-Insertis, F., Schüssler, M.: 1995, *Astrophys. J.* **441**, 886. ADS:1995ApJ...441..886C, doi:10.1086/175410.

Caligari, P., Schüssler, M., Moreno-Insertis, F.: 1998, *Astrophys. J.* **502**, 481. ADS:1998ApJ...502..481C, doi:10.1086/305875.

Cauzzi, G., van Driel-Gesztelyi, L.: 1998, In: Balasubramaniam, K.S., Harvey, J., Rabin, D. (eds.) *Synoptic Solar Physics* **CS-140**, Astron. Soc. Pac, San Francisco, 105. ADS:1998ASPC..140..105C.

Chou, D.-Y., Wang, H.: 1987, *Solar Phys.* **110**, 81. ADS:1987SoPh..110...81C, doi:10.1007/BF00148204.

Cowling, T.G.: 1946, *Mon. Not. Roy. Astron. Soc.* **106**, 218. ADS:1946MNRAS.106..218C.

Du, Z.: 2011, *Solar Phys.* **273**, 231. ADS:2011SoPh..273..231D, doi:10.1007/s11207-011-9849-8.

Fan, Y.: 2009, *Living Rev. Solar Phys.* **6**, 4. ADS:2009LRSP....6....4F, doi:10.12942/lrsp-2009-4.

Fan, Y., Fisher, G.H., DeLuca, E.E.: 1993, *Astrophys. J.* **405**, 390. ADS:1993ApJ...405..390F, doi:10.1086/172370.

Fan, Y., Fisher, G.H., McClymont, A.N.: 1994, *Astrophys. J.* **436**, 907. ADS:1994ApJ...436..907F, doi:10.1086/174967.

Fisher, G.H., Fan, Y., Longcope, D.W., Linton, M.G., Abbett, W.P.: 2000, *Phys. Plasmas* **7**, 2173. ADS:2000PhPl....7.2173F, doi:10.1063/1.874050.

Győri, L., Baranyi, T., Ludmány, A.: 2011, In: Choudhary, D.P., Strassmeier, K.G. (eds.) *The Physics of Sun and Star Spots, IAU Symp.* **P-273**, 403. ADS:2011IAUS..273..403G, doi:10.1017/S174392131101564X.

Gilman, P.A., Howard, R.: 1986, *Astrophys. J.* **303**, 480. ADS:1986ApJ...303..480G, doi:10.1086/164093.

Hathaway, D., Choudhary, D.P.: 2008, *Solar Phys.* **250**, 269 ADS:2008SoPh..250..269H, doi:10.1007/s11207-008-9226-4.

Howard, R.F.: 1989, *Solar Phys.* **123**, 271. ADS:1989SoPh..123..271H, doi:10.1007/BF00149106.

Howard, R.F.: 1992a, *Solar Phys.* **142**, 47. ADS:1992SoPh..142...47H, doi:10.1007/BF00156633.

Howard, R.F.: 1992b, *Solar Phys.* **142**, 233. ADS:1992SoPh..142..233H, doi:10.1007/BF00151452.

Lustig, G., Wöhl, H.: 1994, *Solar Phys.* **152**, 221. ADS:1994SoPh..152..221L, doi:10.1007/BF01473208.

Moreno-Insertis, F., Caligari, P., Schüssler, M.: 1994, *Solar Phys.* **153**, 449. ADS:1994SoPh..153..449M, doi:10.1007/BF00712518.

Muraközy, J., Ludmány, A.: 2012, *Mon. Not. Roy. Astron. Soc.*. **419**, 3624. ADS:2012MNRAS.419.3624M, doi:10.1111/j.1365-2966.2011.20011.x.

Nagy, I., Ludmány, A.: 1980, *Publ. Debr. Heliophys. Obs., Heliogr. Ser.* **4**, 3. ADS:1980PDHO....4.....N.

Petrovay, K., Martínez Pillet, V., van Driel-Gesztelyi, L.: 1999, *Solar Phys.* **188**, 315. ADS:1999SoPh..188..315P, doi:10.1023/A:1005213212336.

Schüssler, M., Rempel, M.: 2005, *Astron. Astrophys.* **441**, 337. ADS:2005A&A...441..337S, doi:10.1051/0004-6361:20052962.

Schüssler, M., Wöhl, H.: 1997, *Astron. Astrophys.* **327**, 361. ADS:1997A&A...327..361S.

Strous, L.H., Scharmer, G., Tarbell, T.D., Title, A.M., Zwaan, C.: 1996, *Astron. Astrophys.* **306**, 947. ADS:1996A&A...306..947S.

Tian, L., Zhang, H., Tong, Y., Jing, H.: 1999, *Solar Phys.* **189**, 305. ADS:1999SoPh..189..305T, doi:10.1023/A:1005252617906.
van Driel-Gesztelyi, L., Petrovay, K.: 1990, *Solar Phys.* **126**, 285 ADS:1990SoPh..126..285V, doi:10.1007/BF00153051.
Yamamoto, T.T.: 2012, *Astron. Astrophys.* **539**, A13. ADS:2012A&A...539A..13Y, doi:10.1051/0004-6361/201014951.

DOI 10.1007/978-1-4939-1182-0_10
Reprinted from *Solar Physics* Journal, DOI 10.1007/s11207-013-0424-3

Migration and Extension of Solar Active Longitudinal Zones

N. Gyenge · T. Baranyi · A. Ludmány

Received: 30 August 2012 / Accepted: 24 September 2013 / Published online: 12 November 2013

Abstract Solar active longitudes show a characteristic migration pattern in the Carrington coordinate system if they can be identified at all. By following this migration, the longitudinal activity distribution around the center of the band can be determined. The half-width of the distribution is found to be varying in Cycles 21 – 23, and in some time intervals it was as narrow as 20 – 30 degrees. It was more extended around a maximum but it was also narrow when the activity jumped to the opposite longitude. Flux emergence exhibited a quasi-periodic variation within the active zone with a period of about 1.3 years. The path of the active-longitude migration does not support the view that it might be associated with the 11-year solar cycle. These results were obtained for a limited time interval of a few solar cycles and, bearing in mind uncertainties of the migration-path definition, are only indicative. For the major fraction of the dataset no systematic active longitudes were found. Sporadic migration of active longitudes was identified only for Cycles 21 – 22 in the northern hemisphere and Cycle 23 in the southern hemisphere.

Keywords Sunspots · Solar activity

1. Introduction

The spatial distribution of active-region emergence has been investigated since the creation of the Carrington coordinate system. The equatorward latitudinal migration of sunspot

Solar Origins of Space Weather and Space Climate
Guest Editors: I. González Hernández, R. Komm, and A. Pevtsov

N. Gyenge (✉) · T. Baranyi · A. Ludmány
Heliophysical Observatory, Research Centre for Astronomy and Earth Sciences, Hungarian Academy of Sciences, P.O. Box 30, 4010 Debrecen, Hungary
e-mail: gyenge.norbert@csfk.mta.hu

T. Baranyi
e-mail: baranyi.tunde@csfk.mta.hu

A. Ludmány
e-mail: ludmany.andras@csfk.mta.hu

group emergence was first observed by Carrington in 1859, a phenomenon later named after Spörer, and Carrington was also the first to observe active-region emergences at the same longitudes (Carrington, 1863). Since then numerous works have been devoted to the longitudinal grouping of activity. Different terminology has been used but the aim was always to reveal whether the emergence is equally probable at any longitude, or if not, what the extent of deviation from the axial symmetry is and where the locations of higher-than-average activity are. This is a much more difficult challenge than the study of latitudinal patterns because one cannot assume that the location of enhanced activity is bound to the Carrington system. The diversity of results obtained can partly be explained by the differences of the methods applied, pre-assumptions, input data, and time intervals.

Several attempts have been made to identify the rotation rate of the frame to which the sources of enhanced activity can be bound. As an example, in the most simple approach Bogart (1982) makes an autocorrelation analysis on a 128-year long dataset of Wolf-numbers. He reports two peaks at 27.5 days and 13.6 days, which allows the interpretation that the non-axisymmetric frame has two opposite maxima and rotates with a period of 27.5 days. The procedure results in different values for the individual cycles. This is a Sun-as-a-star method, which has also been used by Balthasar (2007), who made an FFT analysis on the Wolf-numbers and also found peaks around 27.36 and 27.49 days over the 1848 – 2006 period, and also at 12.07 days (shorter than half of the longer periods), but the period varies from cycle to cycle and also depends on the cycle phase.

This approach using a single daily parameter is of course fairly restricted, and the authors admit that a better understanding requires spatial resolution. Nevertheless, this is the method closest to the possibilities of stellar magnetic-activity research, *i.e.* to see what can be learned from spatially unresolved data. At the same time, Henney and Durney (2005) point out that suggestions of long-term persistency of active longitudes obtained by the method of Bogart (1982) may arise by chance.

The methods considering spatial information are mostly based on sunspot catalogues, primarily on the Greenwich Photoheliographic Results (GPR: Royal Observatory, Greenwich, 1874 – 1976) but also on magnetograms and flare positions. The diversity of the results from different teams is partially caused by the differences in spatial resolution of the data and methods used. Another type of difference is related to the length and date of the chosen time interval. By considering short time intervals one should assume that the rotation rate of the frame containing the active longitudes and the measure of its non-axisymmetry remain constant in time; however, this is not necessarily the case. We will return to the specific results in the discussions by comparing them with the recent findings.

A possible source of controversy is the use of different pre-assumptions about the geometry and dynamics of the frame carrying the active longitudes. The most sophisticated conception is put forward by Berdyugina and Usoskin (2003), Usoskin, Berdyugina, and Poutanen (2005), and Berdyugina *et al.* (2006). They assumed a similar differential-rotation profile as that observed on the surface and searched for its most probable constants, the equatorial angular velocity and the shear of the profile, *i.e.* the constant of the $\sin^2$ term. They reported a century-scale persistence of active longitudes migrating forward with respect to the Carrington frame and containing two preferred activity regions on the opposite sides of the Sun along with a cyclic modulation that originates from differential rotation, *i.e.* from a dependence on the varying latitude. The relative activity levels of the two preferred domains were varying with a mean period of 3.7 years. This variation was interpreted as a solar counterpart of the flip-flop events observed in FK Com (Jetsu, Pelt, and Tuominen, 1993; Oláh *et al.*, 2006).

The method of these works has been criticized by Pelt, Tuominen, and Brooke (2005) and Pelt *et al.* (2006). They pointed out that certain elements of the above procedures and the considered parameter space may result in a bimodal distribution, *i.e.* two persistent preferred longitudes, flip-flop like events, and forward migration even on a computer-generated random dataset. These analyses are a warning for our perspective: we should approach the issue with no pre-assumptions about any internal structures. For instance, the assumption of differential rotation implies a cyclic dependence that cannot be regarded as an *a-priori* fact. On the other hand, according to Bigazzi and Ruzmaikin (2004) the differential rotation makes the non-axisymmetry disappear; it can only remain at the bottom of the convective zone. An apparent signature of the differential rotation can also be detected even in the case of the rigid rotation of a non-axisymmetric magnetic field caused by a stroboscopic effect as demonstrated by Berdyugina *et al.* (2006).

A possible internal non-axisymmetric magnetic structure would have important theoretical consequences. Following Cowling's (1945) idea about a possible relic magnetic field within the radiative zone, Kitchatinov and Olemskoi (2005) and Olemskoy and Kitchatinov (2007) searched for a signature of a period of 28.8 days, which is the rotation period of the radiative zone (Schou *et al.*, 1998). Their significant value was slightly shorter (28.15 days), but it was only detectable in odd cycles. The authors argue that this seems to support the existence of an off-axis relic field, which could alternately be parallel or antiparallel to the varying field in the convective zone. These authors used GPR-data and considered a sector structure with varying rotation rate. Plyusnina (2010) used a similar procedure and obtained different values for the growth and decline phases of the cycles (27.965 and 28.694 days, respectively). In these cases, the longitudinal distribution was unimodal. These procedures used pre-defined sector structures without differential rotation.

There are further results for the synodic rotation period of a rigid non-axisymmetric frame. From flare positions, 26.72 and 26.61 days (Bai, 1988), 27.41 days in Cycles 19–21 (Bai, 2003), and 22.169 days (Jetsu *et al.*, 1997) have been found. The radiation-zone rotation period of 28.8 days has been found on synoptic maps of photospheric magnetic fields with special filtering techniques by Mordvinov and Kitchatinov (2004). However, the different rotation periods do not necessarily suggest mistakes or methodological artifacts; some of the differences may be explained by a varying migration pattern, as demonstrated by Juckett (2006).

2. Data and Analysis

2.1. Activity Maps and Migration Paths

The source of observational data that we use is the Debrecen Photoheliographic Data sunspot catalogue (DPD: Győri, Baranyi, and Ludmány, 2011). This material is the continuation of the classic Greenwich Photoheliographic Results (GPR), the source of numerous works in this field. At present, the DPD covers the entire post-Greenwich era: 1977–2012. The present study covers the time interval 1979–2010, 32 years, 416 Carrington Rotations. There are investigations based on longer datasets, but this is the most detailed sunspot catalogue, and our present aim is to investigate the dynamics of the identified active-longitude zones.

The task is to determine the amount of activity concentration in certain longitudinal belts. The total sunspot area $[A_i]$ of all sunspot groups has been computed within 10° longitudinal bins of the Carrington system for each Carrington Rotation between the years 1977 and 2011 in such a way that the total areas and positions of the individual sunspot groups were taken into account at the time of their largest observed area. Then the data of each bin

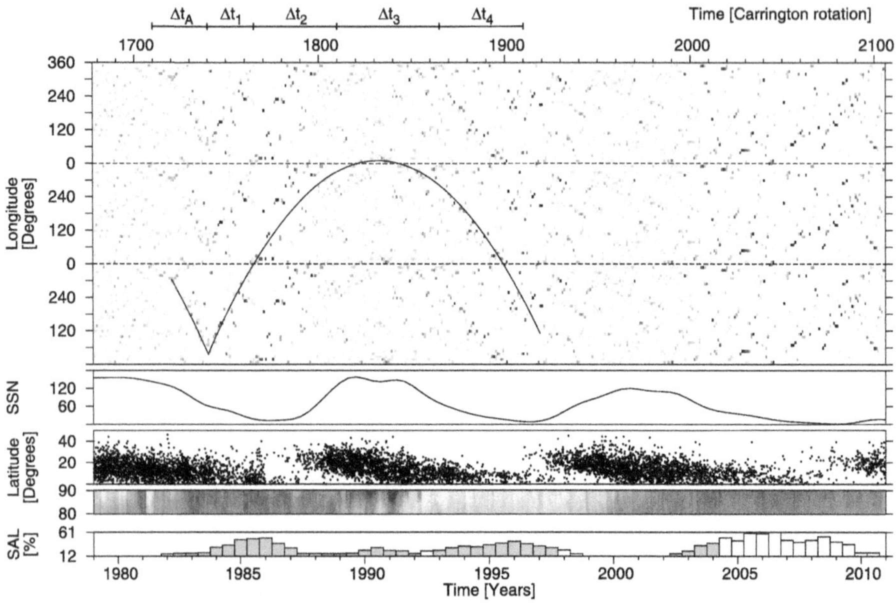

Figure 1 Longitudinal distribution of solar activity between 1997 and 2011 in the northern hemisphere (first row), the time profiles of Cycles 21 – 24 (second row), the hemispheric Spörer diagram (third row), the polarity of the magnetic dipole field at the northern pole (fourth row). The columns in the last row show the Strength of Active Longitude (SAL), the semiannual means of activity concentration expressed in percentage; gray and empty columns mark smooth and chaotic displacements of the active-longitudinal zones, respectively, see Section 2.3.

were divided by the total sum of sunspot areas observed in that rotation in all bins. This normalized quantity represents the activity weight in the given rotation in each bin. The weight of the ith bin is

$$W_i = \frac{A_i}{\sum_{j=1}^{36} A_j} \tag{1}$$

The representation by normalized weights was criticized by Pelt *et al.* (2006) who argued that this amplifies the role of sunspot groups during low-activity intervals. However, in our case the aim is to find a measure for the activity concentration in certain longitudinal regions independently from the particular level of current solar activity. As far as the position is concerned, the location of emergence was used by Usoskin, Berdyugina, and Poutanen (2005) and Zhang *et al.* (2011).

Figures 1 and 2 show the W_i values plotted in each Carrington Rotation (x-axis) and 10° longitudinal bin (y-axis) between 1977 – 2011 coded with the darkness of gray color for the northern and southern hemispheres. In the y-axis, the 360° solar circumference has been repeated three times, similarly to Juckett (2006), in order to follow the occasional shifts of the domains of enhanced sunspot activity in the Carrington system. Below the diagrams, the time profiles of Cycles 21 – 24 are plotted by using the smoothed International Sunspot Number (SIDC-team, 2012) as well as the hemispheric Spörer diagrams by using the sunspot group data of DPD. The next lines of the two figures show the negative and positive magnetic polarities at the poles indicated by dark and light stripes, respectively (Hathaway, 2010). The diagrams of the lowest lines in both figures will be explained in Section 2.3.

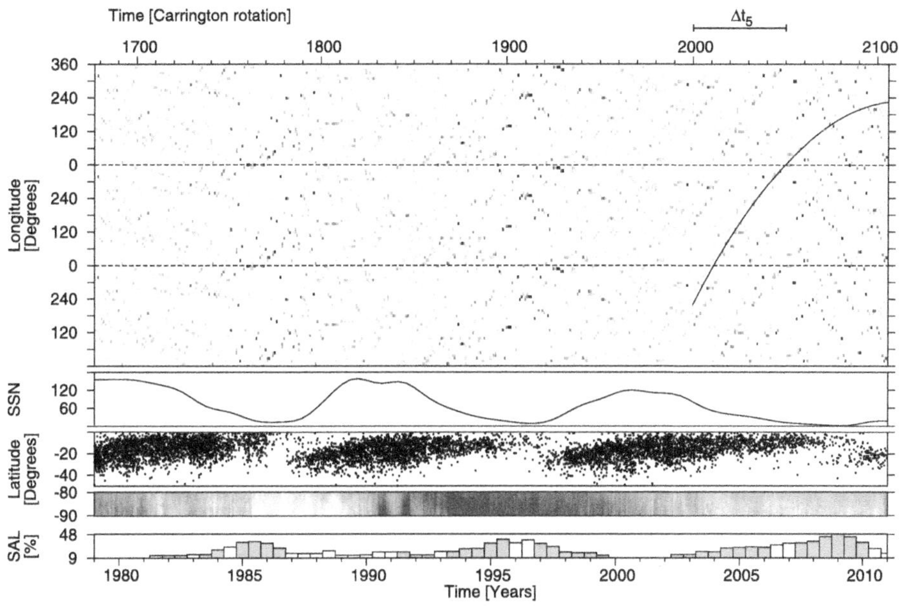

Figure 2 Longitudinal distribution of solar activity between 1997 and 2011 in the southern hemisphere (first row), the time profiles of Cycles 21 – 24 (second row), the hemispheric Spörer diagram (third row), the polarity of the magnetic dipole field at the southern pole (fourth row). The columns in the last row show the Strength of Active Longitude (SAL), the semiannual means of activity concentration expressed in percentage; gray and empty columns mark smooth and chaotic displacements of the active-longitudinal zones, respectively, see Section 2.3.

Remarkable features of Figures 1 and 2 are the migration tracks of enhanced activity toward both increasing and decreasing longitudes. In Cycles 21 – 22 the northern hemisphere exhibits a more evident path; although around the maximum of Cycle 22 the pixels are fainter because the location of activity is more extended, nevertheless the arc of the return can be recognized. In the southern hemisphere the path is more ambiguous in Cycles 21 – 22, it is more evident in Cycles 23 – 24, although no receding part can be seen as yet. The continuous curves indicate parabolas that have been fitted on the paths in the following way. The time intervals of most conspicuous migration tracks were selected, in the northern hemisphere: January 1984 – October 1986 (Carrington Rotation: 1744 – 1781) for the advancing migration, November 1992 – December 1996 (Carrington Rotation: 1862 – 1918) for the receding migration; for the southern hemisphere: April 2003 – January 2007 (Carrington Rotation: 2002 – 2054). The longitudinal center of mass of the sunspot-area data has been determined for each rotation in these intervals, *i.e.* the mean longitude weighted by the sunspot area. To this set of points a parabola has been fitted by using the standard least-squares procedure. The equations of the parabolas are as follows: For the northern hemisphere (Figure 1): $l = -K(r - 1834)^2 + 730$ where the Carrington longitude is $L = l \bmod 360°$ and r is the number of Carrington Rotations; 1834 is the Carrington Rotation number at the position of the symmetry axis of the parabola (September 1990). K is a scale factor between the axis of longitude and the axis of time, $K = 0.082$ degrees rotation^{-1}. For the southern hemisphere (Figure 2) $l = -K(r - 2120)^2 + 948$, and $K = 0.057$ degrees rotation^{-1}.

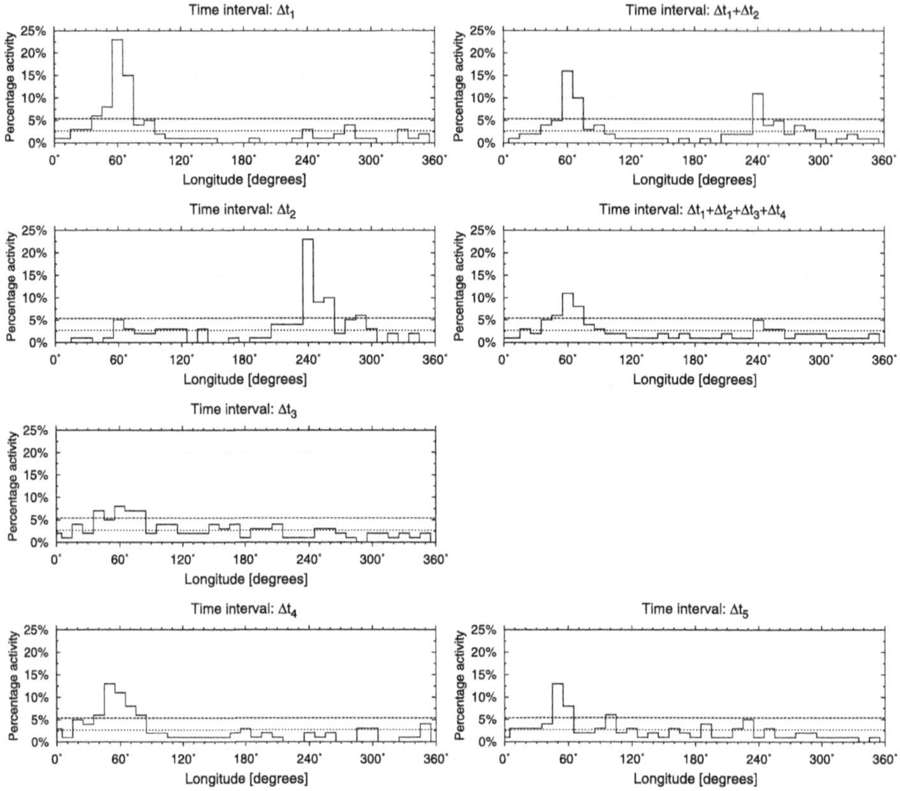

Figure 3 Left panels: Longitudinal distribution of activity along the migration path in the northern hemisphere in the four time intervals indicated in Figure 1. Right column; top panel: distribution in the united 1 – 2 intervals; second panel: distribution for the total length of the path; bottom panel: distribution in interval 5 indicated in the southern hemisphere, see Figure 2. The horizontal dashed lines indicate the levels of 1σ and 2σ, respectively.

2.2. Widths of Active Belts

Taking into account the migration of the active belts, their longitudinal extension can be determined in a moving reference frame that moves along the parabola keeping its longitude at a constant value. In our moving reference frame the parabola is at 60° longitude and the W_i values are calculated in 10° bins for the time interval considered. In Figure 1 time intervals are selected in which the longitudinal distributions are different along the migration paths and therefore they are considered separately. The results are depicted in Figure 3. It can be seen that the width of the active-longitude belt is about 60° at 1σ level (3.5 %), and it is about 30° at 2σ level (7 %) on average.

The separation of the intervals 1 – 4 is not arbitrary. The first and second time intervals of Figure 3 show a flip–flop phenomenon in the northern hemisphere between Cycles 21 and 22. The upper right panel of the figure shows the distribution in the combined 1 – 2 interval. It contains two active longitudes at opposite positions with no activity halfway between them but the separate intervals 1 and 2 contain only one of them. Further consequences of the continuity through the interval 3 will be treated in Section 2.5.

2.3. Another Identification of the Migration Path

The migration path indicated in Figure 1 has been identified by visual inspection as a conspicuous formation in a noisy environment. The sharpness of the active zones presented in Figure 3 seem to corroborate the reality of the choice; nevertheless, different approaches may be necessary to justify this unexpected feature.

The bottom lines of Figures 1 and 2 show the diagrams of a parameter named Strength of Active Longitude (SAL) defined in the following way: The percentage of the total activity in longitudinal bins of 10° has been plotted semiannually, the highest column has been chosen from each histogram, and the semiannual means have been subtracted from them. The differences have been plotted in semiannual resolution and the lowest lines of Figures 1 and 2 show the domains of this histograms above 1σ. The values of SAL are shown between 12 % and 61 % in the northern hemisphere and between 9 % and 48 % in the southern hemisphere. The presence of the columns in consecutive semiannual intervals may be regarded as a necessary condition for the existence of any active longitude, the intervals of missing columns may be disregarded.

The other method investigates the apparently varying rotation rate. As was summarized in the introduction, different investigations found different rotation rates depending on the methods and time intervals. To check the reality of the path marked in Figure 1, the rotation rates have been determined in two-year intervals shifted by steps of one Carrington Rotation along the time axis. In each two-year interval, a time series has been created from the daily total area of sunspots observed within a distance of $\pm 10°$ from the central meridian. The autocorrelations of these time series have been computed in each two-year interval. Figure 4 shows the autocorrelations in five selected intervals. The first one is shorter than two years, corresponding to the backward trend prior to the parabolic path; the rest of them correspond to the intervals 1 – 4 indicated in Figure 1. It can be seen how the rotation rate of the active-longitude zone varies with respect to the Carrington rate during the considered migration, or in other words, how the steepness of the path varies in the diagram of Figure 1. The apparent variation of the rotation rate corresponds to the migration of the active longitude, and this correspondence can be regarded as the sufficient condition of the existence of a migrating active longitude. In the diagrams of the above defined SAL parameter, the columns filled with gray belong to those time intervals in which the autocorrelation values of the relevant rotation rates were significant, otherwise the column is white. This last case means that there are significant activity concentrations at certain longitudes temporally, however, these longitudes do not migrate smoothly, they can have jumps to distant locations.

In the present work those time intervals are considered in which the above-formulated necessary and sufficient conditions were mostly present simultaneously. This sampling principle allowed us to select the time interval between Carrington Rotations 1708 – 1910 in the northern hemisphere, where these criteria were satisfied 84.4 % of the time. In the southern hemisphere the two criteria were satisfied between Carrington Rotations 1990 – 2110 77.7 % of the time. The simultaneous fulfillment of these criteria was restricted to smaller fractions of other intervals. This was the case 23.5 % of the time between Carrington Rotations 1980 – 2110 in the northern hemisphere, and 64.8 % of the time between Carrington Rotations 1700 – 1920 in the southern hemisphere; therefore they were not considered. These fractions are demonstrated in the bottom lines of Figures 1 and 2.

The presented shape of path may appear to be unexpected or even curious, but in fact this article is not the first presentation of this sort of migration. Figure 5 of Juckett (2006) presents practically the same shapes, the forward–backward migration within a couple of

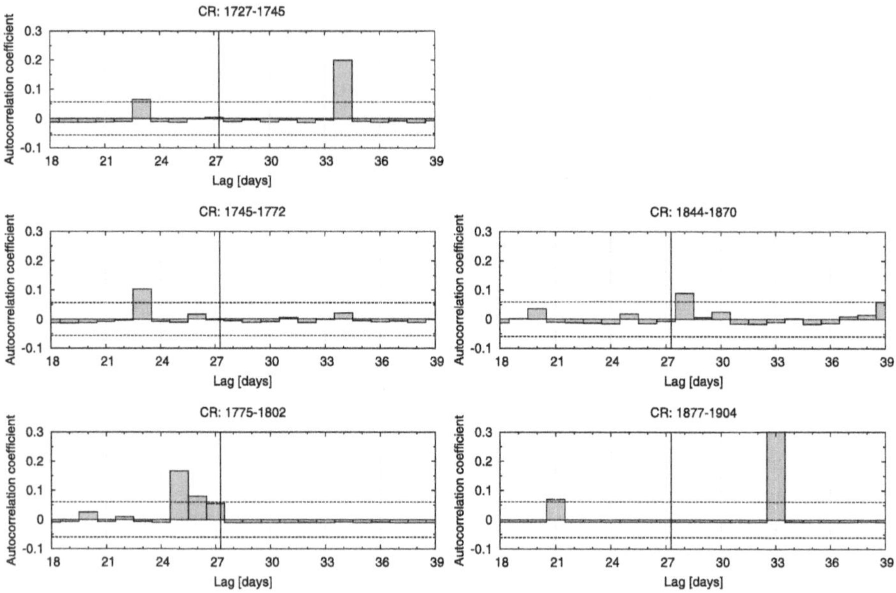

Figure 4 Autocorrelations of five intervals to follow the variation of the rotation rate between the following Carrington Rotations: 1727 – 1745 (upper left, corresponds to interval A in Figure 1) 1745 – 1772 (middle left, interval 1), 1775 – 1802 (bottom left, the flip-flop event), 1844 – 1870 (upper right, interval 3, when the Carrington rate dominates), 1877 – 1904 (lower left, interval 4). The horizontal scale shows the synodic rotation rates, the vertical line marks the Carrington rate.

years and then an abrupt V-shaped turn to move forward again. The similarity is recognizable, *mutatis mutandis*, by taking into account the substantial differences between the two methods.

It should be admitted that the parabolic fitting may be regarded as an assumption about the nature of the migration in spite of our original intention to avoid any pre-assumptions introducing subjective restrictions into the analysis. In the present case, however, the parabola has been chosen as the simplest function for this kind of shape; it does not imply any underlying physics at the moment. A fitting attempt with a section of a sinusoidal curve resulted in higher root-mean-squared error.

2.4. Latitudinal Relations

The flip-flop phenomenon in the northern hemisphere happens between Cycles 21 and 22. This is a jump of dominant activity from the main path to a secondary path at the opposite side, 180° apart and back to the main path in the third interval. To examine the latitudinal relationship of these jumps, the Spörer diagrams of Cycles 21 and 22 have been plotted in the northern hemisphere along with a 60°-wide belt along the path of enhanced activity, *i.e.* in the secondary path during interval 2; see Figure 5. Only the activity within these belts was considered. The activity centers of the main and secondary paths are distinguished by gray circles and black squares, respectively. It can be seen that the flip-flop event happens at the change of the consecutive cycles and both cycles with their distant latitudes are involved in the activity of the secondary path.

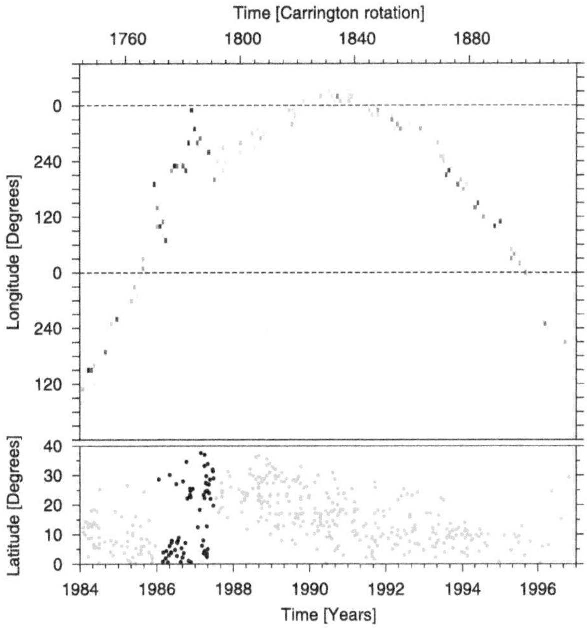

Figure 5 The 60°-wide belts along the main and secondary paths in the time–longitude diagram (top panel). The Spörer diagrams of the active regions within the paths (bottom panel). The main and secondary paths are distinguished by gray and black marks.

2.5. Dynamics of Emergence in the Active Belt

The variability of active-region emergence may also give some information as regards the nature of the belts of enhanced activity; this has been examined by autocorrelation analysis. We considered the 30°-wide band around the main path for Carrington Rotations 1740–1900. The upper panel of Figure 6 shows the fraction of activity in these longitude intervals by rotation (top panel), the plotted values are the sums of the largest observed areas of all sunspot groups reaching their maxima in the given rotations. The lower panel of Figure 6 shows the autocorrelation of this dataset; it has a single significant peak at the 18th rotation which corresponds to about 1.3 years, more precisely 491 days or 1.345 years.

To check whether this peak is an overall feature or it only belongs to the belt of enhanced activity, the same diagrams have been depicted for the sums of maximum sunspot areas on the entire Sun for the same rotations, see Figure 6. The ≈ 1.3-year peak disappeared from the autocorrelogram, so this period can be attributed to the active-longitude belt.

3. Discussion

It should be admitted that active longitudes cannot be identified reliably in a substantial fraction of the examined time period, so the following considerations are only related to those stripes of longitude which exhibit enhanced activity and continuous migration for at least two years according to the presented methods.

To interpret the migration of active longitudes a specific "dynamic reference frame" has been put forward by Usoskin, Berdyugina, and Poutanen (2005). They assumed that the source region of the emerging fields rotates differently and by finding the appropriate constants in the rotation profile, the active longitudes can be localized. Their methodology was criticized by Pelt, Tuominen, and Brooke (2005) and Pelt *et al.* (2006) but they maintained

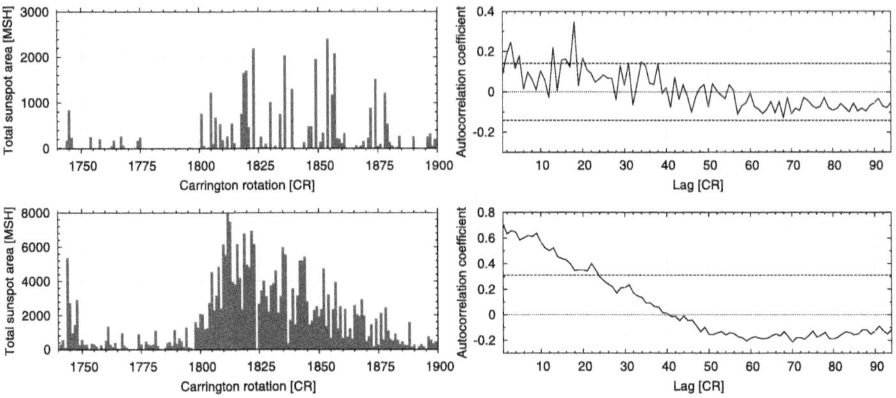

Figure 6 Upper row. Left panel: activity recorded by Carrington Rotations between Rotations 1740 – 1900 within the 30°-wide longitudinal band around the parabola of Figure 1; right panel: autocorrelogram of this data series with the confidence limits, the peak at 18 Carrington Rotations corresponds to 491 days or 1.345 years. Lower row. Left panel: total activity recorded by Carrington Rotations in the same time interval, right panel: autocorrelogram of this data series.

their concept and basic statements after correcting their formula (Usoskin *et al.*, 2007). They, among others, reported a clear cyclic behavior. They evaluated the data by applying numerical criteria.

However, closer scrutiny of Figures 1 and 2 shows that the migration paths of the active belts do not follow the shape of the 11-year cycle. The forward motion in Figure 1 (interval 1) takes place during the declining phase of Cycle 21, as also observed on a restricted interval by Bumba, Garcia, and Klvaňa (2000), while the receding motion coincides with the declining phase of Cycle 22. However, no similar migration pattern can be recognized during Cycle 23. These figures show that differential rotation can hardly be involved in the migration of active longitudes, with neither solar nor with anti-solar profile. Balthasar (2007) did not find signatures of differential rotation either. Juckett (2006, 2007) applied surface spherical harmonics to the surface-activity patterns; the variation of the phase information of this analysis describes the longitudinal migration of the active longitude. The results of these works also did not support the role of differential rotation.

Considering that the arc of the path extends to two cycles, one can only speculate about a possible impact of the Hale cycle because the turn-around point takes place at around the polarity change of the global dipole field. This issue needs more extended studies.

The active-longitude belts also exhibit considerable temporal variation. The width of the belt may be as narrow as 20° – 30° during moderate activity levels but around maximum it is dispersed as can be seen in Figure 3. This narrow extension of the active longitude and the shape of its path in the longitude–time diagram can only be detected with the present resolutions: 10° in longitude and one Carrington Rotation in time. It should be noted that the active longitudes are not always identifiable and the patterns of northern and southern hemispheres are not identical.

The 1.345-year period of spot emergence in the active belt may indicate its depth. Howe *et al.* (2000) detected a radial torsional oscillation at the tachocline zone with this period during 1995 – 2000. The present study makes a distinction between the regions within and out of the active-longitudinal region which is considered in a fairly narrow belt of 30° along the parabola-shaped path in Figure 1. The 1.3-year period is remarkable within the belt but it is absent, or overwhelmed, in the entire material. The presence of the ≈ 1.3-year period within the active belt allows the conjecture that the active longitudes may be connected to a source region close to the tachocline zone. This would be in accordance with Bigazzi and Ruzmaikin (2004) who argued that the non-axisymmetry can only remain at the bottom of the convective zone.

The flip-flop phenomenon was firstly observed by Jetsu, Pelt, and Tuominen (1993) on the active star FK Comae Berenices, further analyzed by Oláh *et al.* (2006). Figure 1 also shows two jumps between the main and secondary belts – there and back. No regular alternation can be observed between the two regions in the considered time interval; similarly, Balthasar (2007) did not find periodicity in the flip-flop events. What is even more important, the activity in the secondary belt takes place at the time of the exchange between Cycles 21 and 22, and furthermore, the active regions of this longitudinal belt are represented in both the ending and beginning cycles, *i.e.* at distant latitudes. This is a further hint that the migration of the active belt cannot be modeled with a differentially rotating frame. Figure 5 gives the impression that the disturbance releasing the field emergence may be either a meridional feature, or a phenomenon affecting the field emergence in a broad latitudinal belt. In other words, the cause of the enhanced activity does not seem to belong to the toroidal field because it affects both of the old and new toruses simultaneously.

The literature provides quite different values for the rotation rate of the source region of active longitudes; some of them were cited in the introduction. In certain cases, some values can be identified as temporally detectable periods influenced by the actual forward or backward motion of the active longitude. Taking into account the steepness of the parabola-shaped path in Figure 1 during the decaying phases of Cycles 21 and 22 the active longitude migrates around the Sun forward and backward during about 34 rotations. This means a virtual synodic rotation period of 26.5 days in Cycle 21 and 28.1 days in Cycle 22. The reported periods between these values may be connected to the migration of the active longitude. Jetsu *et al.* (1997) also reported a 26.722-day period on the northern hemisphere from flare data but their 22.07-day period cannot be identified at the surface, as they also found. Vernova *et al.* (2004) reported a rigid rotation with the Carrington period; this cannot be confirmed with the recent findings.

The concept of a relic magnetic field investigated by Olemskoy and Kitchatinov (2007) depends on the detectability of the 28.8-day period that corresponds to the rotation period of the radiative zone. This has not yet been detected in the surface data, and it is also missing from the present distributions. The relic magnetic-dipole field is conjectured in the radiative zone, and if its axis does not coincide with that of the convection zone magnetic field then their common effect could be a non-axisymmetric activity. This would imply a varying non-axisymmetry from cycle to cycle or the distinction between odd and even cycles because the polarity of the relic field is supposed to be constant.

Another possibility could be a certain disturbance of the meridional flows that could distort the toroidal flux ropes resulting in their buoyancy and emergence. This disturbance could have a counterpart at the opposite side of the Sun, *i.e.* at the secondary belt of active longitude. The interaction of active regions and meridional motions has been investigated by Švanda, Kosovichev, and Zhao (2008) in an opposite causal order; the presence of the active region was considered as an obstacle for the flow. González Hernández *et al.* (2008) also found modified meridional flows at the active regions that remain detectable even after the decay of the active regions. Longitudinal inhomogeneities of the meridional streams have not yet been reported. In the present data, the only intriguing coincidence is the forward migration in the northern hemisphere during negative magnetic polarity of the northern pole and backward migration after the change of polarity.

Dikpati and Gilman (2005) published theoretical results on the active-longitude problem. They concluded that the MHD shallow-water instability could produce bulges in the toroidal field that may be preferred sites to form magnetic loops rising to the surface. The ≈ 1.3-year periodicity of the activity within the active belt proper (Figure 6) could be a hint at this mechanism. To reproduce the migration pattern obtained, this mechanism would also have to imply a wave-like displacement of the bulge along the torus forward and backward depending on the polarity of main dipole field.

4. Summary

i) For the major fraction of our dataset no systematic active longitudes were found. Sporadic migration of active longitudes was identified only for Cycles 21 – 22 in the northern hemisphere and Cycle 23 in the southern hemisphere. The following conclusions are related to a minor fraction of the data, when active longitudes can be identified at all, and not to the entire analyzed time period, they cannot be regarded as general features of sunspot emergence.
ii) The active-longitude regions migrate in longitude forward and backward in the Carrington system independently of the 11-year cycle time profile.
iii) The half-width of the active-longitudinal belt is $20° - 30°$ during moderate activity but it is much less sharp at maximum activity.
iv) Flip-flop variations may occur but not with a regular period.
v) Differential rotation is probably not involved in the migration of active longitudes.
vi) The active-region emergence exhibits a ≈ 1.3-year variation within a 30°-wide belt around the active longitude which is absent in the entire activity.

These results were obtained for a limited time interval of a few solar cycles and, bearing in mind uncertainties of the migration-path definition, are only indicative. Further research based on an extended dataset is needed because the present work is restricted to the DPD era: 1977 – 2011. The study of the same features is under way on a longer dataset; the results will be published in a subsequent article.

Acknowledgements The research leading to these results has received funding from the European Commission's Seventh Framework Programme (FP7/2007 – 2013) under the grant agreement eHEROES (project n^{0} 284461, www.eheroes.eu).

References

Bai, T.: 1988, *Astrophys. J.* **328**, 860 – 878. ADS:1988ApJ...328..860B, doi:10.1086/166344.
Bai, T.: 2003, *Astrophys. J.* **585**, 1114 – 1123. ADS:2003ApJ...585.1114B, doi:10.1086/346152.
Balthasar, H.: 2007, *Astron. Astrophys.* **471**, 281 – 287. ADS:2007A&A...471..281B, doi:10.1051/0004-6361:20077475.
Berdyugina, S.V., Usoskin, I.G.: 2003, *Astron. Astrophys.* **405**, 1121 – 1128. ADS:2003A&A...405.1121B, doi:10.1051/0004-6361:20030748.
Berdyugina, S.V., Moss, D., Sokoloff, D., Usoskin, I.G.: 2006, *Astron. Astrophys.* **445**, 703 – 714. ADS:2006A&A...445..703B, doi:10.1051/0004-6361:20053454.
Bigazzi, A., Ruzmaikin, A.: 2004, *Astrophys. J.* **604**, 944 – 959. ADS:2004ApJ...604..944B, doi:10.1086/381932.
Bogart, R.S.: 1982, *Solar Phys.* **76**, 155 – 165. ADS:1982SoPh...76..155B, doi:10.1007/BF00214137.
Bumba, V., Garcia, A., Klvaňa, M.: 2000, *Solar Phys.* **196**, 403 – 419. ADS:2000SoPh..196..403B, doi:10.1023/A:1005226228739.

Carrington, R.C.: 1863, *Observations of the Spots on the Sun from November 9, 1853, to March 24, 1861 Made at Redhill.* Williams and Norgate, London, 246. ADS:1858MNRAS..19....1C, Section VI.

Cowling, T.G.: 1945, *Mon. Not. Roy. Astron. Soc.* **105**, 166. ADS:1945MNRAS.105..166C.

Dikpati, M., Gilman, P.: 2005, *Astrophys. J. Lett.* **635**, L193 – L196. ADS:2005ApJ...635L.193D, doi:10.1086/499626.

González Hernández, I., Kholikov, S., Hill, F., Howe, R., Komm, R.: 2008, *Solar Phys.* **252**, 235 – 245. ADS:2008SoPh..252..235G, doi:10.1007/s11207-008-9264-y.

Győri, L., Baranyi, T., Ludmány, A.: 2011, In: Choudhary, D.P., Strassmeier, K.G. (eds.) *Physics of Sun and Star Spots, IAU Symp.* **273**, Cambridge University Press, Cambridge, 403 – 407. http://fenyi.solarobs.unideb.hu/DPD/index.html, ADS:2011IAUS..273..403G, doi:10.1017/S174392131101564X.

Hathaway, D.H.: 2010, *Living Rev. Solar Phys.* **7**, 1. ADS:2010LRSP....7....1H, doi:10.12942/lrsp-2010-1.

Henney, C.J., Durney, B.R.: 2005, In: Sankarasubramanian, K., Penn, M., Pevtsov, A. (eds.) *Large-Scale Structures and Their Role in Solar Activity* **CS-346**, Astron. Soc. Pac., San Francisco, 381. ADS:2005ASPC..346..381H.

Howe, R., Christensen-Dalsgaard, J., Hill, F., Komm, R.W., Larsen, R.M., Schou, J., Thompson, M.J., Toomre, J.: 2000, *Science* **287**, 2456 – 2460. ADS:2000Sci...287.2456H, doi:10.1126/science.287.5462.2456.

Jetsu, L., Pelt, J., Tuominen, I.: 1993, *Astron. Astrophys.* **278**, 449 – 462. ADS:1993A&A...278..449J.

Jetsu, L., Pohjolainen, S., Pelt, J., Tuominen, I.: 1997, *Astron. Astrophys.* **318**, 293 – 307. ADS:1997A&A...318..293J.

Juckett, D.A.: 2006, *Solar Phys.* **237**, 351 – 364. ADS:2006SoPh..237..351J, doi:10.1007/s11207-006-0071-z.

Juckett, D.A.: 2007, *Solar Phys.* **245**, 37 – 53. ADS:2007SoPh..245...37J, doi:10.1007/s11207-007-9001-y.

Kitchatinov, L.L., Olemskoi, S.V.: 2005, *Astron. Lett.* **31**, 280 – 284. ADS:2005AstL...31..280K, doi:10.1134/1.1896072.

Mordvinov, A.V., Kitchatinov, L.L.: 2004, *Astron. Rep.* **48**, 254 – 260. ADS:2004ARep...48..254M, doi:10.1134/1.1687019.

Oláh, K., Korhonen, H., Kővári, Zs., Forgács-Dajka, E., Strassmeier, K.G.: 2006, *Astron. Astrophys.* **452**, 303 – 309. ADS:2006A&A...452..303O, doi:10.1051/0004-6361:20054539.

Olemskoy, S.V., Kitchatinov, L.L.: 2007, *Geomagn. Aeron.* **49**, 866 – 870. ADS:2009Ge&Ae..49..866O, doi:10.1134/S001679320907007X.

Pelt, J., Tuominen, I., Brooke, J.: 2005, *Astron. Astrophys.* **429**, 1093 – 1096. ADS:2005A&A...429.1093P, doi:10.1051/0004-6361:20041357.

Pelt, J., Brooke, J.M., Korpi, M.J., Tuominen, I.: 2006, *Astron. Astrophys.* **460**, 875 – 885. ADS:2006A&A...460..875P, doi:10.1051/0004-6361:20065399.

Plyusnina, L.A.: 2010, *Solar Phys.* **261**, 223 – 232. ADS:2010SoPh..261..223P, doi:10.1007/s11207-009-9501-z.

Royal Observatory, Greenwich, Greenwich Photoheliographic Results, 1874 – 1976, in 103 volumes. http://solarscience.msfc.nasa.gov/greenwch.shtml, http://www.ngdc.noaa.gov/stp/solar/greenwich.html.

Schou, J., Antia, H.M., Basu, S., Bogart, R.S., Bush, R.I., Chitre, S.M., *et al.*: 1998, *Astrophys. J.* **505**, 390 – 417. ADS:1998ApJ...505..390S, doi:10.1086/306146.

SIDC-team: 2012, World Data Center for the Sunspot Index, Royal Observatory of Belgium. Monthly Report on the International Sunspot Number. http://www.sidc.be/sunspot-data/.

Švanda, M., Kosovichev, A.G., Zhao, J.: 2008, *Astrophys. J. Lett.* **680**, L161 – L164. ADS:2008ApJ...680L.161S, doi:10.1086/589997.

Usoskin, I.G., Berdyugina, S.V., Poutanen, J.: 2005, *Astron. Astrophys.* **441**, 347 – 352. ADS:2005A&A...441..347U, doi:10.1051/0004-6361:20053201.

Usoskin, I.G., Berdyugina, S.V., Moss, D., Sokoloff, D.D.: 2007, *Adv. Space Res.* **40**, 951 – 958. ADS:2007AdSpR..40..951U, doi:10.1016/j.asr.2006.12.050.

Vernova, E.S., Mursula, K., Tyasto, M.I., Baranov, D.G.: 2004, *Solar Phys.* **221**, 151 – 165. ADS:2004SoPh..221..151V, doi:10.1023/B:SOLA-0000033367-32977-71.

Zhang, L.Y., Mursula, K., Usoskin, I.G., Wang, H.N.: 2011, *Astron. Astrophys.* **529**, 23. ADS:2011A&A...529A..23Z, doi:10.1051/0004-6361/201015255.

DOI 10.1007/978-1-4939-1182-0_11
Reprinted from *Solar Physics* Journal, DOI 10.1007/s11207-012-0220-5

Cyclic and Long-Term Variation of Sunspot Magnetic Fields

Alexei A. Pevtsov · Luca Bertello · Andrey G. Tlatov · Ali Kilcik · Yury A. Nagovitsyn · Edward W. Cliver

Received: 28 August 2012 / Accepted: 22 December 2012 / Published online: 23 January 2013

Abstract Measurements from the Mount Wilson Observatory (MWO) were used to study the long-term variations of sunspot field strengths from 1920 to 1958. Following a modified approach similar to that presented in Pevtsov *et al.* (*Astrophys. J. Lett.* **742**, L36, 2011), we selected the sunspot with the strongest measured field strength for each observing week and computed monthly averages of these weekly maximum field strengths. The data show the solar cycle variation of the peak field strengths with an amplitude of about 500 – 700 gauss (G), but no statistically significant long-term trends. Next, we used the sunspot observations from the Royal Greenwich Observatory (RGO) to establish a relationship between the sunspot ar-

Solar Origins of Space Weather and Space Climate
Guest Editors: I. González Hernández, R. Komm, and A. Pevtsov

A.A. Pevtsov (✉)
National Solar Observatory, Sunspot, NM 88349, USA
e-mail: apevtsov@nso.edu

L. Bertello
National Solar Observatory, 950 N. Cherry Avenue, Tucson, AZ 85719, USA
e-mail: lbertello@nso.edu

A.G. Tlatov
Kislovodsk Solar Station of Pulkovo Observatory, PO Box 145, Gagarina Str., 100, Kislovodsk 357700, Russian Federation
e-mail: tlatov@mail.ru

A. Kilcik
Big Bear Solar Observatory, 40386 Big Bear City, CA, USA
e-mail: kilcik@bbso.njit.edu

Y.A. Nagovitsyn
Pulkovo Astronomical Observatory, Russian Academy of Sciences, Pulkovskoe sh. 65, St. Petersburg 196140, Russian Federation
e-mail: nag@gao.spb.ru

E.W. Cliver
Space Vehicles Directorate, Air Force Research Laboratory, Sunspot, NM 88349, USA
e-mail: ecliver@nso.edu

eas and the sunspot field strengths for cycles 15 – 19. This relationship was used to create a proxy of the peak magnetic field strength based on sunspot areas from the RGO and the USAF/NOAA network for the period from 1874 to early 2012. Over this interval, the magnetic field proxy shows a clear solar cycle variation with an amplitude of 500 – 700 G and a weaker long-term trend. From 1874 to around 1920, the mean value of magnetic field proxy increases by about 300 – 350 G, and, following a broad maximum in 1920 – 1960, it decreases by about 300 G. Using the proxy for the magnetic field strength as the reference, we scaled the MWO field measurements to the measurements of the magnetic fields in Pevtsov *et al.* (2011) to construct a combined data set of maximum sunspot field strengths extending from 1920 to early 2012. This combined data set shows strong solar cycle variations and no significant long-term trend (the linear fit to the data yields a slope of $-0.2 \pm 0.8\ \mathrm{G\,year^{-1}}$). On the other hand, the peak sunspot field strengths observed at the minimum of the solar cycle show a gradual decline over the last three minima (corresponding to cycles 21 – 23) with a mean downward trend of $\approx 15\ \mathrm{G\,year^{-1}}$.

Keywords Magnetic fields · Solar cycle · Sunspots

1. Introduction

Recently, several studies have concentrated on the long-term variations of field strengths in sunspots. The question at the core of these investigations was whether the strength of sunspot magnetic fields has gradually declined over the last two cycles.

Penn and Livingston (2006, 2011) measured the field strength using the separation of two Zeeman components of the magnetically sensitive spectral line Fe I 1564.8 nm over the declining phase of solar cycle 23 and the rising phase of cycle 24. The measurements were taken on a daily basis (in quarterly observing intervals due to telescope scheduling). In early years, only the large sunspots were measured; more recent observations are aimed at including all sunspots and pores that are present on the disk. Monthly averages of these measurements show a gradual decrease in sunspot field strength from the beginning of the project (late 1998) to the present (mid-2012). It is possible that the non-uniformity of the data set (*i.e.*, fewer measurements at the beginning, newer observations include both sunspots and pores) may result in such a decline in average field strengths. However, the most recent study by Livingston, Penn, and Svalgaard (2012) indicates that there is no change in the shape of the distribution of measured field strengths over the observing period; only the mean of the distribution changes. This constancy in the shape of the field distribution appears to rule out the speculation that the decline in field strengths could be explained by the non-uniformity of the data set.

Watson, Fletcher, and Marshall (2011) employed the magnetic flux measurements from the Michelson Doppler Imager (MDI) onboard the *Solar and Heliospheric Observatory* (SOHO), and studied the magnetic flux changes in sunspots over cycle 23 (1996 – 2010). Because MDI only measures the line-of-sight fluxes, the authors reconstructed the vertical flux under the assumption that the magnetic field in sunspots is vertical. The results showed a solar-cycle-like variation and only a minor long-term decrease in vertical magnetic flux of the active regions. Pevtsov *et al.* (2011) employed the manual measurements of magnetic field strengths taken in the Fe I 630.15 – 630.25 nm wavelengths in the period of 1957 – 2010 in seven observing stations that form the solar observatory network across eleven time zones in what is now Azerbaijan, Russian Federation, and in the Ukraine. To mitigate the differences in the atmospheric seeing and the level of the observers' experience, Pevtsov *et al.* (2011) considered only the strongest sunspot measurements for each day of observations. The sunspot field strengths were found to vary strongly with the solar cycle, and no long-term decline was noted. Rezaei, Beck, and Schmidt (2012) used the magnetographic observations from the Tenerife Infrared Polarimeter at the German Vacuum telescope in the period of 1999 – 2011 to confirm the cycle variations of sunspot magnetic fields. Comparing maximum field strengths measured at the rising phases of cycles 23 and 24, the authors had noted a slight reduction in field strengths at the beginning of cycle 24. Still, the main variations in the magnetic field strength were found to be related to the solar cycle.

With the exception of Pevtsov *et al.* (2011), all previous studies based their conclusions on the data from the most recent full cycle 23. In our present article, we extend the analysis to earlier solar cycles. First, we use the data from the Mount Wilson Observatory (MWO) and apply the same technique as in Pevtsov *et al.* (2011). Next, we establish a correlation between the area of a sunspot and its field strength and use this correlation to construct a proxy for the magnetic field strength based on sunspot areas measured by the Royal Greenwich Observatory (RGO) and USAF/NOAA. The proxy for the magnetic field strength allows us to extend the analysis to cycles 11 – 24 and to scale the 1920 – 1958 MWO peak field measurements to those of Pevtsov *et al.* (2011) for the 1957 – 2010 interval. Our analysis is presented in Sections 2 and 3, and our results are summarized and discussed in Section 4.

2. Sunspot Field Strength Measurements from Mount Wilson Observatory (1920 – 1958)

To investigate the properties of the sunspot magnetic fields, we employed the observations from the MWO from May 1920 till December 1958. Specifically, we selected only data published in the *Publications of the Astronomical Society of the Pacific* (PASP). Although the sunspot field strength measurements continue to the present day (with a major interruption between September 2004 to January 2007 due to funding problems), their regular publication was discontinued at the end of 1958. This 1920 – 1958 part of the MWO data set should be considered as the most uniform; in later years there were several major changes to the instrumentation and the observing procedure (*e.g.*, multiple replacements of the spectrograph grating, selection of a new spectral line for measurements, and a different tilt-plate). In addition, the resolution of the field strength measurements changed from 100 gauss (G) through 1958 to 500 G for the later measurements. A summary of these changes can be found in Livingston *et al.* (2006).

We digitized the data summaries in PASP and verified the newly constructed tables against the published record. In a few instances, the MWO had issued corrections to the original tables (also published in PASP). All these corrections were taken into account in the process of data verification. Finally, we found a small number of inconsistencies in the original tables, which were also corrected.

The MWO sunspot summaries provide the MWO sunspot number, the sunspot's field strength (no polarity information), latitude, and an estimated date of the central meridian passage (later data also show the first and last days of an active region on the disk as well as its magnetic classification). In this study, we only use the sunspot field strength, latitude, and the date of the central meridian crossing. For some active regions the maximum field strength is estimated (see explanation in PASP **50**: "When it seems probable that the greatest field-strength observed in any group was not the maximum for that group, an estimated value is given in parentheses"). The field strength was measured manually using a glass tilt-plate. By tilting the plate, the observer co-aligned two Zeeman components of the spectral line, and the tilt angle translated into a linear (wavelength) displacement. Since the stronger field strengths require larger tilt angles, this procedure may introduce a non-linearity in the

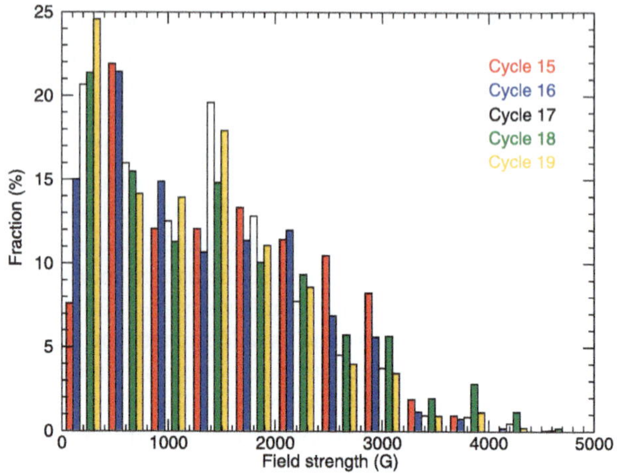

Figure 1 Fractional distribution of sunspot field strengths for different cycles. The bin size is 400 G. Measurements with zero field strength are excluded.

relation between the plate's tilt and the linear displacement. This slight non-linearity was corrected following the procedure specified in Livingston *et al.* (2006). Only measurements exceeding 2400 G were affected by this non-linearity.

A quick examination of the entire data set revealed several systematic effects. First, we found an increased number of values in parentheses (estimated values) in later years of the data set than in earlier years. Therefore, we excluded all estimated data from our investigation. Second, we noted a gradual increase in the fraction of measurements with weak field strengths from earlier to later years. For example, for observations taken during cycle 15 only 4 % of all measurements have field strengths of 100 G. For later cycles, the fraction increases significantly to 9 % (cycle 16), 15 % (cycle 17), 14 % (cycle 18), and 18 % (cycle 19). The tendency for a higher percentage of measurements with weaker fields is quite obvious in the annual number of measurements with zero fields (when observations were taken, but the fields were considered to be weaker than 100 G). Even after normalizing for the level of sunspot activity (using international sunspot numbers), the annual number of measurements with zero fields shows a steady (and significant) increase from 1920 to 1958.

Figure 1 shows the fractional distribution of sunspot field strengths in 400 G bins in the range of 0 – 5000 G. Data for different cycles are shown by different colors. The most significant differences are confined to the first two bins. The field strengths in the range of 100 – 400 G show a systematic increase in the fraction of weak field measurements from about 8 % in cycle 15 to about 25 % in cycle 19. These fractions are given in all measurements taken in a given cycle. The measurements in the next bin (500 – 800 G) show the opposite trend with a decrease in the fraction of measurements from about 22 % in cycles 15 and 16 to about 14 % in cycle 19. Stronger fields do not show any systematic behavior from one cycle to the next.

One can speculate that at least some of these systematic effects could be due to changes in the observing requirements (*e.g.*, increase in the "granularity" of measurements, when measurements are taken in separate umbrae of a single sunspot) and/or a "learning curve" effect (when with increasing experience the observer begins to expand the measurements to smaller sunspots). An increase in the scattered light might have similar effects on the visual measurements of the magnetic fields. These systematic effects may affect the average value of the field strengths. For example, including the larger number of measurements with the weaker field strengths will decrease the average value. To mitigate the negative effects of this possible change in statistical properties of the data sample, we employed an approach similar to the one used by Pevtsov *et al.* (2011), where only the strongest measured field strength is selected for any given day of observations.

Next, we investigated the temporal behavior of the strongest sunspot field strengths. Since the published summaries of the MWO observations do not provide the date of measurements, we adopted the estimated date of the central meridian crossing as a proxy for the day of observations.

Applying the Pevtsov *et al.* (2011) approach to the MWO data is complicated by the fact that MWO measurements are, strictly speaking, not the daily observations. In many cases, there are daily drawings, but no corresponding magnetic field measurements. Also, when measurements do exist, they may exclude some (sometimes even the largest) sunspots that were present on the disk. For example, during a disk passage of a large sunspot (MWO AR 7688) on 17 – 28 November 1944, the magnetic field measurements were taken only on 17 – 19 November and 22 November. On 22 January 1957, the field strengths were measured only in smallest sunspots and pores; the largest sunspots were not measured. From 27 February – 2 March 1942, the measurements alternate between the largest sunspots on the disk (one day) and the smallest sunspots (the following day). Since the published summaries of the MWO observations provide only the estimated date of the central meridian crossing (but not the date of observations), the selection of only one measurement of the strongest field strength for any given day of the observations as in Pevtsov *et al.* (2011) may not work well (*i.e.*, measurements of weaker field strengths are less likely to be excluded even if there are sunspots with stronger fields on the disk). Therefore, we modified the method by selecting only the measurement of the strongest field strength for any given week of observations.

Livingston *et al.* (2006) suggested that, since for field strengths weaker than 1000 G, the Zeeman splitting for Fe 617.33 nm becomes comparable to the Doppler width of the spectral line, the measurements of these fields are unreliable. However, a well-trained observer can consistently measure fields weaker than 1000 G. Livingston *et al.* (2006) relate one example where multiple measurements from the same observer agree within 10 G. We also examined several drawings in more detail and found a good persistence in day-to-day measurements of pores with field strengths below 1000 G. Nevertheless, to be cautious, we chose to exclude field strength measurements below 1000 G in our analysis. Still, including field strength measurements below 1000 G does not change the main results of this article.

Figure 2 (upper panel) shows the latitudinal distribution of active regions in our MWO data set and the monthly averages of the (weekly) strongest field strengths (Figure 2, lower panel). The most prominent trend in the data is a solar cycle variation with a minimum field strength in sunspots around the minima of solar cycles, and a maximum field strength near the maxima of solar cycles (see Table 1). To verify the presence (or absence) of the long-term trend, we fitted the data with linear and quadratic functions. The linear fit suggests a negligible decrease in sunspot field strengths over 40 years of about $0.8 \pm 1.7\ \mathrm{G\,year^{-1}}$.

3. Sunspot Area as Proxy for Magnetic Field Strength

Several studies (*e.g.*, Houtgast and van Sluiters, 1948; Rezaei, Beck, and Schmidt, 2012) found a correlation between the sunspot areas (S, in millionths of solar hemisphere, MSH) and their recorded field strength (H_{MAX}). Ringnes and Jensen (1960) found the strongest correlation between the area logarithm and the field strength. Here, we use this relationship to investigate the changes in sunspot magnetic field strength over a long time-interval by employing the sunspot area as a proxy for the magnetic field strength. Figure 3 shows the relationship between the logarithm of the sunspot area and their field strength. Areas are

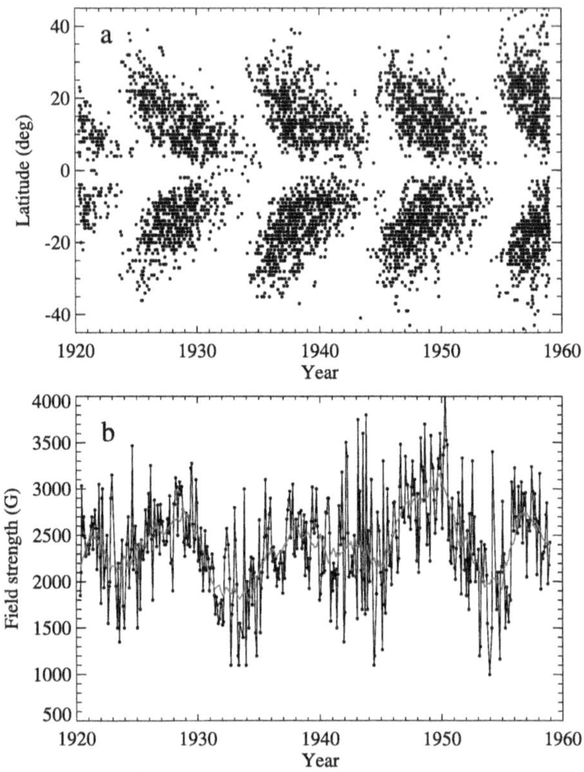

Figure 2 Time-latitude distribution of sunspots in the MWO data set included in our study (upper panel), and monthly average of daily peak sunspot field strengths (lower panel, dots connected by thin line). The thick gray line is the 18-point running average.

Table 1 Years of maxima and minima for solar cycle and sunspot field strengths.

Solar cycle		Sunspot field	
Minimum	Maximum	Minimum	Maximum
1923		1923.6	
	1928		1928.8
1933		1933.3	
	1937		1938.1
1944		1944.7	
	1947		1949.8
1954		1953.9	

taken from the independent data set of the RGO. For this plot, we established a correspondence between the active regions in the MWO and RGO data using the date of the central meridian passage and the latitude of active regions. Only regions with a small difference ($\leq 0.6^\circ$) in latitude and less than 4.8 h (0.2 day) in the time of the central meridian crossing between the two data sets were included. We also excluded RGO areas smaller than 10 MSH. Solar features with area $S \leq 10$ MSH are small spots or pores; their field strengths are more likely to have large measurement errors.

Similar to previous studies, we find a reasonably good correlation between the (logarithm of) sunspot areas and the magnetic field strength (Pearson's correlation coefficient $\rho = 0.756$). Similar (strong) correlation coefficients were found for individual cycles (see

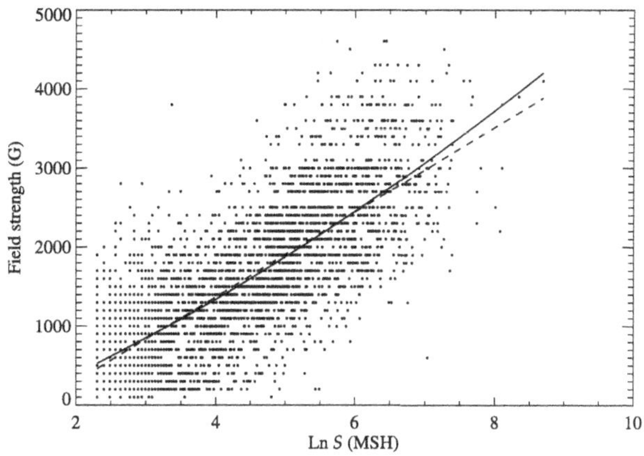

Figure 3 Magnetic field strength (from MWO observations) vs. the natural logarithm of the sunspot area (from RGO observations) for cycles 15 – 19. The dashed line is a first-degree polynomial fit to the data. Fitted coefficients are shown in Table 2, in the entry for "all cycles". For comparison, the solid line shows a fit by a quadratic function.

Table 2 Correlation (ρ) and fitted coefficients for the $H = A + B \times \ln(S)$ dependency between magnetic field strength (H) and the deprojected area (S) of an active region. Student's t-values and maximum sunspot number (SSN) for the $(n+1)$-th cycle are also shown.

Cycle No.	A	B	ρ	t-value	99 %-level	SSN_{n+1}
Cycle 15	-274.1 ± 177.6	507.3 ± 40.2	0.775	12.633	2.623	78.1
Cycle 16	-475.1 ± 63.4	514.9 ± 13.9	0.811	36.947	2.583	119.2
Cycle 17	-771.0 ± 59.9	523.2 ± 13.2	0.781	39.595	2.581	151.8
Cycle 18	-1106.9 ± 78.9	609.2 ± 16.9	0.739	35.966	2.580	201.3
Cycle 19	-800.4 ± 69.5	495.4 ± 14.9	0.784	33.252	2.583	110.6
All cycles	-774.2 ± 35.6	536.0 ± 7.7	0.756	69.170	2.577	
Cycles 16 – 18	-806.3 ± 41.0	551.7 ± 8.9	0.761	61.670	2.578	

Table 2). To verify the statistical significance of the correlations, we used the t-test; the t-values and the cutoff value for the 99 % confidence level are shown in Table 2. Since the t-values are well above the 99 % cutoff values, all correlations are statistically significant. The relationship between the magnetic field strength and the logarithm of the sunspot area is very similar for cycles 15 – 17. For cycle 18 the relation is "steeper" (sunspots with the same areas correspond to stronger magnetic fields than in cycles 15 – 17), and for cycle 19, the field-area relationship is somewhat weaker and more similar to the relationship found in cycle 15. Our data set only partially includes these two cycles. Cycles 16 – 18 are included in their entirety. Overall, we find that the MWO data for cycle 18 contain a slightly higher percentage of stronger field measurements than all other cycles (*e.g.*, see Figure 1). Limiting the data to three complete cycles 16 – 18 does not significantly affect the coefficients for the functional relation between sunspot area and magnetic field strength (see Table 2). Figure 3 indicates a non-linearity between the sunspot area and their magnetic field strength. On the other hand, a quadratic fit to the data (solid line, Figure 3) suggests that this non-linearity is small. For simplicity, in the following discussion we employ a linear relation between the logarithm of sunspot areas and their magnetic field strength.

Using the relation $H_{MAX} = -774.2 + 536.0 \times \ln(S)$ based on cycles 15 – 19 (Table 2), we created a proxy for the magnetic field strength based on the sunspot areas. The data used

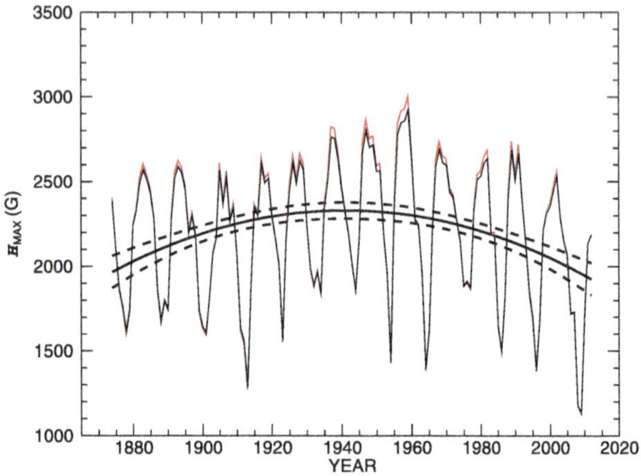

Figure 4 Proxy of the magnetic field strength computed from the deprojected sunspot areas. Annual averages are shown as a thin solid line. The thick solid line is a second-degree polynomial approximation to the data, and the two thick dashed lines represent a one-standard deviation error band for the fit. The red line shows the magnetic field proxy derived using the quadratic dependency shown in Figure 3.

for this exercise represent a combination of the RGO measurements from 1874 – 1977 and the USAF/NOAA data from 1977 – early 2012. The USAF/NOAA data were corrected (by D. Hathaway) by a factor of 1.4 as described in http://solarscience.msfc.nasa.gov/greenwch.shtml. Figure 4 shows annual averages of the proxy of the magnetic field derived from the deprojected active region areas. For consistency with the scaling between the area logarithm and the magnetic field strength derived earlier, Figure 4 does not include areas smaller than 10 MSH. Including all areas does not noticeably change Figure 4. Similar to the direct measurements of the magnetic field, the proxy also traces the solar cycle variations with an approximate amplitude of 500 – 700 G. The data also show a long-term trend (see the parabolic fit to the data), with the mean value of the field proxy increasing from 1874 to around 1920 by about 300 G and then, following a broad relatively flat maximum to around 1960, decreasing by about same amount by early 2012. For comparison, Figure 4 also shows the magnetic field proxy computed using a quadratic fit to the area-field strength relation (red line).

Finally, using the magnetic field proxy as reference, we combined the MWO data and the Russian sunspot field strength observations from Pevtsov *et al.* (2011) into a single data set. Figure 5 shows two data sets on the same scale. The MWO data were re-scaled to the Russian data using the functional dependencies between the H_{MAX} (MWO) and the magnetic field proxy (sunspot areas) and a similar dependency for the Russian data. With the scaling, the mean level of approximately 2500 G in Figure 5 agrees with that observed at NSO/Kitt Peak from 1998 – 2005 by Penn and Livingston (2006).

The most prominent feature of the combined data shown in Figure 5 is the cycle variation of the sunspot field strength. Maxima and minima in the sunspot field strength agree relatively well with the maxima and minima in the sunspot number. The combined data set (Figure 5) shows no statistically significant long-term trend. A linear fit to the data (not shown in Figure 5) has a slope of $\approx -1.9 \pm 0.8$ G year^{-1}. This trend appears to be entirely due to the points corresponding to the deep minimum around 2010, and it disappears once these few points are excluded from the fitting (the fit shown in Figure 5 has a slope of $\approx -0.2 \pm 0.8$ G year^{-1}). In either case, the amplitude of the trend is significantly smaller than the -52 G year^{-1} reported by Penn and Livingston (2006). As the Penn and Livingston (2011) data cover the period of 1998 – 2011, the trend in the field strengths found in their data may be dominated by the declining phase of solar cycle 23.

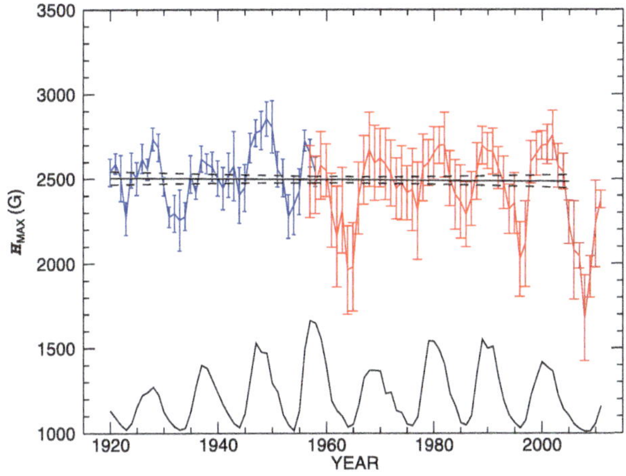

Figure 5 Annual averages of the magnetic field measurements from the MWO (blue color, up to 1958) and the Russian data set from Pevtsov *et al.* (2011, red). The MWO data are scaled to the Russian data set using the proxy of the magnetic field strength in Figure 4 as reference. Error bars show a $\pm 1\sigma$ standard deviation of the mean values. The thick black straight line is a linear approximation to the MWO and the Russian data. Dashed lines indicate the statistical uncertainties of this linear approximation. The black line in the lower part of the figure shows the annual international sunspot numbers.

4. Discussion

We employed observations of the sunspot field strengths from a subset of the MWO data covering solar cycles 15 – 19 (1920 – 1958). The data were analyzed using a modified approach of Pevtsov *et al.* (2011), where the strongest field measurement was selected for each week of observations. This approach allows one to mitigate the effects of some negative properties of the data set, for example, a systematic increase in the number of weak field measurements from the beginning of the data set to its end. Our findings confirm the presence of the 11-year cycle variation in the sunspot daily strongest field strengths similar to that found in the previous studies for solar cycles 19 – 23 (Pevtsov *et al.*, 2011; Watson, Fletcher, and Marshall, 2011; Rezaei, Beck, and Schmidt, 2012). On the other hand, no significant secular trend is found for the period covered by the MWO data (cycles 15 – 19). We do see a weak gradual decrease in average field strengths of $\approx 2.1 \pm 0.9$ G year^{-1}.

A correlation between the area of sunspots and the magnetic field strength allows us to use the sunspot area as a proxy for the magnetic field strength. With this approach, we showed that the cycle variations (in the magnetic field strengths as represented by their proxy) are present during cycles 11 – 24 (1874 – early 2012). The amplitude of these cycle variations is about 1000 G between the solar activity minima and maxima. The magnetic field strength proxy does show a secular trend: between 1874 and $\approx$ 1920, the mean value of the magnetic field proxy increased by about 300 G, and following a broad maximum in 1920 – 1960, it decreased by 300 G. The nature of this trend is unknown, but we note that the broad "maximum" in 1920 – 1960 includes the mid-twentieth century maximum of the current Gleissberg cycle, which began at about 1900. The long-term trend in the magnetic field strength proxy based on the sunspot area (Figure 4) could be subject to several uncertainties. For example, sunspot area estimates from the RGO observations prior to $\approx$ 1910 appear to be systematically lower than later data (Leif Svalgaard, private communication). Moreover, the functional dependency between the sunspot area and the peak magnetic field strength may be non-linear and/or vary from one cycle to another (see the quadratic and linear fits in Figure 3). For example, a long-term variation in the $H_{\mathrm{MAX}} - S$ dependency was found in Ringnes and Jensen (1960). The steepness of the linear regression in cycle n may correlate with the amplitude of cycle $(n+1)$ (compare the B coefficients and the yearly international sunspot number, SSN_{n+1}, in Table 2), although our statistical sample is too small to be more conclusive with respect to this possible dependency. Nevertheless, the H_{MAX} proxies computed using linear and non-linear scaling functions agree reasonably well (compare black and red lines in Figure 4).

Using the proxy of H_{MAX} as a reference allows us to combine the MWO and Russian measurements of the magnetic field into a single data set. Although this data set does not show any statistically significant long-term trend comparable in amplitude with the trend reported in Penn and Livingston (2011), one can notice that the last three (see red curve in Figure 5) minima in the sunspot field strengths progressively decrease. The overall trend derived from this tendency is about 15 G year^{-1}, which could be associated with the Gleissberg cycle variations as derived from other studies (*e.g.*, Mordvinov and Kuklin, 1999). Alternatively, this could be an indication of some other long-term pattern in sunspot field strengths. One can note a similar tendency for progressively lower field strengths in the solar minima of cycles 12 – 14 (Figure 4) and cycles 17 – 19 (Figures 4 and 5). We leave the investigation of these patterns to future studies.

Acknowledgements Y.N. and A.T. acknowledge a partial support from the Federal Program "Scientific and Pedagogical Staff of Innovative Russia", the Russian Foundation for Basic Research (grants Nos. 10-02-00391 and 12-02-00614), and the Programs of the Presidium of Russian Academy of Sciences Nos. 21 and 22. AAP acknowledges funding from NASA's NNH09AL04I inter agency transfer. National Solar Observatory (NSO) is operated by the Association of Universities for Research in Astronomy, AURA Inc under cooperative agreement with the National Science Foundation (NSF).

References

Houtgast, J., van Sluiters, A.: 1948, *Bull. Astron. Inst. Neth.* **10**, 325.
Livingston, W., Harvey, J.W., Malanushenko, O.V., Webster, L.: 2006, *Solar Phys.* **239**, 41.
Livingston, W., Penn, M.J., Svalgaard, L.: 2012, *Astrophys. J. Lett.* **757**, L8.
Mordvinov, A.V., Kuklin, G.V.: 1999, *Solar Phys.* **187**, 223.
Pevtsov, A.A., Nagovitsyn, Y., Tlatov, A., Rybak, A.: 2011, *Astrophys. J. Lett.* **742**, L36.
Penn, M.J., Livingston, W.: 2006, *Astrophys. J. Lett.* **649**, L45.
Penn, M.J., Livingston, W.: 2011, In: Choudhary, D.P., Strassmeier, K.G. (eds.) *The Physics of Sun and Star Spots, IAU Symp.* **273**, 126.
Rezaei, R., Beck, C., Schmidt, W.: 2012, *Astron. Astrophys.* **541**, A60.
Ringnes, T.S., Jensen, E.: 1960, *Astrophys. Nor.* **7**, 99.
Watson, F.T., Fletcher, L., Marshall, S.: 2011, *Astron. Astrophys.* **533**, A14.

DOI 10.1007/978-1-4939-1182-0_12
Reprinted from *Solar Physics* Journal, DOI 10.1007/s11207-013-0337-1

Apparent Solar Tornado-Like Prominences

Olga Panasenco · Sara F. Martin · Marco Velli

Received: 18 September 2012 / Accepted: 25 May 2013 / Published online: 10 July 2013

Abstract Recent high-resolution observations from the *Solar Dynamics Observatory* (SDO) have reawakened interest in the old and fascinating phenomenon of solar tornado-like prominences. This class of prominences was first introduced by Pettit (*Astrophys. J.* **76**, 9, 1932), who studied them over many years. Observations of tornado prominences similar to the ones seen by SDO had already been documented by Secchi (*Le Soleil*, 1877). High-resolution and high-cadence multiwavelength data obtained by SDO reveal that the tornado-like appearance of these prominences is mainly an illusion due to projection effects. We discuss two different cases where prominences on the limb might appear to have a tornado-like behavior. One case of apparent vortical motions in prominence spines and barbs arises from the (mostly) 2D counterstreaming plasma motion along the prominence spine and barbs together with oscillations along individual threads. The other case of apparent rotational motion is observed in a prominence cavity and results from the 3D plasma motion along the writhed magnetic fields inside and along the prominence cavity as seen projected on the limb. Thus, the "tornado" impression results either from counterstreaming and oscillations or from the projection on the plane of the sky of plasma motion along magnetic-field lines, rather than from a true vortical motion around an (apparent) vertical or horizontal axis. We discuss the link between tornado-like prominences, filament barbs, and photospheric vortices at their base.

Solar Origins of Space Weather and Space Climate
Guest Editors: J. González Hernández, R. Komm, and A. Pevtsov

Electronic supplementary material The online version of this article (doi:10.1007/s11207-013-0337-1) contains supplementary material, which is available to authorized users.

O. Panasenco (✉)
Advanced Heliophysics, Pasadena, CA, USA
e-mail: panasenco.olga@gmail.com

S.F. Martin
Helio Research, La Crescenta, CA, USA

M. Velli
Jet Propulsion Laboratory, California Institute of Technology, Pasadena, CA, USA

Keywords Coronal mass ejections, low coronal signatures · Coronal mass ejections, initiation and propagation · Magnetic fields, corona · Coronal holes, prominences, formation and evolution · Filaments

1. Introduction

Prominences closely resembling terrestrial tornadoes in form, when projected on the solar limb, have been observed spectroscopically since 1868. Probably the first published drawings, similar to the tornado-like prominences recently observed by the *Transition Region and Coronal Explorer* (TRACE: Handy *et al.*, 1999) and the *Atmospheric Imaging Assembly* (AIA: Lemen *et al.*, 2012) onboard the *Solar Dynamics Observatory* (SDO), can be found in Secchi (1877) under the type "Flammes." Young (1896) described these transient structures as resembling "whirling waterspouts, capped by a great cloud." The first successful photographs of tornado-like prominences, made by Slocum in 1910 and by Pettit in 1919, were published by Pettit (1925). Pettit preserved his interest in this subject and periodically came back to its study (Pettit, 1932, 1941, 1943, 1946, 1950). Pettit summarized the appearance of tornado-like prominences as "Vertical spirals or tightly twisted ropes" (Pettit, 1932), and stated: "In silhouette, tornado prominences are ... columnar, usually with a small smoke-like streamer issuing from the top, often bent over, even touching the chromosphere" (Pettit, 1950). Most prominences of this type are from 5000 to 22 000 km in width and 25 000 to 100 000 km in height (Pettit, 1943). The tornado-like prominences described so far are transient objects, often appearing in groups.

Recent high-resolution observations from SDO have reawakened interest in apparent solar tornado-like prominences. The higher spatial resolution and greater temporal cadence of the SDO observations together with their unprecedented duration and continuity shed new light on the possible origin, formation, and evolution of solar tornado-like prominences. Recent works in this area have described the SDO observations of apparent tornado-like prominences as helical structures with rotational motions (Li *et al.*, 2012; Su *et al.*, 2012; Orozco Suárez, Asensio Ramos, and Trujillo Bueno, 2012; Panesar *et al.*, 2013). We question this geometry and describe alternative models and mechanisms of formation. We also discuss the role that photospheric vortices may play in their dynamics (Velli and Liewer, 1999; Attie, Innes, and Potts, 2009; Wedemeyer-Böhm *et al.*, 2012; Rappazzo, Velli, and Einaudi, 2013; Kitiashvili *et al.*, 2013).

The first spacecraft observations of this phenomenon were done with TRACE (Figure 1). The TRACE movie from 27 November 1999 (see supplementary materials) seems to show a tornado-like motion at the limb, which can also be interpreted in terms of the counterstreaming of prominence plasma, with velocities up to 50 km s^{-1}, along the local magnetic-field lines comprising the prominence spine and barbs (Zirker, Engvold, and Martin, 1998; Lin, Engvold, and Wiik, 2003; Lin *et al.*, 2007; Lin, Engvold, and Rouppe van der Voort, 2012). Both the filament spine and barbs are composed of thin threads (Lin *et al.*, 2005a), but the barbs are threads or groups of threads that branch from the axis of the filament to the chromosphere on each side of the filament. Barbs are not a ubiquitous feature of prominences, as there is a continuous spectrum of filaments: from smaller, short-lived active-region filaments with no barbs, up to huge long-lasting quiescent filaments that have very large barbs (Martin, Lin, and Engvold, 2008).

Barbs are threads that no longer run the full length of the spine because they have reconnected to other fields beneath or to the side of the spine. They are secondary to the spine in that they form after the spine and do not seem to prevent the spine from erupting (Martin

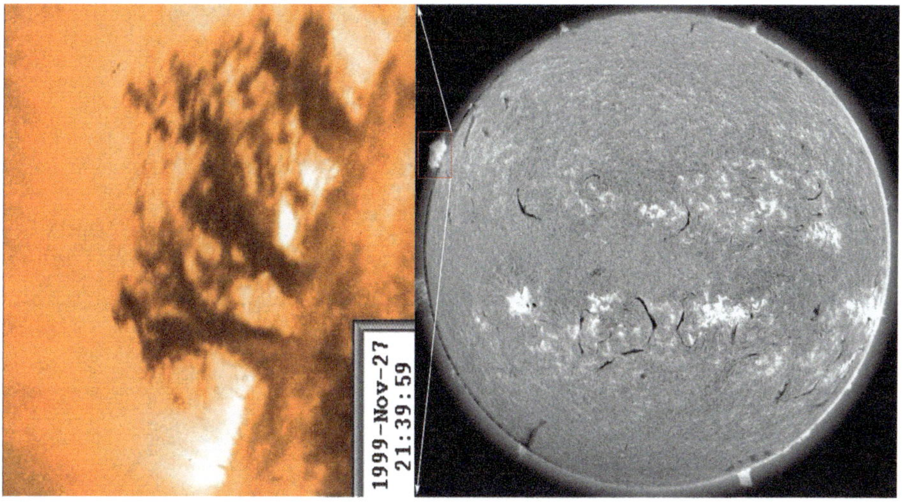

Figure 1 A group of tornado-like prominences observed by TRACE in the 171 Å spectral line on 27 November 1999 (left). The corresponding prominence structure on the limb as seen in a full-disk Hα image from BBSO (right). The orange box around the prominence in the full-disk image corresponds to the TRACE field of view and image in the left panel.

and Echols, 1994; Martin, Bilimoria, and Tracadas, 1994; Gaizauskas *et al.*, 1997; Engvold, 1998; Wang and Muglach, 2007; Martin, Lin, and Engvold, 2008). They readily detach from the spine during eruption, presumably by magnetic reconnection, and cease to exist (Martin and McAllister, 1997). The part of the barb above the reconnection site in the corona appears to shrink upward and merge into the expanding spine. Counterstreaming plasma is observed along the threads of the filament spine and barbs. At the solar limb, the counterstreaming plasma motion along the prominence spine and barbs will create, in specific circumstances, an illusion of rotational motion around the prominence barbs, which also have a vertical geometry and connect the mostly horizontal prominence spine to the chromosphere. We will call this limb appearance of the mostly two-dimensional (2D) motion along the prominence spine and barbs the "vortical illusion" of apparent tornado-like prominences.

Another tornado-like motion observed only at the solar limb is caused by the sporadic plasma flow along the prominence cavity: the 3D volume between the prominence spine and overlying coronal arcade. This phenomenon is a pure 2D projection on the plane of the sky of the 3D motion along the prominence cavity magnetic-field lines. Implied cavity magnetic-field lines generally possess writhe (see Figures 3 and 4 of Török, Berger, and Kliem, 2010). The field-line writhe gradually increases with height, starting from the top of the ribbon-shaped prominence spine and becoming very strong as one moves upwards through the cavity cross-section to the low part of the overlying arcade (Panasenco and Martin, 2008; Martin *et al.*, 2012). When observed from the appropriate angle, the writhed field will appear as a loop (black lines in Figure 2), and the corresponding motion along the writhed lines will be observed as a writhing motion, creating an illusion of a tornado.

We now discuss observational and 3D modeling aspects of the two different plasma motions and their relationships to magnetic topology.

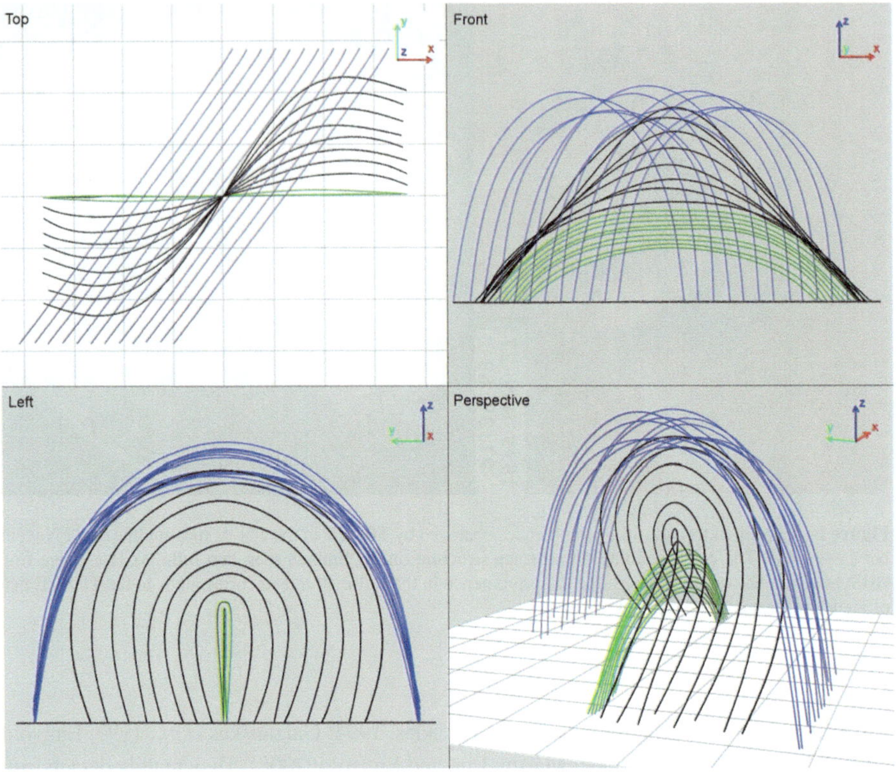

Figure 2 A 3D representation of the filament-channel topological system illustrating the three main components: filament spine (green), coronal loops (blue), and filament cavity (black: the space between the filament and overlying coronal loops). Cavity field lines have a strong writhe (top view), which appears as a loop in the perspective view. This writhe is the result of the interaction of the cavity field with the overlying filament arcade. The filament spine is modeled as a ribbon-like structure, and the overlying filament channel coronal arcade shows a left skew in this model for dextral filaments (see review by Martin, 1998).

2. Apparent Vortical Motions in Prominence Spines and Barbs

What produces the illusion of vortical motion in apparent tornado-like prominences? It is mainly the fact that we can see plasma appearing and disappearing on the upper sides of a vertical column, which looks wider at the top. This illusion is easily produced by counterstreaming flows: indeed, the apparent rotational motion is only observed in 2D projection on the limb in the plane of the sky and never on the disk; the constant counterstreaming motion of the prominence plasma along the thin threads, especially when they connect the vertical parts of the prominence (legs and barbs) to the much more horizontal spine, creates an effect that the eye associates with rotation. But the overall prominence topology and magnetic structure are very important in understanding the true nature of such observed motions at the solar limb (Panasenco and Martin, 2008; Lin, Martin, and Engvold, 2008; Martin *et al.*, 2012; Liewer, Panasenco, and Hall, 2013).

Limb observations of a group of four tornado-like prominences in 171 Å taken by TRACE on 27 November 1999 are shown in Figure 1 together with the position of these prominences on the limb in a full-disk image. Figure 3 shows a sequence of full-disk Hα images of this group from the Big Bear Solar Observatory (BBSO) as the group rotates onto

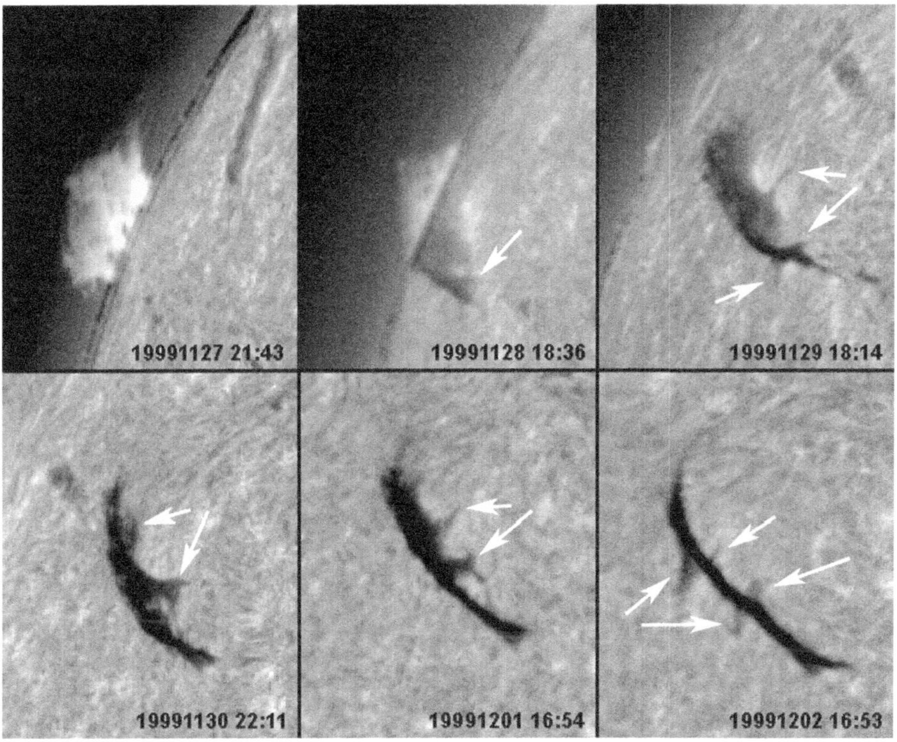

Figure 3 Apparent tornado-like prominences gradually become visible against the disk – as shown in this sequence of images from BBSO – first from the side and, in the last image shown here, from the top. The vertical trunk-like parts of the "tornadoes," which connect the horizontal coronal part of the limb prominence with the low chromosphere, become visible as filament barbs – easily observed on the disk. Scales on the limb: height of the prominence $\approx 50-60$ Mm; width of the individual "tornado" trunk at the bottom $\approx 6-10$ Mm; at the top ≈ 100 Mm. On the disk: length of the filament ≈ 200 Mm, width of the filament spine $\approx 5-9$ Mm or less; projected length of the barbs on the disk ≈ 40 Mm; projected width of the barbs on the disk 5 – 18 Mm.

the disk. Even these low-resolution images, taken at relatively low cadence, establish a direct connection between apparent tornado-like prominences on the limb and filament structures known as barbs (Martin, 1990, 1998). If one does assume that tornadoes at the limb are the result of a vortex-like motion around a conical structure, the top view of this cone would then be described by a circle with diameter equal to the width of the cone top as viewed from the side, *i.e.* on the limb. Measurements show that the width of the tornado-like cone near the top, where the strongest motion occurs, is ≈ 100 Mm as observed by TRACE (Figure 1). However, the top view of the same structure seen on the disk has a width of $\approx 5-10$ Mm, corresponding to the filament spine width (Figure 3). The vortical interpretation therefore cannot stand, and we conclude that observed illusion of the 3D rotation comes from the mostly 2D counterstreaming motion along the prominence/filament spine and barb threads. The four tornado-like prominences observed at the limb by TRACE on 27 November 1999 are simply four filament barbs on the two sides of a filament spine that had a length of ≈ 200 Mm on 2 December 1999 (Figure 3).

To further support these conclusions based on TRACE and BBSO 1999 data, we present here two examples of more recent high-resolution and high-cadence observations obtained by the SDO/AIA instrument in February and March 2012. Figure 4 shows an apparent

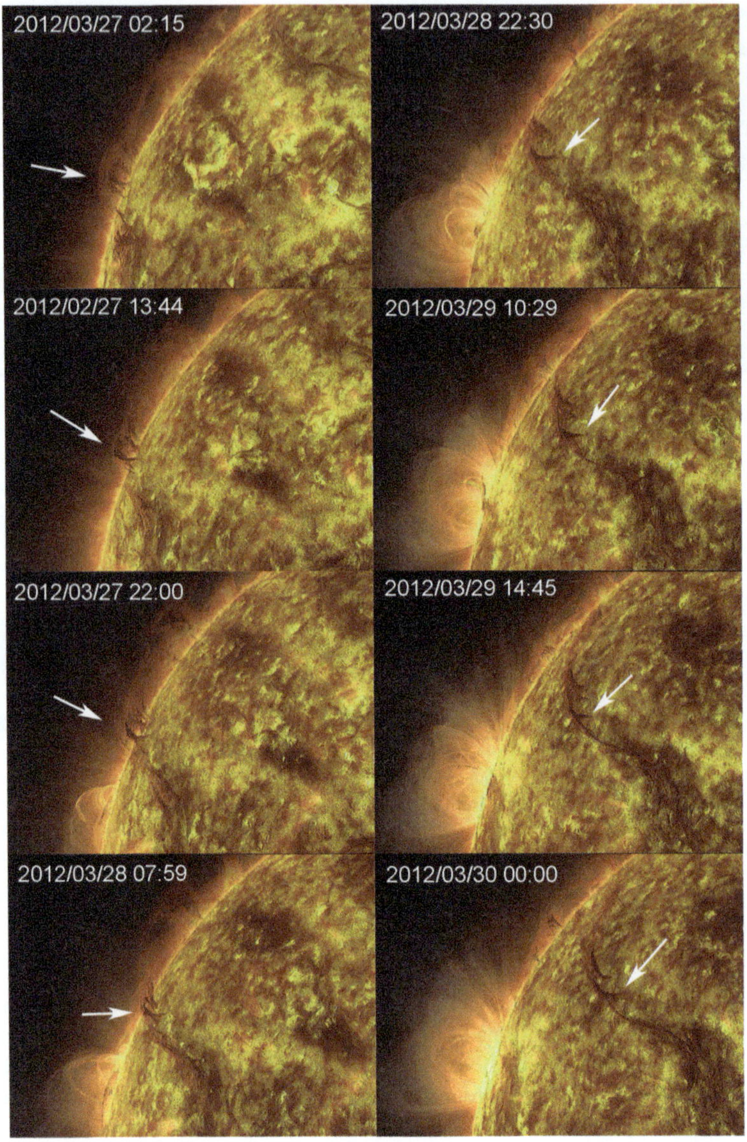

Figure 4 Composite images from SDO/AIA 193 Å, 171 Å, and 304 Å. The width of the apparent "tornado" at the limb near its top is ≈ 40 Mm; the width of the filament spine at the bending point in the last image is ≈ 2 Mm in 193 Å and ≈ 3 Mm in 304 Å. White arrows point to the prominence/filament barb – the source of the tornado illusion. As the Sun rotates, the apparent "tornado" gradually becomes just a barb.

tornado-like prominence on the limb and its subsequent appearance on the disk. The width of the tornado observed at the limb is ≈ 40 Mm on 27 March, and the width of the same area viewed from the top against the disk is $\approx 2-3$ Mm on 30 March. The difference between these two measurements is an order of magnitude, substantiating the interpretation of the apparent tornado-like structure as 2D counterstreaming. An observation supporting the narrow width of the filament channel and spine in the corona comes from coronal cells:

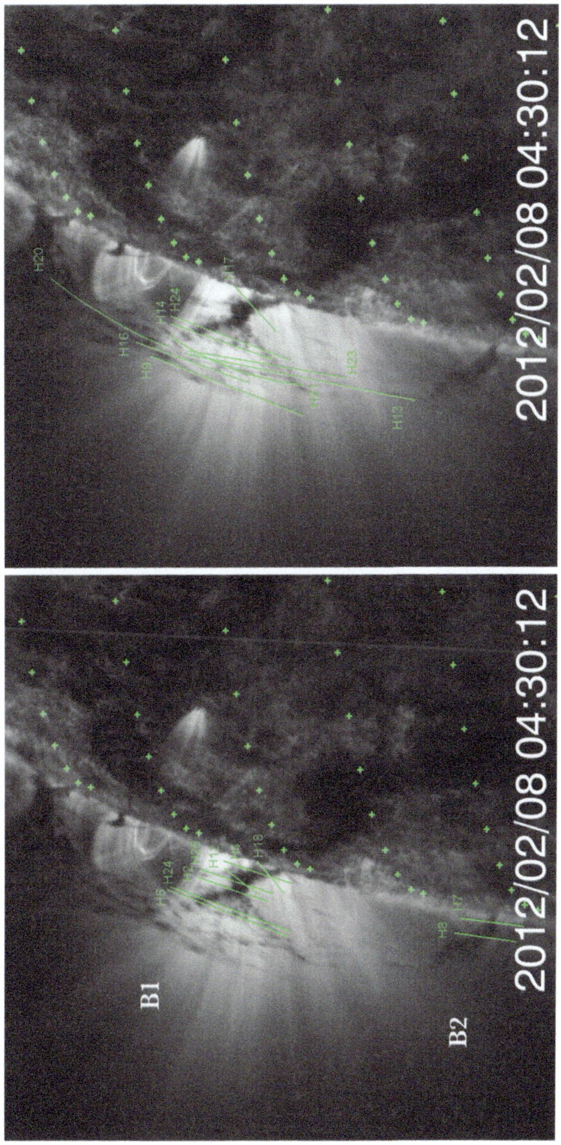

Figure 5 SDO/AIA image of the prominence spine and barbs on the limb taken on 8 February 2012, in 171 Å. The left panel shows cuts used for the subsequent time-slice diagram in the lower barb regions, while the right panel displays the cuts used for time-slice diagrams of the upper barb region and spine. Crosses mark a longitude–latitude grid with five-degree separation on the Sun. The width of each cut is 3 pix ≈ 1.3 Mm across. Each cut has its origin at the northern point, and each cut's length is measured southward.

cellular features in Fe XII 193 Å images of the 1.2 MK corona first observed and modeled by Sheeley and Warren (2012). Panasenco *et al.* (2012) found that coronal cells do not cross the polarity-reversal boundary within a filament channel at heights below the filament spine top. Coronal cells originate from network-field concentrations and show the same pattern of chromospheric fibrils (which align along the filament channel axis) because they follow the same filament-channel magnetic topology, with a (presumably) strong horizontal component of the field. The distance between coronal cells on opposite sides of the filament channel is very narrow and does not exceed 1 – 15 Mm, even though coronal cells reach up to 100 Mm and more in height. These coronal observations support our chromospheric measurements of the filament channels, but much higher up into the corona.

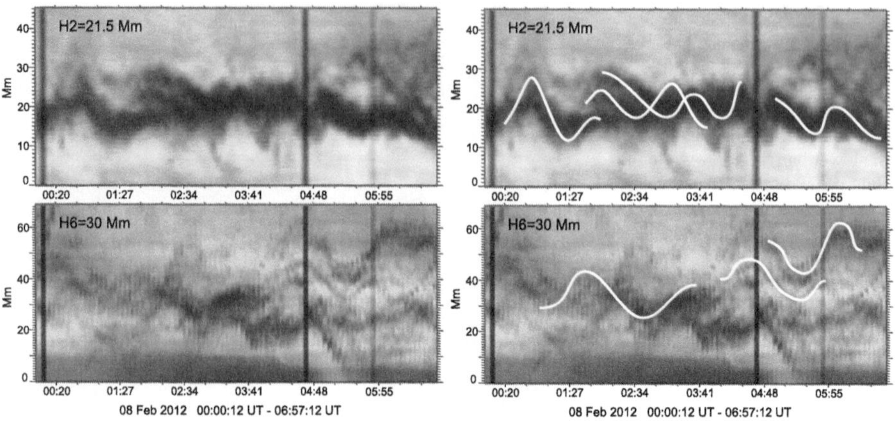

Figure 6 Left panel: time-slice diagrams for cuts H2 and H6 in prominence barb B1 from Figure 5 taken by SDO/AIA in 171 Å with five-minute cadence over ≈ seven hours on 8 February 2012. Right panel: same time-slice diagrams with lines showing barb substructure as an aid to illustrate calculation of oscillation periodicities, ranging from 45 to 70 minutes for the substructures.

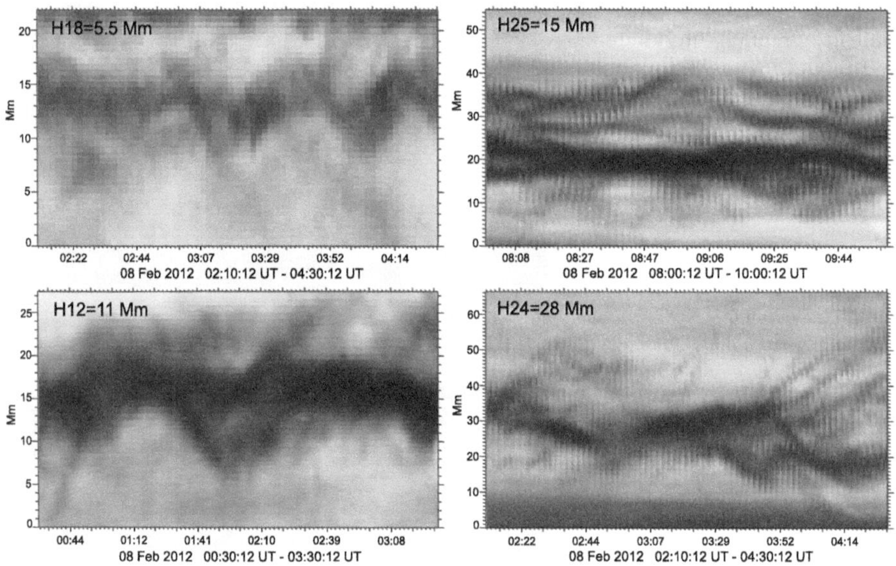

Figure 7 Time-slice diagrams for cuts at different heights (H18, H12, H25, H24 in ascending order) in prominence barb B1 from Figure 5 taken by SDO/AIA in the 171 Å spectral line with two-minute cadence on 8 February 2012. The overall duration is 120 minutes for panel H25, 140 minutes for panels H18 and H24, and 180 minutes for panel H12. Oscillation periods vary from 40 minutes for H18 to 65 minutes for H12.

The speeds of counterstreaming plasma motions in apparent tornado-like prominences seen on the limb (along the prominence threads observed in the 304 Å spectral line on 27 March 2012) were estimated to be up to $45 \pm 5\ \mathrm{km\ s^{-1}}$ in the plane of the sky. Oppositely directed plasma motion along the threads at different heights contributes to the illusion of rotational tornado-like motion. A similar apparent tornado-like prominence was observed on 7 – 8 February 2012 (see supplementary movies), but it erupted on 9 February, so we were

Table 1 Period [T] of oscillations *versus* height [H] of the cuts across barb 1 and barb 2 on 8 February 2012.

	Barb 1										Barb 2	
H [Mm]	5	5.5	11	14	15	21	21.5	24.5	28	30	7.5	15
T [min]	70	40	65	66 – 80	50 – 55	70 – 90	45 – 66	60 – 66	40 – 50	66 – 70	25 – 40	35 – 45

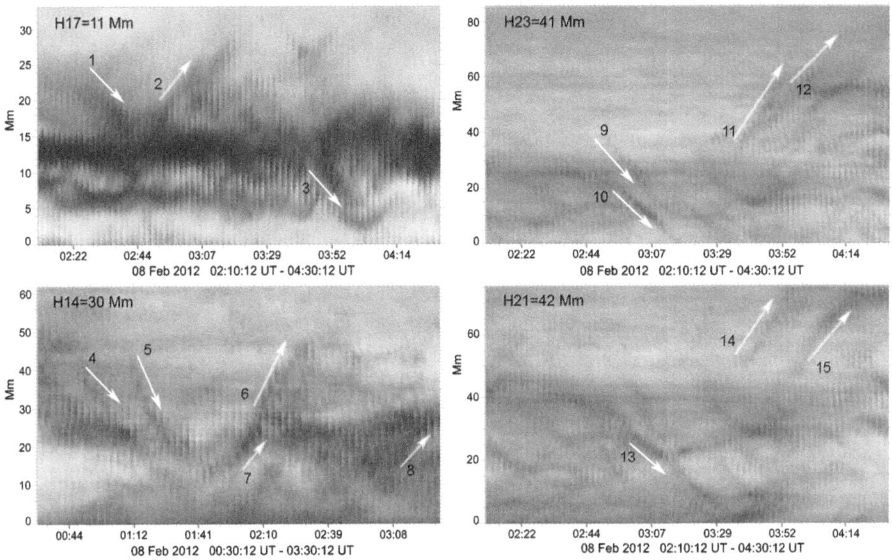

Figure 8 Time-slice diagrams for cuts at different heights (H17, H14, H23, H21 in ascending order) in prominence barb B1 from Figure 4 taken by SDO/AIA in 171 Å with two-minute cadence on 8 February 2012. The overall duration is 140 min for panels H17, H23, and H21, and 180 min for panel H14. White arrows outline counterstreaming signatures along the slices and corresponding velocities: $V_1 = -6 \pm 2$ km s^{-1}; $V_2 = 10 \pm 2$ km s^{-1}; $V_3 = -10 \pm 2$ km s^{-1}; $V_4 = -17 \pm 2$ km s^{-1}; $V_5 = -27 \pm 3$ km s^{-1}; $V_6 = 27 \pm 3$ km s^{-1}; $V_7 = 15 \pm 2$ km s^{-1}; $V_8 = 13 \pm 2$ km s^{-1}; $V_9 = -21 \pm 3$ km s^{-1}; $V_{10} = -16 \pm 2$ km s^{-1}; $V_{11} = 26 \pm 3$ km s^{-1}; $V_{12} = 14 \pm 2$ km s^{-1}; $V_{13} = -14 \pm 2$ km s^{-1}; $V_{14} = 25 \pm 3$ km s^{-1}; $V_{15} = 21 \pm 3$ km s^{-1}.

not able to trace it to the disk as we did for the 27 March prominence. The region of interest of the filament channel was nearly parallel to the limb, allowing us to make measurements of the plasma speeds along the prominence spine and across barbs in more preferable observational conditions. Figure 5 shows the prominence with barbs observed on the limb in the 171 Å spectral line by SDO on 8 February 2012. We used 28 cuts to create time-slice diagrams for the lower barb regions, upper barb regions, and prominence spine. The time-slice diagrams for the low barb regions are shown in Figures 6 and 7. Because barbs connect the prominence spine in the corona with the photosphere, the plasma motion along these lower parts of the barbs is mostly vertical; we do not observe continuous horizontal motions here, but sporadic ejections of plasma. However, it is interesting to notice that the trunks of the prominence barbs show oscillations with an average period ≈ 40 – 70 minutes. These oscillations could also cause an illusion of rotational motion. Figure 6 shows two time-slice diagrams for cuts H2 and H6 in prominence barb B1 (see Figure 5) taken at heights 21.5 Mm and 30 Mm, respectively. The right panel of Figure 6 shows the same time-slice diagrams

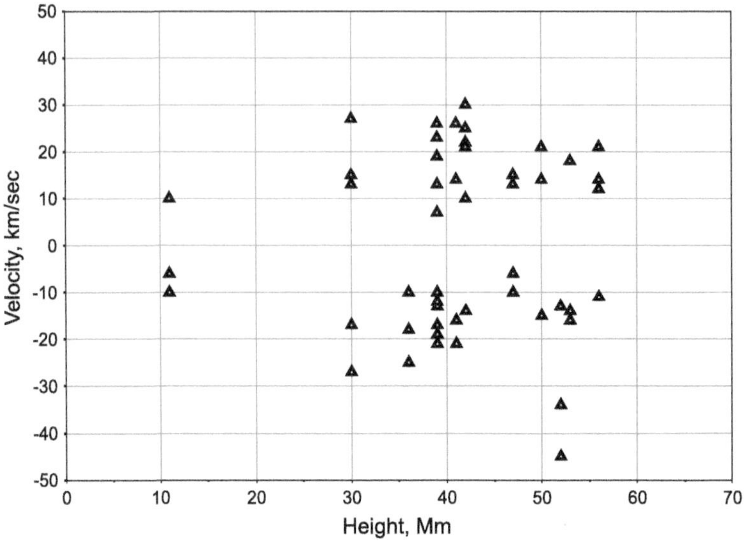

Figure 9 Distribution of counterstreaming velocities measured along cuts at different heights.

with added white lines showing the barb substructure as an aid to illustrate the calculation of oscillation periods, ranging from 45 to 70 minutes for the substructures. Figure 7 shows four time-slice diagrams for cuts at different heights (H18, H12, H25, H24 in ascending order) in prominence barb B1, with oscillation periods varying from 40 minutes for H18 to 65 minutes for H12. Table 1 summarizes all observed oscillations in the low parts of the prominence barbs. The filament spine and barbs are composed of thin threads, and the widths of the filament threads in Hα are ≤ 200 km (Lin *et al.*, 2005a). The oscillations in the individual Hα threads have been found to be $\approx 10-20$ minutes (Lin *et al.*, 2007, 2009; Lin, 2011), but the oscillations of groups of prominence threads have been found to be ≈ 40 minutes (Berger *et al.*, 2008; Panasenco and Velli, 2009). The time-slice diagrams in Figures 6 and 7 show that barbs have a substructure and are often composed of two to five bundles of threads resolved in the 171 Å spectral line. Each such bundle has an oscillation period of $\approx 40-70$ minutes. We will review and discuss the possible sources of these oscillations in the Discussion section.

The time-slice diagram in the bottom right panel in Figure 7 shows plasma motion along cut H24 at a height of 28 Mm. One can see that approximately at 03:52 UT one of the relatively solid bundles of the barb B1 splits into many thin separate subbundles where plasma is moving with a speed of 10 ± 2 km s^{-1} in opposite directions. A similar situation has been observed for cuts H2, H3, H5, and H6 with corresponding heights 21.5, 24.5, 21, and 30 Mm. This allows us to conclude that at heights above ≈ 21 Mm the plasma motion along the vertical part of the barbs became more and more horizontal and parallel to the solar surface. Consequently, the horizontal component of the plasma speed gradually increases from 5 km s^{-1} at a height of 21 Mm up to 45 km s^{-1} at a height of 54 Mm.

The next step in our study is to measure the plasma speed along the upper barbs and spine of the prominence on 8 February 2012. These cuts correspond to areas where the plasma motion is less vertical and can be nicely estimated from time-slice diagrams. We made 14 different cuts across the upper part of the barb B1, and along the prominence spine between the barbs B1 and B2 (see labels on Figure 5). The width of each cut is three pixels

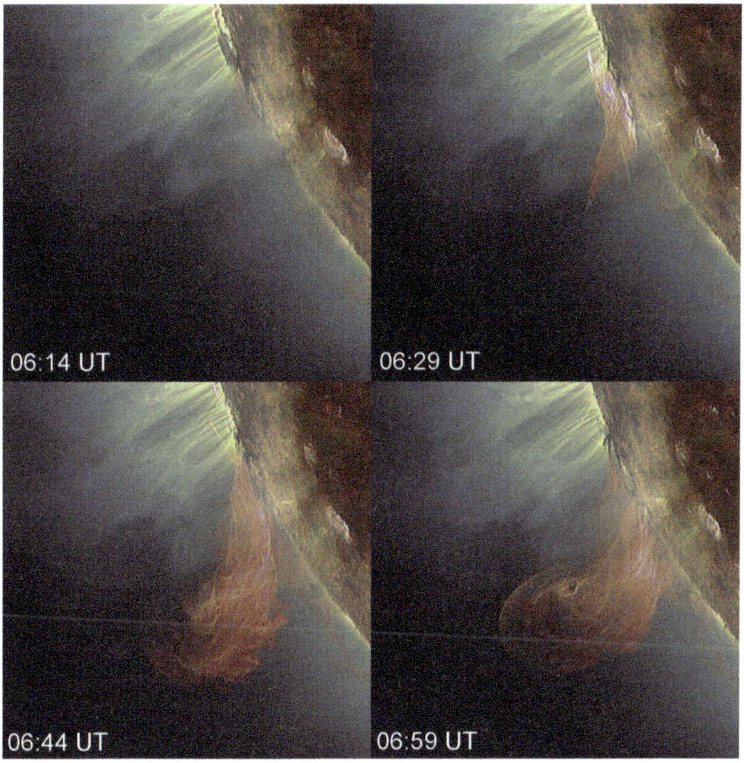

Figure 10 The plasma motion along the filament cavity when the filament channel is crossing the limb on 16 June 2011. Superposition of 171 Å, 193 Å, and 304 Å from SDO/AIA. The field of view is 390×390 Mm.

≈ 1.3 Mm. This width can accommodate at most $\approx 6-7$ thin filament threads along which plasma is moving. The observations from SDO/AIA have a resolution that does not allow us to resolve all threads. We will estimate in our measurements the speed of plasma pieces moving in and out of our cuts at different heights.

The time-slice diagram in the upper left panel in Figure 8 shows an exceptional horizontal motion of plasma from the barb B1 at a low height of 11 Mm along cut H17. The sporadic ejections of plasma from the barbs along the low threads occur relatively rarely, and usually have a low horizontal component of the speed, which is $6-10\ \mathrm{km\,s^{-1}}$ in opposite directions in this case. The other three panels in Figure 8 correspond to heights of 30, 41, and 42 Mm. Time-slice diagrams for these heights show a wide spectrum of plasma speeds in the plane of the sky, ranging from 13 to $27\ \mathrm{km\,s^{-1}}$. The distribution of the plasma speed along cuts at different heights is shown in Figure 9. Here we combine all measurements along all 14 cuts across the upper part of barb B1, along the prominence spine above barb B1, and between barbs B1 and B2. One can see that the plasma moves in opposite directions with speeds that have a similar distribution. These oppositely directed, mostly 2D motions at different heights create a perfect illusion of rotating vortices. The counterstreaming is easily observed in Doppler measurements and might be interpreted, incorrectly, as a rotational motion when observed on the limb (Orozco Suárez, Asensio Ramos, and Trujillo Bueno, 2012). Note that Orozco Suárez, Asensio Ramos, and Trujillo Bueno (2012) estimated this rotational velocity from Doppler measurements as $\pm 2-6\ \mathrm{km\,s^{-1}}$, which is within the noise range for

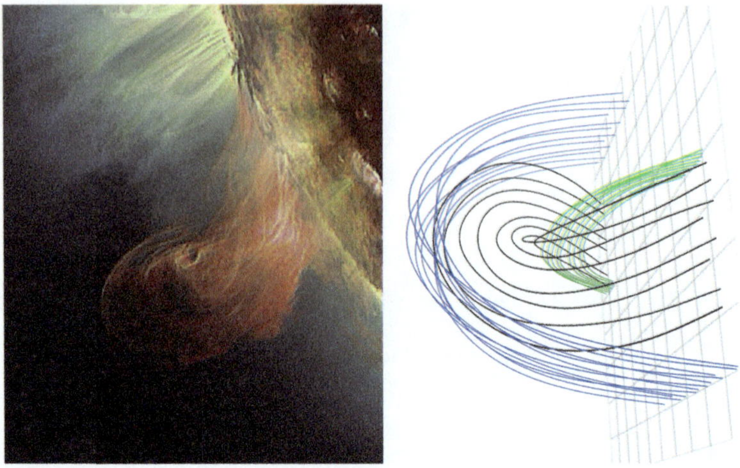

Figure 11 The similarity between the observed trajectory of the plasma moving along the filament cavity (16 June 2011, 06:59 UT, 171 Å, 193 Å, and 304 Å SDO/AIA) and the modeled cavity with a strong writhe (black lines in the right panel) when viewed mostly along the filament channel. The field of view of the left image is 390 × 390 Mm.

counterstreaming plasma motions along prominence threads. The filament-channel orientation relative to the limb plays a crucial role in Doppler measurements. Any deviation from the parallel orientation to a more perpendicular one with respect to the limb will increase the Doppler velocities due to the same counterstreaming motions.

3. Apparent Rotational Motions in Prominence Cavities Due to Writhe

Apparent tornado-like motions on the limb can also be observed when rapidly moving plasma from an outside source propagates into and along a filament channel cavity. In this case the plasma motion does have a 3D geometry, but it does not develop along the ribbon-like filament spine or barbs, rather along the filament channel cavity magnetic-field lines, which have a strong writhe (black lines in Figure 2). Usually the source of this plasma is from a neighboring active region. For quiescent filaments, the injections of plasma inside the filament cavity occur during flare-like activity. Sporadic injections of plasma inside the filament channel cavity create an illusion of rotational plasma motion on the limb, when observed along the filament channel axis as the filament channel rotates through the limb. This illusion is created by the projection of the writhed magnetic-field lines onto the limb plane. The mystery is easily resolved if limb observations from L_1 or the Earth point of view are compared with simultaneous observations by the *Sun Earth Connection Coronal and Heliospheric Investigation* (SECCHI: Howard *et al.*, 2008) onboard the *Solar TErrestial RElations Observatory* (STEREO: Kaiser *et al.*, 2008).

Figure 10 shows multispectral observations from SDO/AIA of the plasma motion with speed up to 200 km s^{-1} along the filament cavity when the filament was crossing the limb on 16 June 2011. The speed was measured using the 195 Å STEREO-B images of this plasma motion when observed against the disk. The separation angle between STEREO-B and SDO for 16 June was ≈ 93 degrees. The position of the active region – the source of sporadic plasma injections inside the filament cavity – is S16W7 for STEREO-B, and exactly at the

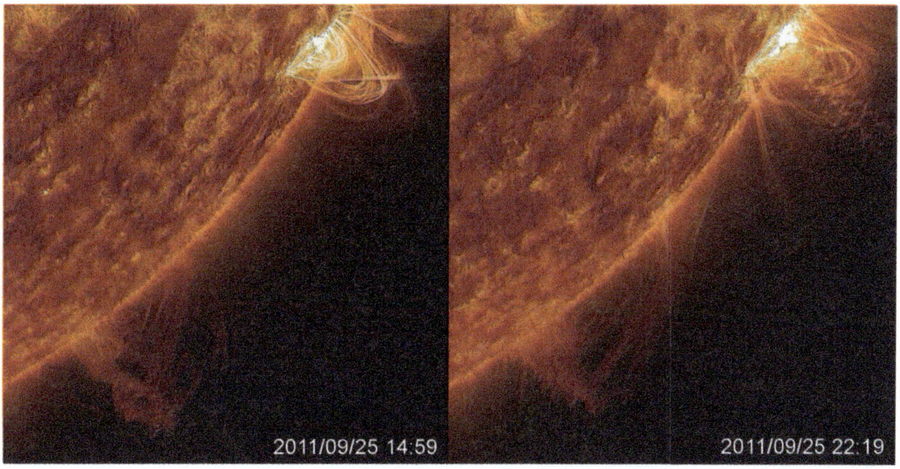

Figure 12 The plasma motion along the prominence spine and cavity from the eastern to western prominence footpoints after the flare of class M1.5 in the active region NOAA 11303 (09:25), which was located $\approx 20^{\circ}$ northward from the eastern end of the polar crown filament. The field of view is 380 × 380 Mm.

limb for SDO. Figure 11 shows the similarity between the observed trajectory of the plasma moving along the filament cavity (16 June 2011 06:59 UT, SDO/AIA) and a simple cavity model including fields with a strong writhe (black lines in the right panel) when viewed mostly along the filament channel. The black lines of the modeled cavity show the same loop-like geometry in the upper regions, close to the overlying coronal arcade (blue lines), clearly a projection effect of the lines with writhe.

Another writhing motion in a prominence cavity was observed on 25 September 2011 and has been described in detail by Li *et al.* (2012) and Panesar *et al.* (2013). Figure 12 shows the eastern end of a polar crown filament as observed at the limb on 25 September 2011. Two bundles of thin threads were visible on 24 September: two barbs, one visible on the disk and at the limb, and another with its footpoint completely behind the limb. With solar rotation, these barbs coalign and superpose, so only one barb can be observed on 25 September. The active region NOAA 11303, located ≈ 20 degrees northward from the eastern filament end, produced multiple flares. The strongest flare (class M1.5) occurred on 25 September, 09:25 UT. This last flare created coronal disturbances with subsequent strong oscillations of the whole prominence system, triggering magnetic reconnection of the fields in the prominence barbs and footpoints. As a result of these reconnections, plasma was injected inside the prominence system from the east end of the channel, and later propagated along the prominence spine and cavity in opposite directions towards both filament footpoints (Figure 12). The illusion of a tornado was created by the projection effect together with the oscillatory disturbances caused by the active region flaring only 20 heliographic degrees from the filament. Some possible reasons for these oscillatory disturbances are described by Panesar *et al.* (2013) and include the Hudson effect (Hudson, 2000).

When the position of the filament channel is nearly parallel to the limb with a very small acute angle, the plasma injected inside the filament cavity produces a spectacular picture as it traces the magnetic topology of the cavity (where the plasma β is arguably less than unity). The observed lines exhibit clear writhe. Figure 13 shows a sequence of images of the plasma moving inside the filament cavity on 23 July 2012 from 06:15 to 08:16 UT. Panels (g) and (h) in Figure 13 show the cavity maximally filled with plasma tracing the field, which can be best described as bundles with writhe.

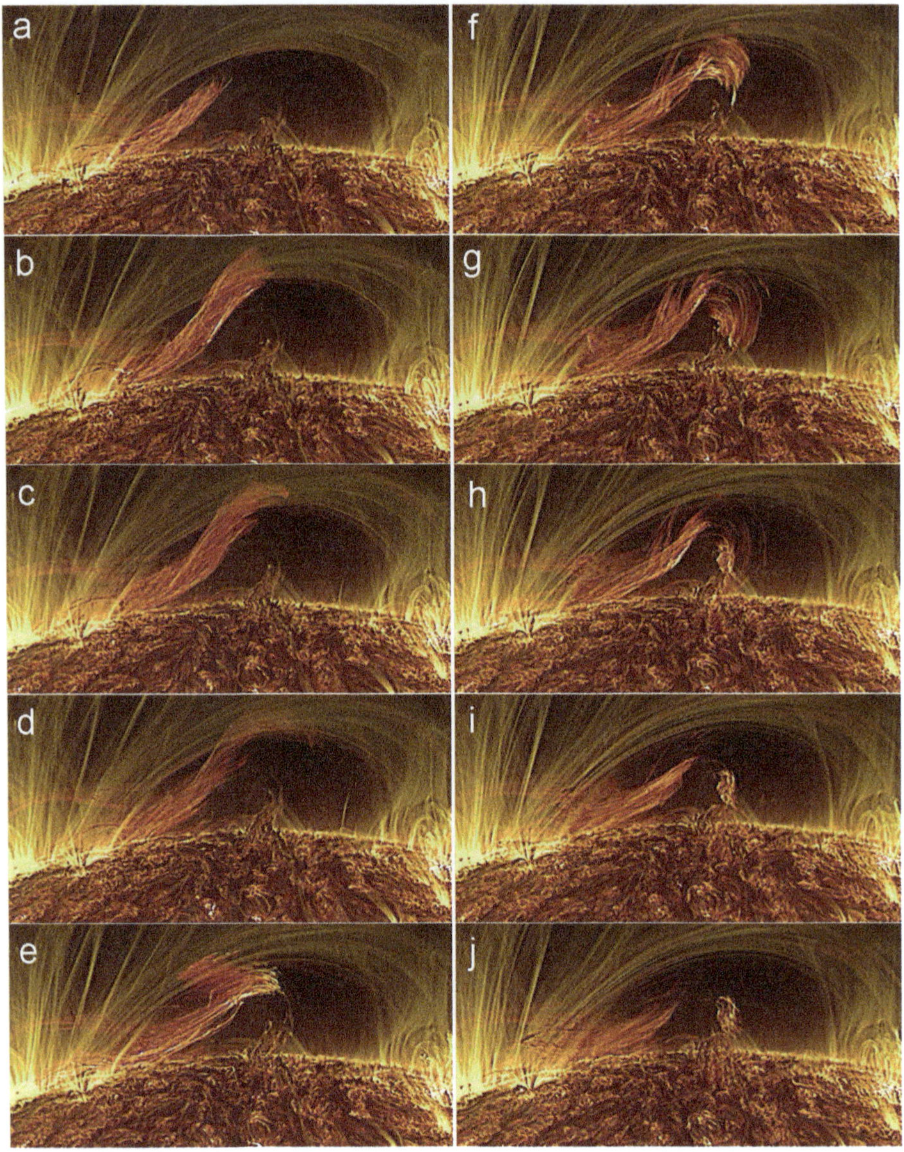

Figure 13 Plasma motion inside the filament cavity for the case when the filament is mostly parallel to the limb. Superposition of 171 Å and 304 Å SDO/AIA (images rotated 90° clockwise). The time sequence of images is from 06:15 UT to 08:16 UT on 23 July 2012. The field of view is 180 × 360 Mm.

The sporadic propagation of the externally injected plasma inside the filament cavity is not a rare phenomenon. It is a very important tool in our understanding of the magnetic topology of the filament system, which also includes the filament-channel-cavity field lines. These lines usually are not traceable due to the very low plasma density of the cavity compared with the filament/prominence or surrounding corona. The sporadic injections of the plasma inside the filament channel system work as perfect tracers to uncover the cavity magnetic topology.

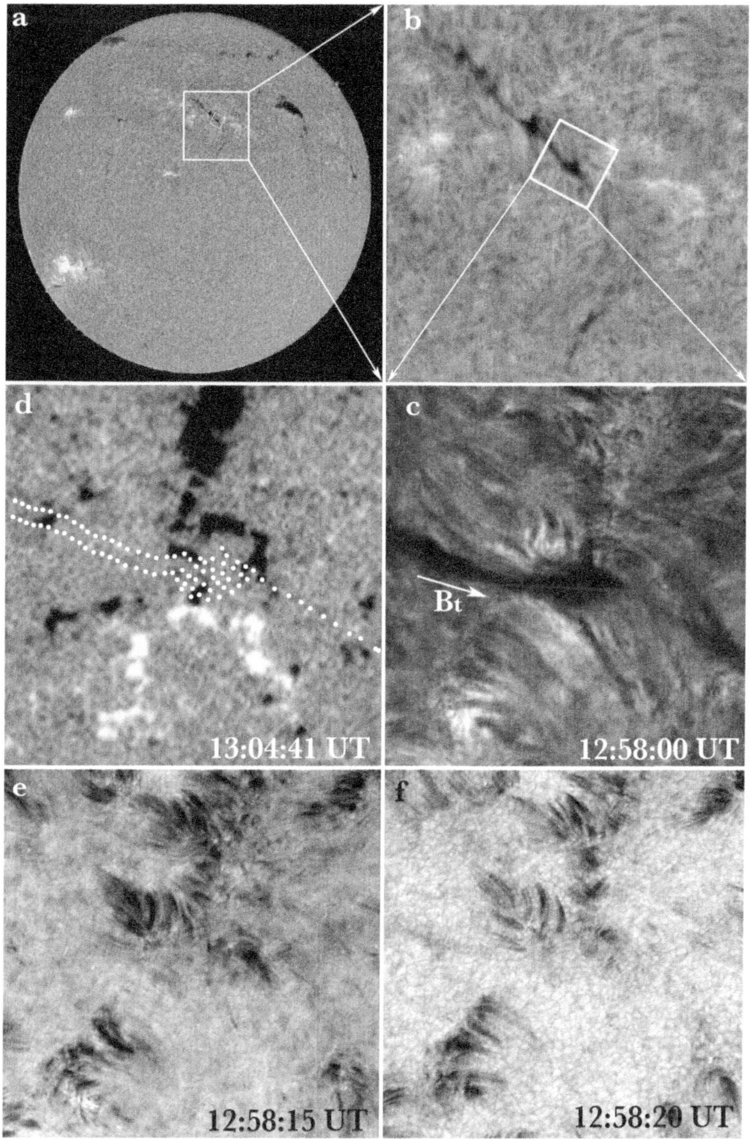

Figure 14 The magnetic boundary at the source of a barb. (a) A full-disk Hα image taken by the BBSO on 13 September 2010. The white square is centered around a dextral filament which is displayed in panel (b); (b) the white square shows part of the filament as observed by the *Dutch Open Telescope* in the core of the Hα line and shown, rotated, in panel (c). The direction of the horizontal component of the magnetic field B_t along this dextral filament channel is given by the white arrow. Panel (d) is the corresponding magnetogram from SDO/HMI. The dotted line superimposed on the magnetogram outlines the filament barb anchored at the junction of four supergranular cells. Panels (e) and (f) are images in the blue and red wings of Hα ($\mp$ 0.07 nm), respectively. The field of view spans 45.5 × 45.5 Mm for panels (c) – (f).

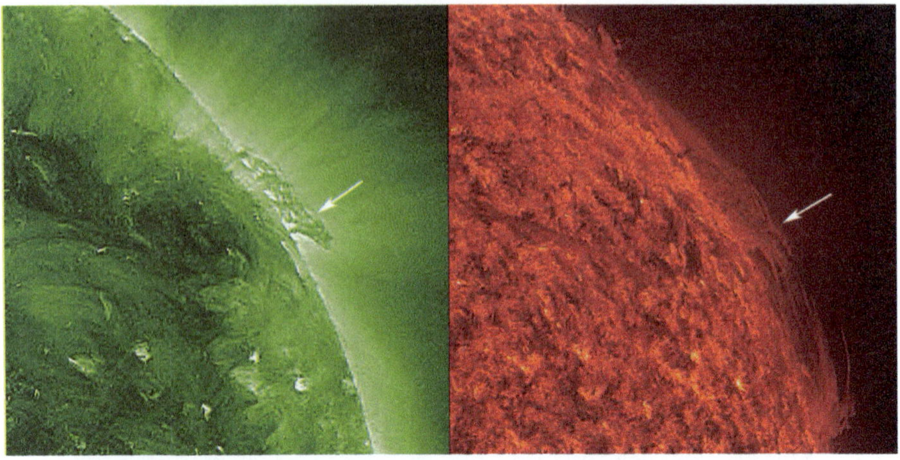

Figure 15 "Tornado prominences" observed by SDO/AIA 193 Å at the west limb on 20 September 2010 (left) and its appearance in 304 Å as a regular prominence (right). One of these "tornadoes," indicated with a white arrow, corresponds to the barb of the filament observed at the disk on 13 September (Figure 14). The field of view is 386 × 392 Mm.

4. Discussion

Lin *et al.* (2005b) have shown that close to 65 % of the observed end points of barbs fall within photospheric network boundaries. Recent high-resolution observations at the *Dutch Open Telescope* (DOT) reveal that filament barbs are anchored at the intersections of supergranular cells (Figure 14). We trace the filament observed by BBSO and the part of its spine and barb observed by DOT on 13 September 2010 to the west limb, where the corresponding prominence was observed by SDO/AIA in 304 Å (Figure 15). When observed in 171 Å this prominence turns out to be a group of apparent tornado-like prominences. One is the barb observed in high resolution by the DOT. The fact that the barb is rooted in the junction of four supergranular cells creates conditions for magnetic field canceling (Litvinenko, 1999, 2010), a necessary requisite for filament formation, since supergranular intersections are perfect locations where opposite polarities can be pushed together and cancel. It is interesting to note that the DOT observations in the far-blue and far-red wings (shown in panels (e) and (f) in Figure 14) demonstrate the pattern of fibrils in the filament channel. They are nearly parallel to the filament spine and do not cross it. The same pattern was observed in coronal cells (Panasenco *et al.*, 2012) and provides additional observational evidence of how narrow a filament spine at the chromospheric level is – the distance between two oppositely directed fibrils being $\approx 4.5-6$ Mm (Figure 14 (e), (f)) – the same order as the width of filament spines (2 – 3 Mm) and the distance between oppositely directed coronal cells (10 – 15 Mm).

Supergranular cell intersections are also natural locations where photospheric convective motions turn into downdrafts, and downdrafts are also locations where fluid vortical motions can develop (Attie, Innes, and Potts, 2009). A number of recent articles have linked the apparent vortical motions seen in the outer solar atmosphere, at different heights in the solar chromosphere or corona, to vortices coming from the unwinding of magnetic fields entrained by convection in the much higher density photosphere (Su *et al.*, 2012; Wedemeyer-Böhm *et al.*, 2012). That photospheric vortices might be a source of the large-amplitude Alfvénic turbulence observed in the solar wind had already been suggested by Velli and

Liewer (1999). Indeed, photospheric supergranulation flows concentrate magnetic flux at the network boundaries, where the resulting flux tubes exceed dynamic pressure equipartition and are close to evacuation pressure balance. Vorticity obeys the same equation as magnetic induction, so in regions where the field does not dominate the dynamics, vorticity and magnetic flux will concentrate in the same regions of space. Vorticity filaments are the natural dissipative structures of 3D hydrodynamic turbulence and are observed to form in simulations of the solar convection zone (see, *e.g.*, Brandenburg *et al.*, 1996).

Simon and Weiss (1997) gave several examples of vorticity sinks, associated with photospheric downdrafts, at mesogranular scales. They fit the observed vorticity with a profile

$$\omega(r) = (V/R)\exp\left(-r^2/R^2\right),$$

where V is the characteristic rotational velocity associated with the vortices, and R their characteristic radius. A typical photospheric vortex lasts several hours and takes about two hours to develop. The strongest vortex that they observed had $|\omega| \approx 1.4 \times 10^{-3}$ rads s^{-1}, a size $R \approx 2.5 \times 10^3$ km, and a maximum azimuthal velocity of order 0.5 km s^{-1}. Coincidentally, Su *et al.* (2012) associate the apparent tornado-like motion of barbs with photospheric vortices of approximately the same dimensions and vorticity. A vortex of this type is associated with periods of hours, and not tens of minutes as the barb oscillations, seen by Su *et al.* (2012) and in this article, display.

Photospheric vortices on open field lines would produce an Alfvén wave packet, as the magnetic-field lines are entrained by the rotational motion, whose frequency is given by the vorticity itself, and duration coincides with the life of the vortex. Such Alfvénic-like motions would propagate upwards into the solar wind, evolving due to nonlinear interactions in the process, and amplifying and decaying with height over several solar radii. However, on closed field lines, such as barbs, the twist injected by a photospheric vortex would lead to the formation of tangential discontinuities: it is fundamental to remark here that photospheric tornadoes with periods of hours would, at coronal heights, appear as quasi-steady motions (the travel time along a barb over a distance of, say, 50 000 km would require only a fraction of a minute for a coronal Alfvén speed of 1000 km s^{-1}). Rappazzo, Velli, and Einaudi (2013) have studied the evolution of a coronal volume subject to such vortical forcing at the photosphere and showed that the resulting field lines would first develop twist and then become unstable to an internal kink mode, releasing most of the stored magnetic energy and removing twist from the field lines, over periods of tens of Alfvén crossing times (*i.e.* tens of minutes). This being the case, it is difficult to interpret barb motions as tornadoes: a more realistic suggestion, given that barb strands are indeed formed at intersections of supergranules, is that reconnection releasing the twisted magnetic-field component generated from the vortices produces mass motions along the barb itself, which leads to the observed oscillations. In other words, though the photospheric vortical motions might contribute to barb formation and dynamics, this would more likely be indirectly, through the dissipation and reconnection of the vortical magnetic field.

A reliable estimate of the contribution of energy in photospheric vorticity sinks to coronal dynamics and heating, as channeled by the magnetic field and rotational motions, would really require a long-term statistical analysis of the distribution of the number, intensity, and duration of such vorticity tubes in the photosphere. We hope that this becomes possible with the extended SDO/HMI observations. However, as the present discussion has shown, the interpretation of the motions in prominence barbs and spines as direct transmission of such vortical motions seems unlikely.

5. Conclusions

Observations of solar apparent tornado-like prominences have been interpreted as extraordinary vortical motions in the solar atmosphere. However, a careful analysis of older events seen on the limb and then on the disk from space and ground-based instruments, together with the multiwavelength images from SDO and the multiple viewpoints provided by STEREO, reveal solar apparent tornado-like prominences to be due to two different types of illusions involving the motions of the plasma in the solar atmosphere as projected onto the plane of the sky.

We have shown that apparent vortical-like motions in prominences are really the projection above the solar limb of the mostly 2D counterstreaming plasma motion and oscillations along the prominence spine and barbs. A writhing motion, creating an illusion of a tornado, on the other hand, consists in the limb projection of the 3D plasma motion following the magnetic fields inside and along the prominence cavity.

The impression of tornado-like rotational motion results in both cases from plasma motion and oscillations along magnetic-field lines observed on the plane of the sky, rather than from a true vortical motion around an apparent vertical or radial axis. Apparent tornado-like prominences, once understood, provide a tool to understand the magnetic structure of filament channels, filaments, and prominences, whose role in space weather is very important. Most coronal mass ejections (CMEs) originate from coronal-loop systems surrounding filament channels. Understanding the correct magnetic structure of the filament channel system will shed light on the formation of the CMEs and their propagation in the corona.

This interpretation of apparent tornado-like prominences does not mean that rotational motions or swirls in the chromosphere and corona are absent (especially at smaller scales and times: Wedemeyer-Böhm *et al.*, 2012). Indeed, such motions may be present and participate, as large-amplitude Alfvénic motions, in the heating of the solar corona and acceleration of the solar wind along open field lines (Verdini *et al.*, 2010). However, in closed-field regions, it is much more likely that such twists must relax in tangential discontinuities and current sheets as Parker nanoflares (Rappazzo, Velli, and Einaudi, 2013), and in injected plasma flows into the barbs and up into the prominence spines (Cirtain *et al.*, 2013).

Acknowledgements OP and SM are supported in this research under the National Aeronautics and Space Administration (NASA) grant NNX09AG27G and NSF SHINE grant 0852249. The work of MV was conducted at the Jet Propulsion Laboratory, California Institute of Technology, under a contract from NASA. We are thankful to Aram Panasenco for contributions in image processing. The SECCHI data are produced by an international consortium of the NRL, LMSAL and NASA GSFC (USA), RAL and University of Birmingham (UK), MPS (Germany), CSL (Belgium), IOTA, and IAS (France). The AIA data used here are courtesy of SDO (NASA) and the AIA consortium. The *Dutch Open Telescope* (DOT) is located at Observatorio del Roque de los Muchachos (ORM) on La Palma. The DOT was designed and built by Rob H. Hammerschlag. We thank the referee for interesting comments and suggestions.

References

Attie, R., Innes, D.E., Potts, H.E.: 2009, *Astron. Astrophys.* **493**, L13. doi:10.1051/0004-6361:200811258.

Berger, T.E., Shine, R.A., Slater, G.L., Tarbell, T.D., Title, A.M., Okamoto, T.J., *et al.*: 2008, *Astrophys. J. Lett.* **676**, L89. doi:10.1086/587171.

Brandenburg, A., Jennings, R.L., Nordlund, Å., Rieutord, M., Stein, R.F., Tuominen, I.: 1996, *J. Fluid Mech.* **306**, 325. doi:10.1017/S0022112096001322.

Cirtain, J.W., Golub, L., Winebarger, A.R., de Pontieu, B., Kobayashi, K., Moore, R.L., *et al.*: 2013, *Nature* **493**, 501. doi:10.1038/nature11772.

Engvold, O.: 1998, In: Webb, D.F., Schmieder, B., Rust, D.M. (eds.) *IAU Colloq. 167: New Perspectives on Solar Prominences* **CS-150**, Astron. Soc. Pac., San Francisco, 23.

Gaizauskas, V., Zirker, J.B., Sweetland, C., Kovacs, A.: 1997, *Astrophys. J.* **479**, 448. doi:10.1086/512788.
Handy, B.N., Acton, L.W., Kankelborg, C.C., Wolfson, C.J., Akin, D.J., Bruner, M.E., *et al.*: 1999, *Solar Phys.* **187**, 229. doi:10.1023/A:1005166902804.
Howard, R.A., Moses, J.D., Vourlidas, A., Newmark, J.S., Socker, D.G., Plunkett, S.P., *et al.*: 2008, *Space Sci. Rev.* **136**, 67. doi:10.1007/s11214-008-9341-4.
Hudson, H.S.: 2000, *Astrophys. J. Lett.* **531**, L75. doi:10.1086/312516.
Kaiser, M.L., Kucera, T.A., Davila, J.M., St. Cyr, O.C., Guhathakurta, M., Christian, E.: 2008, *Space Sci. Rev.* **136**, 5. doi:10.1007/s11214-007-9277-0.
Kitiashvili, I.N., Kosovichev, A.G., Lele, S.K., Mansour, N.N., Wray, A.A.: 2013, arXiv:1301.0018.
Lemen, J.R., Title, A.M., Akin, D.J., Boerner, P.F., Chou, C., Drake, J.F., *et al.*: 2012, *Solar Phys.* **275**, 17. doi:10.1007/s11207-011-9776-8.
Li, X., Morgan, H., Leonard, D., Jeska, L.: 2012, *Astrophys. J. Lett.* **752**, L22. doi:10.1088/2041-8205/752/2/L22.
Liewer, P.C., Panasenco, O., Hall, J.R.: 2013, *Solar Phys.* **282**, 201. doi:10.1007/s11207-012-0145-z.
Lin, Y.: 2011, *Space Sci. Rev.* **158**, 237. doi:10.1007/s11214-010-9672-9.
Lin, Y., Engvold, O., Rouppe van der Voort, L.H.M.: 2012, *Astrophys. J.* **747**, 129. doi:10.1088/0004-637X/747/2/129.
Lin, Y., Engvold, O.R., Wiik, J.E.: 2003, *Solar Phys.* **216**, 109. doi:10.1023/A:1026150809598.
Lin, Y., Martin, S.F., Engvold, O.: 2008, In: Howe, R., Komm, R.W., Balasubramaniam, K.S., Petrie, G.J.D. (eds.) *Subsurface and Atmospheric Influences on Solar Activity* **CS-383**, Astron. Soc. Pac., San Francisco, 235.
Lin, Y., Engvold, O., Rouppe van der Voort, L., Wiik, J.E., Berger, T.E.: 2005a, *Solar Phys.* **226**, 239. doi:10.1007/s11207-005-6876-3.
Lin, Y., Wiik, J.E., Engvold, O., Rouppe van der Voort, L., Frank, Z.A.: 2005b, *Solar Phys.* **227**, 283. doi:10.1007/s11207-005-1111-9.
Lin, Y., Engvold, O., Rouppe van der Voort, L.H.M., van Noort, M.: 2007, *Solar Phys.* **246**, 65. doi:10.1007/s11207-007-0402-8.
Lin, Y., Soler, R., Engvold, O., Ballester, J.L., Langangen, Ø., Oliver, R., *et al.*: 2009, *Astrophys. J.* **704**, 870. doi:10.1088/0004-637X/704/1/870.
Litvinenko, Y.E.: 1999, *Astrophys. J.* **515**, 435. doi:10.1086/307001.
Litvinenko, Y.E.: 2010, *Astrophys. J.* **720**, 948. doi:10.1088/0004-637X/720/1/948.
Martin, S.F.: 1990, In: Rudzjak, V., Tandberg-Hanssen, E. (eds.) *Dynamics of Quiescent Prominences, IAU Colloq, Lecture Notes in Physics* **117**, Springer, Berlin, 1. doi:10.1007/BFb0025641.
Martin, S.F.: 1998, *Solar Phys.* **182**, 107. doi:10.1023/A:1005026814076.
Martin, S.F., Bilimoria, R., Tracadas, P.W.: 1994, In: Rutten, R.J., Schrijver, C.J. (eds.) *Solar Surface Magnetism*, Kluwer Academic, Dordrecht, 303.
Martin, S.F., Echols, C.R.: 1994, In: Rutten, R.J., Schrijver, C.J. (eds.) *Solar Surface Magnetism*, Kluwer Academic, Dordrecht, 339.
Martin, S.F., Lin, Y., Engvold, O.: 2008, *Solar Phys.* **250**, 31. doi:10.1007/s11207-008-9194-8.
Martin, S.F., McAllister, A.H.: 1997, In: Crooker, N., Joselyn, J.A., Feynman, J. (eds.) *Geophys. Monogr. Ser.* **99**, 127, AGU, Washington.
Martin, S.F., Panasenco, O., Berger, M.A., Engvold, O., Lin, Y., Pevtsov, A.A., *et al.*: 2012, In: Rimmele, T.R., Tritschler, A., Wöger, F., Collados Vera, M., Socas-Navarro, H., Schlichenmaier, R., *et al.* (eds.) *Second ATST-EAST Meeting: Magnetic Fields from the Photosphere to the Corona* **CS-463**, Astron. Soc. Pac., San Francisco, 157.
Orozco Suárez, D., Asensio Ramos, A., Trujillo Bueno, J.: 2012, *Astrophys. J. Lett.* **761**, L25. doi:10.1088/2041-8205/761/2/L25.
Panasenco, O., Martin, S.F.: 2008, In: Howe, R., Komm, R.W., Balasubramaniam, K.S., Petrie, G.J.D. (eds.) *Subsurface and Atmospheric Influences on Solar Activity* **CS-383**, Astron. Soc. Pac, San Francisco, 243.
Panasenco, O., Velli, M.: 2009, In: Lites, B., Cheung, M., Magara, T., Mariska, J., Reeves, K. (eds.) *The Second Hinode Science Meeting: Beyond Discovery-Toward Understanding* **CS-415**, Astron. Soc. Pac, San Francisco, 196.
Panasenco, O., Martin, S.F., Velli, M., Vourlidas, A.: 2012, *Solar Phys.* **321**. doi:10.1007/s11207-012-0194-3.
Panesar, N.K., Innes, D.E., Tiwari, S.K., Low, B.C.: 2013, *Astron. Astrophys.* **549**, A105. doi:10.1051/0004-6361/201220503.
Pettit, E.: 1925, *Publ. Yerkes Obs.* **3**, 4.
Pettit, E.: 1932, *Astrophys. J.* **76**, 9.
Pettit, E.: 1941, *Publ. Astron. Soc. Pac.* **53**, 289.
Pettit, E.: 1943, *Astrophys. J.* **98**, 6.
Pettit, E.: 1946, *Publ. Astron. Soc. Pac.* **58**, 150.
Pettit, E.: 1950, *Publ. Astron. Soc. Pac.* **62**, 144.

Rappazzo, A.F., Velli, M., Einaudi, G.: 2013, *Astrophys. J.* **771**, 76. doi:10.1088/0004-637X/771/2/76.
Secchi, P.A.: 1877, *Le Soleil* **1**, Gauthier-Villars, Paris.
Sheeley, N.R. Jr., Warren, H.P.: 2012, *Astrophys. J.* **749**, 40. doi:10.1088/0004-637X/749/1/40.
Simon, G.W., Weiss, N.O.: 1997, *Astrophys. J.* **489**, 960. doi:10.1086/304800.
Su, Y., Wang, T., Veronig, A., Temmer, M., Gan, W.: 2012, *Astrophys. J. Lett.* **756**, L41. doi:10.1088/2041-8205/756/2/L41.
Török, T., Berger, M.A., Kliem, B.: 2010, *Astron. Astrophys.* **516**, A49. doi:10.1051/0004-6361/200913578.
Velli, M., Liewer, P.: 1999, *Space Sci. Rev.* **87**, 339. doi:10.1023/A:1005110315988.
Verdini, A., Velli, M., Matthaeus, W.H., Oughton, S., Dmitruk, P.: 2010, *Astrophys. J. Lett.* **708**, L116. doi:10.1088/2041-8205/708/2/L116.
Wang, Y.-M., Muglach, K.: 2007, *Astrophys. J.* **666**, 1284. doi:10.1086/520623.
Wedemeyer-Böhm, S., Scullion, E., Steiner, O., Rouppe van der Voort, L., de La Cruz Rodriguez, J., Fedun, V., *et al.*: 2012, *Nature* **486**, 505. doi:10.1038/nature11202.
Young, C.A.: 1896, *The Sun*, 2nd edn. D. Appleton & Company, New York, 224.
Zirker, J.B., Engvold, O., Martin, S.F.: 1998, *Nature* **396**, 440. doi:10.1038/24798.

DOI 10.1007/978-1-4939-1182-0_13
Reprinted from *Solar Physics* Journal, DOI 10.1007/s11207-012-0216-1

Forecasting the Maxima of Solar Cycle 24 with Coronal Fe XIV Emission

Richard C. Altrock

Received: 30 August 2012 / Accepted: 11 December 2012 / Published online: 15 January 2013

Abstract The onset of the "Rush to the Poles" of polar-crown prominences and their associated coronal emission is a harbinger of solar maximum. Altrock (*Solar Phys.* **216**, 343, 2003) showed that the "Rush" was well observed at 1.15 $R_{\odot}$ in the Fe XIV corona at the Sacramento Peak site of the National Solar Observatory prior to the maxima of Cycles 21 to 23. The data show that solar maximum in those cycles occurred when the center line of the Rush reached a critical latitude of $76° \pm 2°$. Furthermore, in the previous three cycles solar maximum occurred when the highest number of Fe XIV emission features per day (averaged over 365 days and both hemispheres) first reached latitudes $20° \pm 1.7°$. Applying the above conclusions to Cycle 24 is difficult due to the unusual nature of this cycle. Cycle 24 displays an intermittent Rush that is only well-defined in the northern hemisphere. In 2009 an initial slope of $4.6°\,\mathrm{year}^{-1}$ was found in the north, compared to an average of $9.4 \pm 1.7°\,\mathrm{year}^{-1}$ in the previous cycles. An early fit to the Rush would have reached 76° at 2014.6. However, in 2010 the slope increased to $7.5°\,\mathrm{year}^{-1}$ (an increase did not occur in the previous three cycles). Extending that rate to $76° \pm 2°$ indicates that the solar maximum in the northern hemisphere already occurred at 2011.6 ± 0.3. In the southern hemisphere the Rush to the Poles, if it exists, is very poorly defined. A linear fit to several maxima would reach 76° in the south at 2014.2. In 1999, persistent Fe XIV coronal emission known as the "extended solar cycle" appeared near 70° in the North and began migrating towards the equator at a rate 40 % slower than the previous two solar cycles. However, in 2009 and 2010 an acceleration occurred. Currently the greatest number of emission features is at 21° in the North and 24° in the South. This indicates that solar maximum is occurring now in the North but not yet in the South.

Keywords Corona · Solar cycle

Solar Origins of Space Weather and Space Climate
Guest Editors: I. González Hernández, R. Komm, and A. Pevtsov

R.C. Altrock (✉)
Air Force Research Laboratory, Space Weather Center of Excellence, PO Box 62, Sunspot, NM 88349, USA
e-mail: altrock@nso.edu

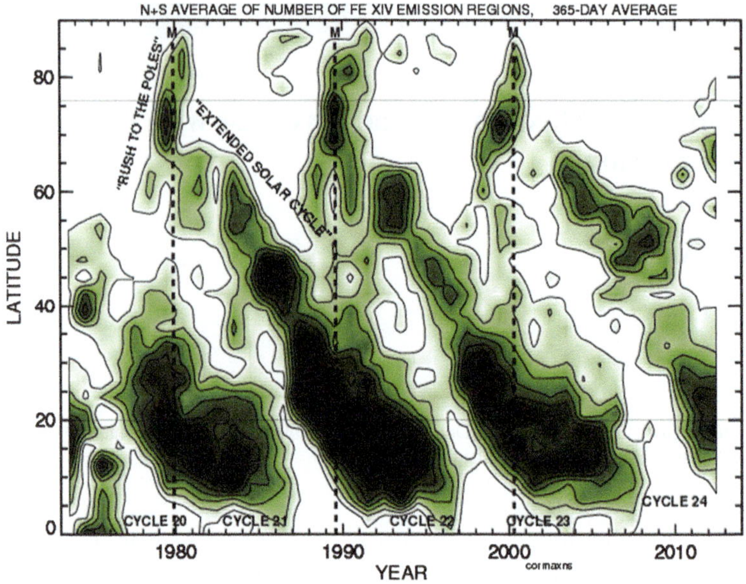

Figure 1 Annual northern-plus-southern-hemisphere averages of the number of Fe XIV emission features from 1973 to 2012. The Rush to the Poles beginning near 1978 and the extended solar cycle beginning near 1980 (see text) are indicated. Vertical dashed lines indicate the time of solar maxima. Contours are drawn at 0.065, 0.085, 0.105, ... emission features per day. Shading darkens with each change of contour. See text for description of other features.

1. Introduction

Altrock (2011) discussed in detail the methods by which observations of Fe XIV 530.3 nm obtained with the *Photoelectric Coronal Photometer* and 40 cm coronagraph at the John W. Evans Solar Facility of the National Solar Observatory at Sacramento Peak (Fisher 1973, 1974; Smartt, 1982) can be used to forecast solar maxima at varying times prior to the occurrence of the maxima. For further discussion of the observations see Altrock (1997) as well as earlier observations discussed in Altrock (2011).

Altrock (2011) found that a synoptic map of the location in latitude of local intensity maxima in the Fe XIV scans at 1.15 solar radii from the center of the disk ($R_{\rm o}$) clearly showed the progress of the emission from high to low latitudes, known as the extended solar cycle, in Cycles 22 – 23 and the Rush to the Poles overlying polar-crown prominences preceding solar maxima. We will hereafter refer to these intensity maxima as emission features.

As discussed in Altrock (2011), high-latitude emission features are situated above the high-latitude neutral line of the large-scale photospheric magnetic field seen in Wilcox Solar Observatory synoptic maps. This is also the locus of polar-crown prominences. These features are therefore likely parts of streamers overlying the polar-crown prominences, although the low resolution of the observations does not allow a rigorous connection to be made. At lower latitudes the emission features may also overly active regions, prominences, other large-scale magnetic field boundaries, *etc.*

As in Altrock (2011), we average the number of points at each latitude over a given time interval. This process allows us to correct the data for missing days, which is an essential step for correctly interpreting the data. Figure 1 shows annual averages of the number of emission features, also averaged over the northern and southern hemispheres.

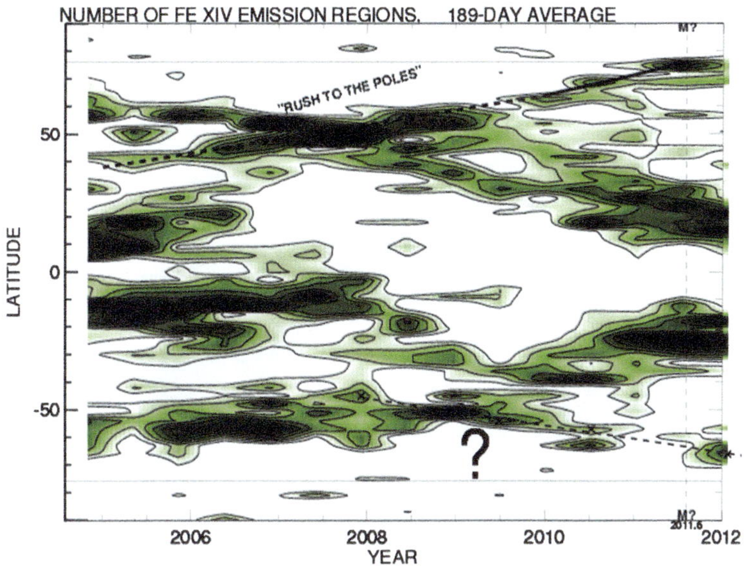

Figure 2 Seven-rotation (approximately semiannual) averages of the number of Fe XIV emission features from 2005 to 2012 plotted separately for each hemisphere (−90° to +90° latitude). See text for further information.

2. Discussion

In Figure 1, we see that extended Solar Cycles 22 and 23 begin near 70° latitude and end near the equator about 18 years later. The initial rate of migration towards the equator for Cycle 24 was 40 % slower than the previous cycles (see Altrock, 2011). More recently, emission took a more rapid rate down to near 20° latitude, and there is a suggestion that a Rush may have developed.

Also in Figure 1, there are fine lines drawn at 76° and 20°. As discussed in Altrock (2011), during the last three cycles, solar maximum occurred when a linear fit to the Rush reached 76° ± 2° latitude. This may be seen in Figure 1. Altrock (2011) also found that, in the previous three cycles, solar maximum occurred when the greatest number of Fe XIV emission features, averaged over 365 days and both hemispheres (as in Figure 1), were at latitudes 20° ± 1.7°. Suggestions of this may also be seen in Figure 1.

In order to attempt to apply these conclusions to Cycle 24, let us examine a higher-resolution (if noisier) graphic.

2.1. The Rush to the Poles

Figure 2 shows the data plotted separately for each hemisphere from 2005 – 2012 (for earlier years, see Altrock (2011)). Fine lines mark 76° latitude. In 2005 there is an appearance of a weak Rush in the northern hemisphere, marked by a label, "Rush to the Poles", and a linear fit from 2005 – 2010 (dashed line). This early fit to the Rush resulted in more of a stroll than a Rush, which would have reached 76° at 2014.6. Between 2009 and 2010 an acceleration occurred (solid line). The new, higher rate of the Rush in the northern hemisphere indicates it reached 76° ± 2° latitude at 2011.6 ± 0.3. In other words, in the northern hemisphere, solar maximum has already occurred at about 2011.6. This is indicated by a fine vertical dashed line.

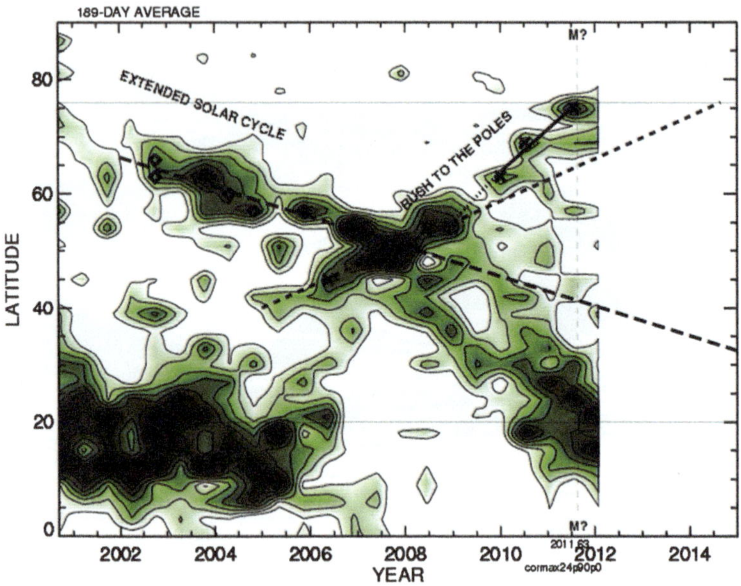

Figure 3 Seven-rotation averages of Cycle 24 northern-hemisphere emission features.

In the southern hemisphere, the Rush to the Poles, if it exists, is very poorly defined. I show a linear fit to several maxima with a dashed line underlain by a large interrogative. This fit would reach 76° at 2014.2, which gives a best guess to solar maximum in the southern hemisphere.

2.2. Low-Latitude Emission

Figure 3 shows the northern-hemisphere emission from 2000.5 to 2012. The northern-hemisphere fits to the Rush discussed in Section 2.1 are shown, as well as the initial fit to the extended solar cycle, which is extrapolated to 2015. This initial fit (long-dashed line) would not reach 20° until 2019.8.

However, the initial rate of migration towards the equator of emission features was suddenly interrupted around 2009. The higher-latitude emission is ending, and a new lower-latitude band has developed. This low-latitude band may be used to infer when solar maximum will occur. Currently the greatest number of emission features is at 21° in the North and 24° in the South. This indicates that solar maximum is occurring now in the North but not yet in the South.

Note, from Sections 2.1 and 2.2, that something remarkable happened on the Sun around 2009, resulting in a dramatic increase in the latitude-time rates in the northern hemisphere of the extended solar cycle and the Rush to the Poles.

2.3. Results from Other Authors and Other Data Sets

Gopalswamy *et al.* (2012) come to similar conclusions from a study of the microwave butterfly diagram and prominence eruption latitudes and show that the northern high-latitude magnetic field changes sign in 2012, which is an indicator that solar maximum has occurred in the North. Rušin, Minarovjech, and Saniga (2009), using similar Fe XIV data but a different technique, predicted two solar maxima: one in the time frame 2010 – 2011 and one in 2012.

Other authors have discussed North-South asymmetries. Gopalswamy *et al.* (2003) showed that in Cycle 23 the magnetic polarity of the north pole reversed before that of the south pole. This is consistent with the observations that the northern polar-crown prominences disappeared in October of 2000, while those in the South disappeared in early 2002, and northern high-latitude activity disappeared in November 2000, while that in the South disappeared in May 2002.

Svalgaard and Kamide (2012) note that in Cycle 19 the south pole reversed polarity first followed by the north pole more than a year later. Since then polar field reversals have happened first in the North. They show that North-South asymmetries are normal in the sunspot number (each hemisphere may have two or more maxima), and the Rush to the Poles for Cycle 20 happened in the North well before the South.

Gopalswamy (2012) studied solar energetic particle (SEP) events for Cycle 24 and found that all but one of the 15 large SEP events during the first 4.5 years occurred in the northern hemisphere, and SEP events did not appear in the southern hemisphere until June 2012.

Cycle 24 therefore resembles the rise and maximum phases of Cycle 21, when most of the events occurred in the northern hemisphere until after the maximum phase.

Altrock (2011) discusses the use of simulated coronal emission *vs.* observed coronal emission. He concludes that there is no reason to use simulated emission in studies of the properties of coronal emission. Robbrecht *et al.* (2010) used potential-field source-surface simulated emission to conclude that the extended solar cycle does not exist. However, other authors (*cf.* Sandman and Aschwanden, 2011) doubt the validity of this simple model.

The usual measure of solar maximum, the global smoothed sunspot numbers (ftp://ftp.ngdc.noaa.gov/STP/SOLAR_DATA/SUNSPOT_NUMBERS/INTERNATIONAL/smoothed/SMOOTHED), show an inflection point in late 2011, which could represent solar maximum in the northern hemisphere.

Figure 4 shows North and South sunspot areas as compiled at NASA Marshall SFC by David Hathaway (http://solarscience.msfc.nasa.gov/greenwch.shtml) from 2007 to 2012. These monthly values have been smoothed in the same manner as the global smoothed sunspot numbers but are extended up to the current time by use of IDL edge truncation. The last value unaffected by edge truncation is January 2012. However, it seems all but certain that solar maximum has occurred in the North but not yet in the South, albeit in the sunspot *area*, not in the sunspot *number*. The Solar Influences Data Analysis Center (SIDC) (http://www.sidc.be/) has northern and southern sunspot numbers. Their data show a maximum in the northern hemisphere near 2012.0 but no maximum yet in the South.

3. Conclusions

The location of Fe XIV emission features in time-latitude space displays an 18-year progression from near 70° to the equator, which has been referred to as the extended solar cycle. Cycle 24 emission features began migrating towards the equator similarly to previous cycles, although at a 40 % slower rate. In addition, in approximately 2005 the northern-hemisphere Rush to the Poles began at a 50 % slower rate than in recent cycles. It accelerated beginning in 2010.

Analysis of the northern-hemisphere Rush to the Poles indicates that the northern-hemisphere maximum already occurred in 2011.6 ± 0.3. This conclusion is different from

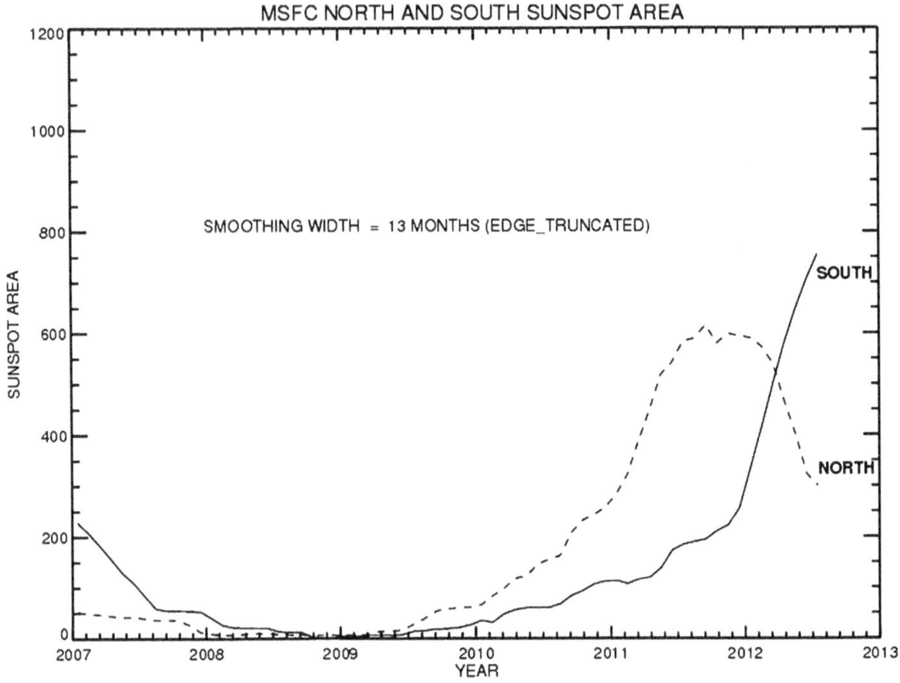

Figure 4 Smoothed NASA Marshall SFC northern and southern sunspot areas.

that in Altrock (2011), because the acceleration beginning in 2010 had not yet been recognized.

A weak Rush may be occurring in the southern hemisphere, and an analysis of its properties yields an estimate for the southern-hemisphere maximum of 2014.2.

Low-latitude emission features are migrating towards the equator in both hemispheres. In previous cycles, solar maximum occurred when the greatest concentration of 1.15 R_o Fe XIV emission features reached $20° \pm 1.7°$ latitude. Currently the greatest number of emission features is at 21° in the North and 24° in the South. This indicates that solar maximum is occurring now in the North but not yet in the South.

A recent study of microwave brightness and prominence eruptions by Gopalswamy *et al.* (2012) comes to similar conclusions. In addition, there is strong evidence that the northern-hemisphere sunspot *area* reached a maximum around the end of 2011. The southern-hemisphere area appears to still be increasing. An inflection point occurred in the global smoothed sunspot *number* late in 2011, which could be evidence that maximum occurred in one hemisphere.

This analysis does not indicate the strength of the maximum.

Acknowledgements The observations used herein are the result of a cooperative program of the Air Force Research Laboratory and the National Solar Observatory. I am grateful for the assistance of NSO personnel, especially John Cornett, Timothy Henry, Lou Gilliam, and Wayne Jones, for observing and data-reduction and -analysis services and maintenance of the Evans Solar Facility and its instrumentation and to Raymond N. Smartt, who completely redesigned the Sacramento Peak *Photoelectric Coronal Photometer* filters in 1982, making it the excellent instrument it is today. I wish to also thank the referee for pointing out several deficiencies in the paper, which I have endeavored to correct.

References

Altrock, R.C.: 1997, *Solar Phys.* **170**, 411.
Altrock, R.C.: 2003, *Solar Phys.* **216**, 343.
Altrock, R.C.: 2011, *Solar Phys.* **274**, 251.
Fisher, R.R.: 1973, A photoelectric photometer for the Fe XIV solar corona, AFCRL-TR-73-0696. 15 pp.
Fisher, R.R.: 1974, *Solar Phys.* **36**, 343.
Gopalswamy, N.: 2012, In: *Proc. 11th Annual Astrophysical Conference, Palm Springs, CA, 19–23 March 2012*. arXiv:1208.3951.
Gopalswamy, N., Lara, A., Yashiro, S., Howard, R.A.: 2003, *Astrophys. J. Lett.* **598**, L63.
Gopalswamy, N., Yashiro, S., Makela, P., Michalek, G., Shibasaki, K., Hathaway, D.H.: 2012, *Astrophys. J. Lett.* **750**, L42.
Robbrecht, E., Wang, Y.-M., Sheeley, N.R. Jr., Rich, N.B: 2010, *Astrophys. J.* **716**, 693.
Rušin, V., Minarovjech, M., Saniga, M.: 2009, *Contrib. Astron. Obs. Skaln. Pleso* **39**, 71.
Sandman, A.W., Aschwanden, M.J.: 2011, *Solar Phys.* **270**, 503.
Smartt, R.N.: 1982, In: *Instrumentation in Astronomy IV, Proc. SPIE* **331**, 442.
Svalgaard, L., Kamide, Y.: 2012, arXiv:1207.2077.

DOI 10.1007/978-1-4939-1182-0_14
Reprinted from *Solar Physics* Journal, DOI 10.1007/s11207-013-0301-0

Coronal Magnetic Field Evolution from 1996 to 2012: Continuous Non-potential Simulations

A.R. Yeates

Received: 23 August 2012 / Accepted: 4 April 2013 / Published online: 25 April 2013

Abstract Coupled flux transport and magneto-frictional simulations are extended to simulate the continuous magnetic-field evolution in the global solar corona for over 15 years, from the start of Solar Cycle 23 in 1996. By simplifying the dynamics, our model follows the build-up and transport of electric currents and free magnetic energy in the corona, offering an insight into the magnetic structure and topology that extrapolation-based models cannot. To enable these extended simulations, we have implemented a more efficient numerical grid, and have carefully calibrated the surface flux-transport model to reproduce the observed large-scale photospheric radial magnetic field, using emerging active regions determined from observed line-of-sight magnetograms. This calibration is described in some detail. In agreement with previous authors, we find that the standard flux-transport model is insufficient to simultaneously reproduce the observed polar fields and butterfly diagram during Cycle 23, and that additional effects must be added. For the best-fit model, we use automated techniques to detect the latitude–time profile of flux ropes and their ejections over the full solar cycle. Overall, flux ropes are more prevalent outside of active latitudes but those at active latitudes are more frequently ejected. Future possibilities for space-weather prediction with this approach are briefly assessed.

Keywords Coronal mass ejections, theory · Magnetic fields, corona · Magnetic fields, models · Magnetic fields, photosphere · Solar cycle, models

1. Introduction

Modelling the Sun's coronal magnetic field over the full solar cycle is important because it acts as a driver of space-weather events, and changes significantly over the cycle, as well as from one cycle to the next. It is fundamentally time-dependent, as manifested in the

Solar Origins of Space Weather and Space Climate
Guest Editors: I. González Hermández, R. Komm, and A. Pevtsov

A.R. Yeates (✉)
Department of Mathematical Sciences, Durham University, Durham, DH1 3LE, UK
e-mail: anthony.yeates@durham.ac.uk

variations of almost all observed properties of the Sun, including sunspot number, magnetic flux, rates of flares and coronal mass ejections (CMEs), and even the total solar irradiance (Willson and Hudson, 1991). The coronal magnetic field is driven by activity in the solar interior, including large-scale flows and convection in the photosphere as well as the periodic emergence of new active regions.

Previous studies of the coronal magnetic evolution over months to years have mostly used potential-field extrapolations (Altschuler and Newkirk, 1969; Schatten, Wilcox, and Ness, 1969). From the space-weather viewpoint, these models give a first approximation of the Sun's open flux (Wang and Sheeley, 2002) and the heliospheric current sheet (Hoeksema, Wilcox, and Scherrer, 1983), and have been coupled to models of the heliosphere (Arge and Pizzo, 2000; Luhmann *et al.*, 2002; Pizzo *et al.*, 2011). They have also been applied to predict CME rates using the topology of magnetic nulls and "breakout" configurations (Cook, Mackay, and Nandy, 2009). However, their value for predicting flares and CMEs is limited since they allow no free energy, and the lack of electric currents leads them to underestimate the open flux, particularly during active periods (Riley, 2007; Yeates *et al.*, 2010a).

Attempts to move beyond potential-field models and extrapolate more general magnetic equilibria suffer from the problem of non-uniqueness of magnetic topology for a given distribution of observed magnetic field in the photosphere. For example, the current-sheet source-surface (CSSS) model can better match the observed open flux by allowing certain forms of electric currents (Zhao and Hoeksema, 1995; Jiang *et al.*, 2010a), but these are chosen for mathematical convenience rather than any particular physical basis. On the other hand, when trying to extrapolate nonlinear force-free fields from photospheric vector magnetograms, different numerical extrapolation codes tend to produce different results (DeRosa *et al.*, 2009). These difficulties have led to a more pragmatic approach for nonlinear force-free modelling of local structures such as coronal cavities or sigmoids, whereby the computation is initialised with a flux-rope structure in the corona (van Ballegooijen, 2004; Su *et al.*, 2009; Savcheva, van Ballegooijen, and DeLuca, 2012).

In recent years, global magnetohydrodynamic (MHD) models that more realistically account for thermodynamic properties of the plasma have become practical, allowing for comparison with observed emission at various wavelengths (Lionello, Linker, and Mikić, 2009; Rušin *et al.*, 2010; Downs *et al.*, 2010). However, it is not yet practical to simulate the temporal evolution of the corona for many months, and MHD models are generally limited to finding individual equilibria by relaxing from an initial condition appropriate for a given day. So far, the initial conditions for the magnetic field have been potential-field extrapolations. Thus the MHD models inherit the topology of the potential field, and do not account for the gradual build up of the magnetic topology over time, or any long-term memory of previous interactions that remains imprinted in the coronal field.

The approach of our model is to gain new insight into the magnetic structure and topology by not just extrapolating from photospheric data at a single time, but simulating the coronal field in a time-dependent way, using a simplified approximation to the real coronal evolution. The technique was originally introduced to study the formation of filaments (van Ballegooijen, Priest, and Mackay, 2000; Mackay, Gaizauskas, and van Ballegooijen, 2000; Mackay and van Ballegooijen, 2001, 2005, 2006), whose magnetic structure is non-potential and conjectured to depend on the build-up and transport of coronal magnetic helicity over time (van Ballegooijen and Martens, 1989). It was extended to model the global corona by Yeates, Mackay, and van Ballegooijen (2008). The model effectively produces a continuous sequence of nonlinear force-free fields, in response to flux emergence and shearing by photospheric footpoint motions. As such, a particular magnetic topology is chosen automatically at each time step by the evolutionary history, thus providing a physically motivated solution to the non-uniqueness problem. Moreover, the model – henceforth the NP ("non-potential") model – offers interesting possibilities for modelling and predicting space weather, because it allows for the build up and transport of free magnetic energy, electric currents, and magnetic helicity. In the model – as is hypothesized in the real corona – helicity tends to concentrate in flux-rope structures overlying photospheric polarity-inversion lines. When too much helicity accumulates, the flux ropes "erupt" and are ejected through the outer boundary of the simulation domain (Mackay and van Ballegooijen, 2006; Yeates and Mackay, 2009).

The developments in this article are threefold. Firstly, we extend the simulations to a continuous 15-year evolution (Section 2). Our previous study of solar-cycle variations in the NP model was limited to six separate six-month runs (Yeates *et al.*, 2010b). The new simulation allows for longer-term magnetic memory to take effect; our initial finding that this affects the chirality of high-latitude filaments has been described elsewhere (Yeates and Mackay, 2012). Secondly, we describe how the surface flux-transport component of the model – the lower boundary condition – must be carefully calibrated to observations for such long simulations (Section 3). Thirdly, we consider the distribution of flux ropes and flux-rope ejections over the 15-year simulation (Section 4). We conclude in Section 5 with a brief discussion of the future prospects for space-weather forecasting using this approach.

2. Coronal Magnetic Model

The non-potential (NP) model couples surface flux transport to magneto-frictional relaxation in the overlying corona (van Ballegooijen, Priest, and Mackay, 2000).

2.1. Formulation

The large-scale mean coronal magnetic field [$\mathbf{B}_0 = \nabla \times \mathbf{A}_0$] is evolved by the induction equation

$$\frac{\partial \mathbf{A}_0}{\partial t} = \mathbf{v}_0 \times \mathbf{B}_0 - \mathbf{E}_0, \tag{1}$$

where we neglect ohmic diffusion and the mean electromotive force [$\mathbf{E}_0$] describes the effect of unresolved small-scale fluctuations. Following van Ballegooijen and Cranmer (2008), we apply a hyperdiffusion

$$\mathbf{E}_0 = -\frac{\mathbf{B}_0}{B_0^2} \nabla \cdot \left(\eta_4 B_0^2 \nabla \alpha_0\right), \tag{2}$$

where

$$\alpha_0 = \frac{\mathbf{B}_0 \cdot \mathbf{j}_0}{B_0^2} \tag{3}$$

is the current helicity density, with $\mathbf{j}_0 = \nabla \times \mathbf{B}_0$ the current density, and $\eta_4 = 10^{11}\ \mathrm{km^4\,s^{-1}}$. This form of hyperdiffusion preserves magnetic-helicity density [$\mathbf{A}_0 \cdot \mathbf{B}_0$] in the volume and describes the tendency of the magnetic field to relax to a state of constant α_0 (Boozer, 1986; Bhattacharjee and Hameiri, 1986), although such a state is never reached in the global simulation. The velocity is determined by the magneto-frictional technique (Yang, Sturrock, and Antiochos, 1986; Craig and Sneyd, 1986) as

$$\mathbf{v}_0 = \frac{1}{\nu}\frac{\mathbf{j}_0 \times \mathbf{B}_0}{B_0^2} + v_{\mathrm{out}}(r)\mathbf{e}_r. \tag{4}$$

This replaces the full momentum equation and allows numerical solution of the model over months and years. The first term enforces relaxation towards a force-free equilibrium [$\mathbf{j}_0 \times \mathbf{B}_0 = 0$]. The second term is a radial outflow imposed only near the outer boundary [$r = 2.5\mathrm{R}_\odot$] to represent the effect of the solar wind radially distending magnetic-field lines (Mackay and van Ballegooijen, 2006).

2.2. Photospheric Boundary Condition

On the photospheric boundary $r = \mathrm{R}_\odot$, the magneto-frictional velocity is not applied, and instead the radial magnetic field [B_{0r}] is evolved by the surface flux-transport model (Sheeley, 2005; Mackay and Yeates, 2012). In spherical polar coordinates (r, θ, ϕ), the vector potential evolves according to

$$\frac{\partial A_{0\theta}}{\partial t} = u_\phi B_{0r} - \frac{D}{\mathrm{R}_\odot \sin\theta}\frac{\partial B_{0r}}{\partial \phi} + S_\theta(\theta, \phi, t), \tag{5}$$

$$\frac{\partial A_{0\phi}}{\partial t} = -u_\theta B_{0r} + \frac{D}{\mathrm{R}_\odot}\frac{\partial B_{0r}}{\partial \theta} + S_\phi(\theta, \phi, t). \tag{6}$$

Here D is a (constant) diffusivity modelling the random walk of magnetic flux owing to the changing supergranular convection pattern (Leighton, 1964). In Section 3, we experiment with an additional exponential decay of B_{0r} (Schrijver, DeRosa, and Title, 2002). The differential rotation velocity $u_\phi = \Omega(\phi)\mathrm{R}_\odot \sin\theta$ uses the observationally determined Snodgrass (1983) profile

$$\Omega(\theta) = 0.18 - 2.3\cos^2\theta - 1.62\cos^4\theta \ \mathrm{deg\,day}^{-1}, \tag{7}$$

written in the Carrington frame. For the basic meridional flow we assume the form of Schüssler and Baumann (2006), namely

$$u_\theta(\theta) = u_0 \frac{16}{110}\sin(2\lambda)\exp\bigl(\pi - 2|\lambda|\bigr), \tag{8}$$

where $\lambda = \pi/2 - \theta$ is latitude and u_0 is a constant controlling the flow amplitude. In Section 3, we experiment with activity-dependent perturbations of this flow profile.

The source terms S_θ and S_ϕ in Equations (5) and (6) represent the emergence of new active regions, and are necessary to maintain an accurate description of the observed surface B_{0r} over the continuous 15-year simulation. Rather than specify a functional form for these source terms, we insert individual bipolar magnetic regions with properties chosen to match those in observed synoptic magnetograms (Yeates, Mackay, and van Ballegooijen, 2007). The inserted bipolar regions take an idealised three-dimensional form (Yeates, Mackay, and van Ballegooijen, 2008). In the future, we hope to incorporate more detailed models of the structure of individual active regions, built up in a time-dependent manner. For the simulations described here, we use synoptic normal-component magnetograms from US National Solar Observatory, Kitt Peak. Until 2003, these were taken with the older *Vacuum Telescope*, and from 2003 onward with *Synoptic Optical Long-term Investigations of the Sun* (SOLIS). We insert a total of 2040 bipoles between Carrington Rotation CR1911 (June 1996) and CR2122 (April 2012). We do not insert any bipoles to replace those missed during the three data gaps (CR2015 – 16, CR2040 – 41, CR2091).

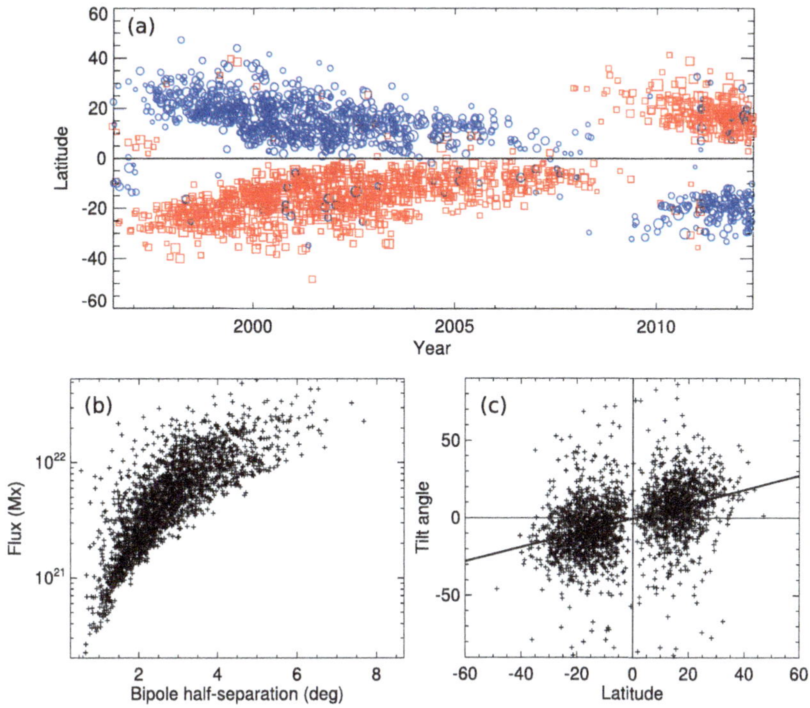

Figure 1 Properties of the 2040 magnetic bipoles determined from NSO/KP synoptic magnetograms: (a) locations in time and latitude, with colour/symbol showing polarity and symbol size proportional to flux; (b) bipole flux against bipole size; (c) tilt angles against latitude. In (c), the solid line is a linear fit to the measured tilt angles.

The 2040 bipolar regions are summarised in Figure 1. Figure 1(a) shows their latitude–time distribution, along with their leading/following polarity. The majority polarity reverses with each 11-year cycle, as it should. Figure 1(b) shows the range of sizes and fluxes of bipoles in our dataset. Our semi-automated technique identifies bipoles using only flux above 50 gauss, thus imposing a lower size cut-off. Figure 1(c) shows the latitude distribution of bipole tilt angles [γ] (the angle with respect to the Equator of the line connecting leading and following polarity centroids). The solid line shows a linear fit $\gamma = -0.41^\circ + 0.459\lambda$, which is consistent with magnetogram observations of Wang and Sheeley (1989), who used NSO/KP data for Cycle 21, or Stenflo and Kosovichev (2012), who used SOHO/MDI magnetograms. White-light studies of tilt angles return a rather lower slope (Dasi-Espuig *et al.*, 2010). We feel that the distribution derived from magnetograms should be more appropriate for our application, although we find in Section 3 that reducing the tilt angles by 20 % is a straightforward way to calibrate the flux-transport simulation without the need for more complex effects (Cameron *et al.*, 2010). For the 20 % reduced tilt angles, the linear fit is $\gamma = -0.33^\circ + 0.367\lambda$.

2.3. Numerical Methods

The equations are solved for the vector potential [$\mathbf{A}_0$] on a staggered grid, using a flux-conserving (finite-volume) scheme for the advection terms. To avoid the problem of grid convergence near the poles, we have newly incorporated a variable grid (see Appendix A for details). This greatly reduces the computational time. For the simulations described in this article, we use a resolution of 192 cells in longitude at the Equator, corresponding to an angular resolution of 1.875°, and 28 cells in radius. At the polar grid boundary ($\theta \approx 0.67°$), there are only 12 cells in longitude, owing to the variable grid. Boundary conditions are described in Appendix B. The coronal magnetic field is initialised using a potential-field extrapolation for 15 June 1996, derived from the synoptic magnetogram for CR1910 (Yeates, Mackay, and van Ballegooijen, 2007). The code is parallelised with OpenMPI and the full 5822-day run took approximately five days with 48 cores.

3. Observational Calibration

Since we do not reset the photospheric magnetic field to observed magnetograms once the simulation has begun, it is important to calibrate the surface flux-transport model to reproduce the observed long-term evolution on the Sun. This is particularly true for simulations as long as 15 years, so we look at this issue in detail here. Unfortunately, it is a delicate balance between the properties of newly emerging active regions and the chosen profiles for the transport processes of meridional flow and supergranular diffusion. Differential rotation is better constrained by observations and we do not consider its variation. Previous parameter studies of the flux-transport model have been undertaken by Baumann *et al.* (2004), and for Cycle 23 by Schrijver and Liu (2008) and Jiang *et al.* (2011), although here we are additionally constrained by the individual measured properties of the bipolar regions. Since there is no feedback of the coronal (magneto-frictional) component of the model on the surface flux-transport component, the calibration can be done by running only the latter.

To calibrate the surface simulations we use several numerical indicators for the photospheric magnetic field:

i) Cross-correlation of total (unsigned) magnetic flux [$\Phi_\odot$] with the observed time series.
ii) Least value of $\Phi_\odot$ around the Cycle 23 Minimum (ca. 2009).
iii) Cross-correlation of axial dipole strength $B_{\mathrm{ax}} \equiv b_{1,0}$ with the observed time series.
iv) Time of sign-reversal in the axial dipole strength.
v) Peak negative value of the axial dipole strength during Cycle 23.
vi) Cross-correlation of equatorial dipole strength $B_{\mathrm{eq}} \equiv \sqrt{|b_{1,1}|^2 + |b_{1,-1}|^2} \equiv \sqrt{2}|b_{1,1}|$ with the observed time series.
vii) Cross-correlation of longitude-averaged field $\langle B_{0r} \rangle \equiv \int_0^{2\pi} B_{0r} \mathrm{R}_\odot \sin\theta \, \mathrm{d}\phi$ with the observed butterfly diagram.

Here $b_{l,m}$ are the complex-valued spherical-harmonic components of B_{0r}. Because of the smoothing effect of supergranular diffusion, the total simulated flux [B_{0r}] tends to be lower than the total magnetogram flux. To account for this, our cross-correlations i) and vii) use a smoothed version of the observed magnetogram, obtained by removing spherical harmonic components above order $\ell = 128$. Higher-resolution simulations in the future will need to incorporate a more realistic random-walk model for discrete flux concentrations (*e.g.* Worden and Harvey, 2000; Schrijver, 2001). To facilitate cross-correlation of time series in i), iii), and vi), the simulation data are first interpolated to the same times as the magnetogram observations. For cross-correlation of the two-dimensional butterfly diagrams, the simulation data are first interpolated to the same times and latitudes as the observations.

Table 1 lists values of the numerical indicators for a representative set of test runs, which are described in more detail below. For comparison, the total unsigned flux [$\Phi_\odot$], axial dipole strength [B_{ax}], and equatorial dipole strength [B_{eq}] for each run are shown in Figure 2. Butterfly diagrams of $\langle B_{0r} \rangle$ in each run are shown in Figure 3.

Table 1 Performance of surface flux-transport calibration runs.

Run	Indicators						
	i)	ii) [10^{23} Mx]	iii)	iv) [years]	v) [gauss]	vi)	vii)
Observed	–	1.520	–	2000.04	−5.47	–	–
Smoothed	–	0.895	–	–	–	–	0.945
0	0.941	1.314	0.934	1999.71	−14.49	0.688	0.693
A35	0.965	0.392	0.938	2001.19	−6.47	0.570	0.590
B200	0.968	1.146	0.951	1999.93	−9.85	0.684	0.709
C80	0.960	1.119	0.933	2000.08	−10.32	0.709	0.711
D80	0.954	1.147	0.931	2000.08	−10.88	0.687	0.708
E	0.967	0.898	0.898	2001.04	−8.36	0.673	0.695
F5	0.966	0.553	0.913	1999.19	−9.74	0.682	0.706
C80F10	0.967	0.702	0.966	1999.64	−8.65	0.706	0.726

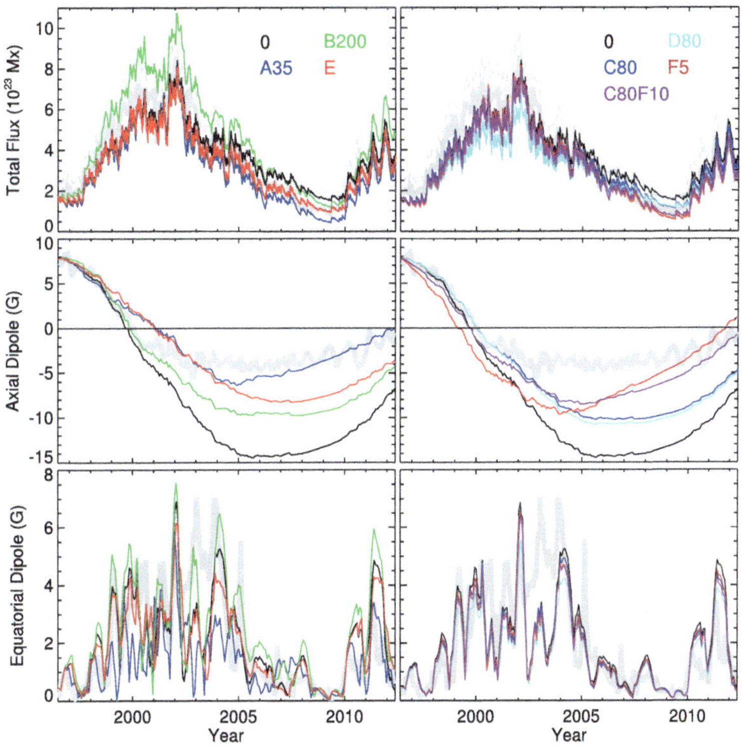

Figure 2 Time series of total unsigned photospheric flux [$\Phi_\odot$], axial dipole component [B_{ax}], and equatorial dipole component [B_{eq}] for the various surface calibration runs. Observed (NSO/KP) values are shown by thick-grey lines (in the top row, the dashed-grey line shows value from the original magnetograms and the solid-grey line that from smoothed magnetograms).

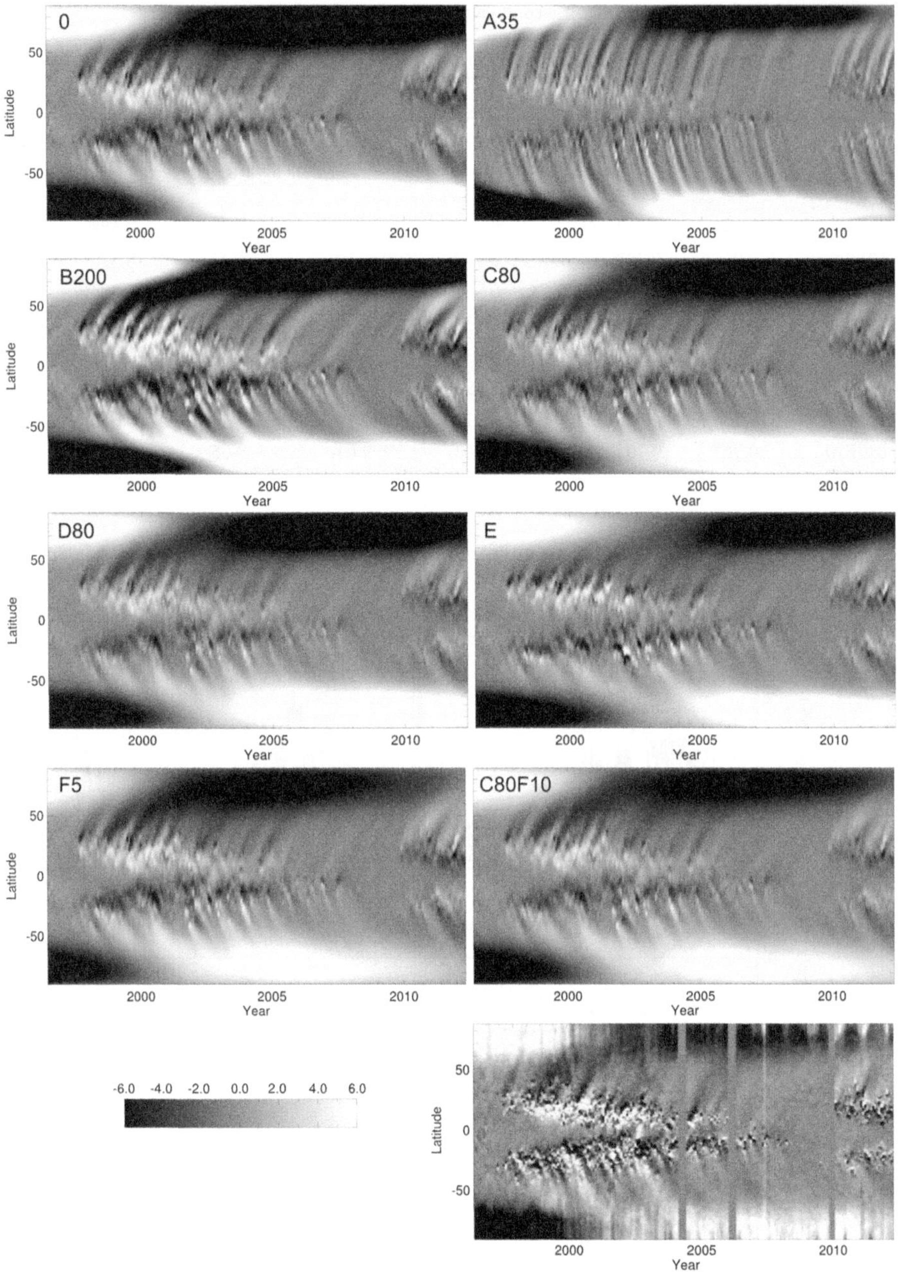

Figure 3 Magnetic butterfly diagrams showing the longitude-average $\langle B_{0r} \rangle$ for each calibration run, and for the observed KP magnetograms (bottom right). In each case the grey scale saturates at ± 6 G. Data gaps and problems with high-latitude measurements are evident in the observed butterfly diagram.

3.1. Reference Case

Consider first the reference case Run 0, with peak meridional flow $u_0 = 11\ \mathrm{m\,s^{-1}}$ (consistent with Cycle 23 observations by Hathaway and Rightmire, 2010) and diffusivity $D = 450\ \mathrm{km^2\,s^{-1}}$. The time series for Run 0 are shown by black curves in all panels of Figure 2, while corresponding values for the observed magnetograms are shown by thick grey lines. It is clear that $\Phi_\odot$ is too high from about 2004 onwards, during the declining phase. This is clearly caused by an excess in B_{ax} rather than B_{eq}. Indeed, the butterfly diagram (Figure 3) shows that Run 0 builds up unrealistically large polar caps. The low polar field of Cycle 23 is not reproduced by this standard model. A second inconsistency in Run 0 is a deficit in the equatorial dipole strength [B_{eq}] around 2002 – 2004. This is present in all of the calibration runs and appears to be due to an underestimate of emerged flux in a handful of particular active regions near the Equator, to which B_{eq} is sensitive. This is likely caused by our simplified method of extracting bipolar regions, but fortunately does not have a significant effect on $\Phi_\odot$ or on the later polar field produced.

3.2. Meridional Flow Speed

A possible solution to the excess-polar-field problem is to retain the standard flux-transport model but to alter the meridional flow speed [u_0] or diffusivity [D]. Differential rotation is better constrained and does not significantly affect the dipole strengths (Baumann *et al.*, 2004). In order to reduce the polar-field production, one must increase u_0, so as to reduce cancellation at the Equator (DeVore, Sheeley, and Boris, 1984). In Run A35, the required speed of $u_0 = 35\ \mathrm{m\,s^{-1}}$ is used to bring B_{ax} down to the observed level (Figure 2). But in addition to being inconsistent with observations (Hathaway and Rightmire, 2010), this causes a very late reversal of B_{ax}, and too low a level of B_{eq} over much of the cycle. The resulting butterfly diagram is also poorly correlated with observations (Figure 3).

Schrijver and Liu (2008) and Jiang *et al.* (2011) were able to improve the match of their simulations to observed dipole moments/polar fields by changing the meridional flow. However, when we run a simulation with our measured bipole properties and the diffusivity ($250\ \mathrm{km^2\,s^{-1}}$) and meridional-flow profile of Jiang *et al.* (2011), we find that the axial dipole strength is close to that of Run 0, even with their 55 % enhanced flow speed ($17\ \mathrm{m\,s^{-1}}$). Moreover, correlation with the observed butterfly diagram is again poor. If we implement the meridional-flow profile of Schrijver and Liu (2008), with a stronger latitudinal gradient in u_θ at the Equator, and $u_0 = 15\ \mathrm{m\,s^{-1}}$, we still obtain similar results. Note that neither of those studies used observations of individual bipolar magnetic regions, as we do here.

3.3. Supergranular Diffusivity

In Run B200, we instead reduce D to $200\ \mathrm{km^2\,s^{-1}}$. This produces slightly high $\Phi_\odot$ compared with the observations throughout the cycle, although the butterfly diagram and B_{eq} are a better fit to the observations. In the declining phase, B_{ax} is still rather high. Here we note a discrepancy with the parameter study by Baumann *et al.* (2004) for our u_θ profile and bipole properties: as we increase D, the peak axial dipole B_{ax} *increases*, whereas they found the maximum polar field strength to *decrease* over the same range of D.

3.4. Bipole Properties

Although our bipolar regions are constrained individually by observations, we also consider systematically modifying their properties. One can reduce the polar field either by reducing the bipole fluxes, or by reducing the average tilt angle so that each region contributes less to the axial dipole (Wang, Sheeley, and Lean, 2000). In Run C80, all tilt angles were reduced by 20 %, while in Run D80 all bipole fluxes were reduced by 20 %. Both runs produce a comparable improvement in B_{ax}, with accurate reversal times. Overall, Run C80 produces better results than Run D80, because the latter causes too much reduction in B_{eq} and consequently in $\Phi_{\odot}$, particularly during active periods. Jiang *et al.* (2011) found that a 28 % decrease in their tilt angles (as compared to previous cycles) produced a reasonable polar field in Cycle 23. However, even in Run C80 the peak of B_{ax} remains too strong compared to the observations, so we are led to consider an additional change to the model.

3.5. Exponential Flux Decay

In run F5, we try adding additional exponential decay terms of the form $-A_{0\theta}/\tau$ and $-A_{0\phi}/\tau$ to Equations (5) and (6), respectively, leading to an exponential decay on a timescale $\tau = 5$ years. Such a decay has previously been introduced by Schrijver, DeRosa, and Title (2002) in flux-transport simulations of the past 340 years, in order to maintain regular polar-field reversals when cycles vary in strength from one to the next. Baumann, Schmitt, and Schüssler (2006) have introduced a similar enhanced decay, proposing the physical explanation to be volume diffusion of the surface magnetic field in the three-dimensional solar interior. Figure 2 shows that this decay is able to bring B_{ax} down to the observed level in 2010 while maintaining a good correlation with the observed butterfly diagram, although it has the side-effect of bringing the reversal time of B_{ax} too early. By combining a weaker exponential decay term ($\tau = 10$ years) with 20 % reduced tilt angles (as in Run C80), Run C80F10 produces a better compromise and an even stronger correlation with the observed butterfly diagram. This is the run chosen for driving the coronal simulations in Section 4, and used by Yeates and Mackay (2012).

3.6. Time-Varying Meridional Flow

An alternative improvement to the flux-transport model is to introduce temporal and spatial fluctuations in the meridional flow. Either a high-latitude countercell (Jiang *et al.*, 2009) or converging flows towards active regions (Jiang *et al.*, 2010b) can reduce the resulting polar field. Although we do not carry out an exhaustive investigation here, we illustrate (Run E) the effect of such converging flows using the method of Cameron and Schüssler (2010). An axisymmetric perturbation is added to the steady flow u_θ to model the net effect of non-axisymmetric perturbations (DeRosa and Schrijver, 2006), and takes the form

$$u'(\lambda, t) = c_0\left(\frac{\mathrm{d}}{\mathrm{d}\lambda}\langle|B_{0r}|\rangle(\lambda, t)\right), \qquad (9)$$

where $\langle|B_{0r}|\rangle(\lambda, t)$ is the longitudinal average of $|B_{0r}|$, to which we also apply a Gaussian smoothing in latitude at each time. Following Cameron and Schüssler (2010) we set $c_0 = 10\ \mathrm{m\,s^{-1}\,gauss^{-1}\,deg}$. The Gaussian smoothing is implemented (at each time step of the main simulation) by taking 80 one-dimensional (in λ) diffusive steps with diffusion coefficient $2.4 \times 10^7\ \mathrm{km^2\,s^{-1}}$. Cameron and Schüssler (2010) show that this form of $u'(\lambda, t)$ leads naturally to variation of the $P_1^2(\cos\theta)$ component of the total meridional flow over the cycle. It thus reproduces both the observations of Hathaway and Rightmire (2010), who found a variation of the P_1^2 component using MDI tracking, and Basu and Antia (2010), who both substantiated the Hathaway and Rightmire result and found evidence for inflows towards active regions, using MDI helioseismology. For our bipole properties, Figure 2 shows that these flow perturbations lead to reasonable $\Phi_\odot$, and reasonable B_{ax} later in the cycle. But the reversal of B_{ax} is much later than observed, because poleward transport is reduced during the rising phase of the cycle. This leads also to poorer correlation with the observed butterfly diagram. Hence we do not pursue this route here, although temporal variations in meridional flow will be important to develop in the future, and particularly to take into account variations between different cycles (Wang, Lean, and Sheeley, 2002; Schrijver and Liu, 2008).

4. Flux Ropes and Eruptions

As an application of the NP model, we focus here on the formation and ejection of magnetic-flux ropes over the 15-year simulation, driven by the optimal flux transport Run C80F10. Twisted flux ropes form naturally in the simulation when surface motions concentrate magnetic helicity above polarity inversion lines in the photospheric field. Previously, we have developed automated techniques to detect them, and have analysed the effect of the various coronal-simulation parameters on their formation and ejection (Yeates and Mackay, 2009). We have also undertaken a detailed comparison with CME source regions observed in the extreme ultraviolet (Yeates *et al.*, 2010b). Yeates, Constable, and Martens (2010) looked at how the flux-rope statistics differed in different phases of Cycle 23, using six-month "snapshots". Here, we present the distribution of flux ropes and flux-rope ejections over the continuous 15-year simulation. This is intended to be an illustration of the model capabilities, rather than a detailed parameter study. In particular, the model is not yet suitable for prediction of individual space-weather events (see Section 5).

4.1. Definition and Automated Identification

Identifying magnetic-flux ropes in complex three-dimensional magnetic fields is a non-trivial task. We do this once per day during the course of the simulation, using a slight modification of the automated technique described by Yeates and Mackay (2009). The technique works by first identifying flux-rope *points* on the numerical grid satisfying certain criteria, before grouping these points into flux *ropes* using a clustering algorithm. We have settled on the following criteria for defining flux-rope points. At each point on the computational grid, we compute the normalised vertical magnetic-tension force and pressure gradient,

$$T_r = \frac{\mathrm{R}_\odot}{B_0^2}\mathbf{B}_0 \cdot \nabla\left(\frac{B_{0r}}{\mu_0}\right), \qquad P_r = -\frac{\mathrm{R}_\odot}{B_0^2}\frac{\partial}{\partial r}\left(\frac{B_0^2}{2\mu_0}\right). \tag{10}$$

A grid point (r_i, θ_j, ϕ_k) is then selected if it satisfies the following five conditions:

$$P_r(r_{i-1}, \theta_j, \phi_k) < -0.4, \tag{11}$$

$$P_r(r_{i+1}, \theta_j, \phi_k) > 0.4, \tag{12}$$

$$T_r(r_{i-1}, \theta_j, \phi_k) > 0.4, \tag{13}$$

$$T_r(r_{i+1}, \theta_j, \phi_k) < -0.4, \tag{14}$$

$$|\mathbf{j}_0 \cdot \mathbf{B}_0| > \alpha^* B_0^2, \tag{15}$$

where $\alpha^* = 0.7 \times 10^{-8}\ \mathrm{m}^{-1}$. Thus a flux rope is defined as a twisted structure where magnetic pressure acts outward from the flux-rope axis and magnetic tension acts inward (Figure 4). As an illustration, Figure 5 shows four snapshots of the corona in Run C80F10, with red/orange magnetic-field lines traced from identified flux-rope points.

Figure 4 Flux-rope points are identified by an inward magnetic tension force $[T_r]$ and an outward magnetic pressure force $[P_r]$.

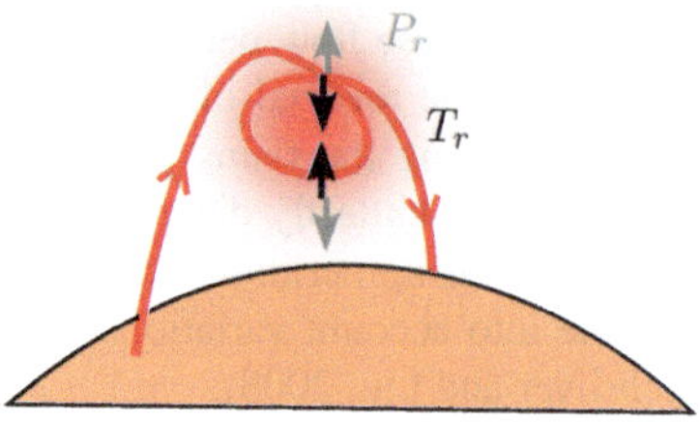

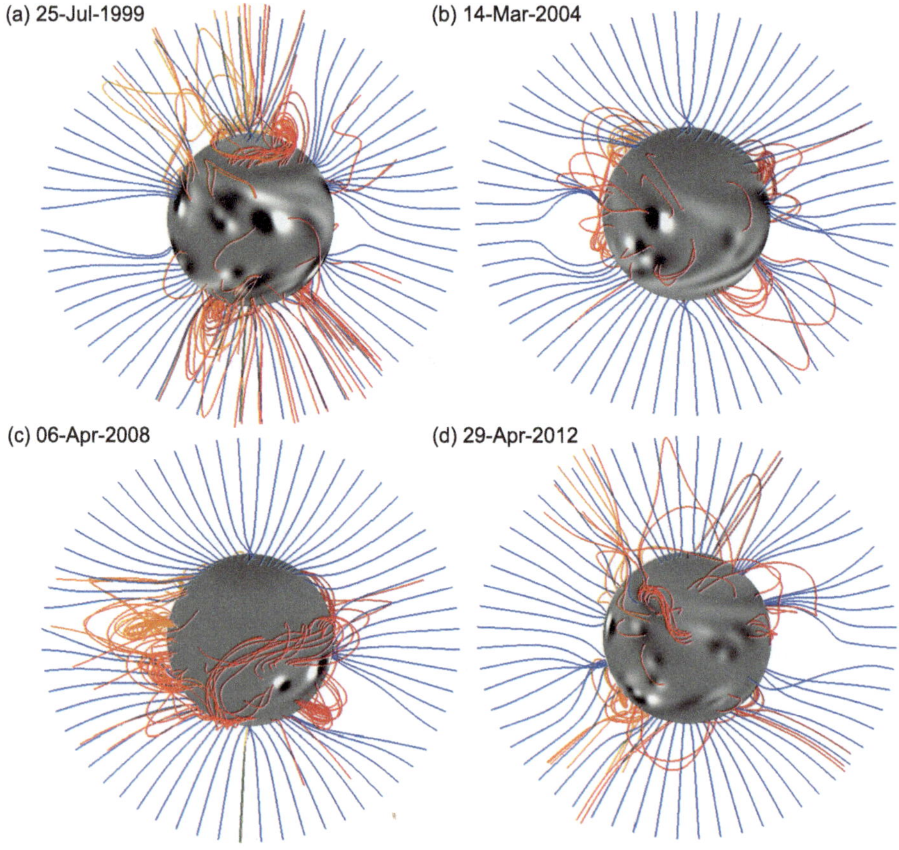

Figure 5 Snapshots of the NP model at a sequence of times during Run C80F10, including (a) Cycle 23 Maximum prior to polar reversal, (b) the declining phase, (c) Cycle 23 Minimum, and (d) Cycle 24. In each case, blue field lines are traced down from the source surface $r = 2.5\mathrm{R}_\odot$, while red field lines are traced from (a subset of) flux-rope points.

To identify flux-rope ejections, we use the automated post-processing procedure of Yeates and Mackay (2009). This selects those flux-rope points with $v_{0r} > 0.5\ \mathrm{km\,s^{-1}}$ in the magneto-frictional code and clusters them both spatially and temporally into separate ejection events. To be classed as separate events, clusters must be separated by at least five days in time and have at least eight points (on the grid resolution used here). This yields a list of times, locations, and sizes of flux-rope ejections during the simulation.

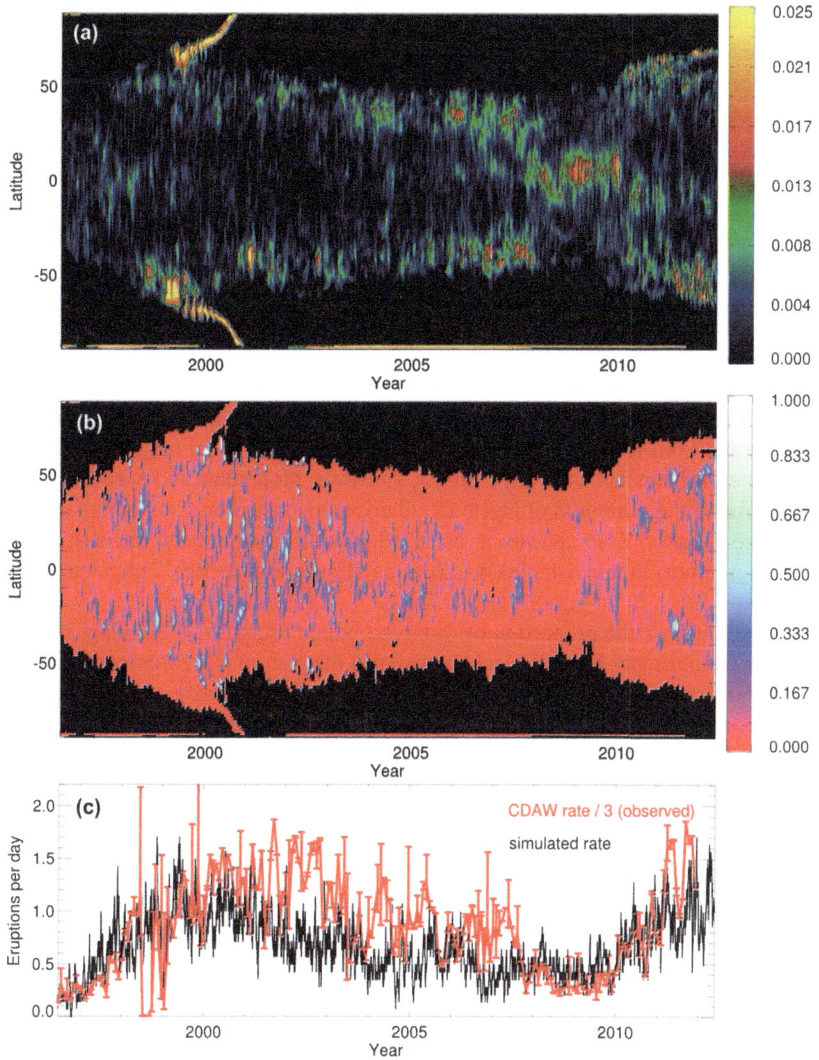

Figure 6 Latitude–time distribution of (a) flux ropes and (b) flux-rope eruptions, and (c) time series of the flux-rope eruption rate. In (a), the colour scale shows flux rope filling factor, while in (b), the colour scale shows the fraction of flux-rope points erupting in a given latitude–time bin. Bins with no flux-rope points are coloured black. In (c), the observed CME rate from the CDAW catalogue divided by three is plotted in red (see text).

4.2. Latitude–Time Distribution

At any one time, there are multiple flux ropes, of varying sizes, present in the simulation. To show how these vary over latitude and time, Figure 6 summarises the results of the automated detection routines applied to Run C80F10.

Figure 6(a) shows the distribution of flux ropes in time and latitude. The quantity plotted is the *filling factor* of flux ropes, namely the proportion of grid points at each latitude and time that are identified as flux-rope points. There are two features to notice about the flux-rope distribution. Firstly, the latitude range of flux ropes reflects the extent of polarity inversion lines (PILs) on the solar surface, and varies over the cycle. The distribution of PILs reaches its broadest extent during the "rush-to-the-poles" in the run up to polar field reversal in 2000 – 2001, after which high-latitude flux ropes disappear again. The second feature to notice is that the filling factor is greater outside of active regions, either at high latitudes or during the Minimum period from 2007 – 2010 when there were few active regions present. This anti-correlation with the magnetic butterfly diagram fits the picture of flux ropes forming as a result of gradual transport and build-up of helicity by surface motions. In fact, the number of individual flux ropes does increase by a factor of about 1.5 during solar maximum (Yeates, Constable, and Martens, 2010), but this is because a greater number of smaller flux ropes are present in and around active regions. Larger ropes are found in decaying flux regions where helicity has had time to concentrate.

This observed distribution of flux ropes shares some similarities with observed butterfly diagrams of solar filaments, which also overlie PILs (d'Azambuja and d'Azambuja, 1948; Mouradian and Soru-Escaut, 1994). Namely, the overall latitudinal range and the increase in the number of filaments with solar activity. However, Mouradian and Soru-Escaut (1994) find that, around the minimum of Cycle 21, filaments persisted only at higher latitudes and not near the Equator. This appears to be at odds with the simulated distribution of flux ropes in Figure 6(a) in the years 2008 – 2010, as seen in Figure 5(c). This requires further investigation.

Figure 6(b) shows the proportion of flux-rope points found to be erupting, within each latitude–time bin. Empty bins are black. Notice that the pattern is predominantly the inverse of Figure 6(a), and now correlates with the magnetic butterfly diagram. So we find that, although less space is filled with flux ropes at active latitudes, the flux ropes that are present at active latitudes are more likely to erupt. Hence the eruption rate is modulated by solar activity (black curve in Figure 6c). This agrees with the findings of Yeates, Constable, and Martens (2010), who found that the flux-rope eruption rate increased by a factor of eight between 1996 and 1999, comparable to the relative increase of the bipole emergence rate, or of the total magnetic energy.

Note that the rate in 1996 may be underestimated – both in Figure 6 and in the earlier simulation – due to a lack of helicity stored in the initial configuration. Yeates and Mackay (2012) show how the helicity on high-latitude PILs depends not only on *in-situ* generation by differential rotation, but also on the gradual poleward transport of helicity from lower latitudes. Thus there is a time delay for the build up of non-potential structure in the corona. Indeed the subsequent Cycle 23 minimum (2007 – 2009) shows a higher floor in the eruption rate of roughly 0.4 per day. So if the simulation were initialised earlier, we might expect a higher eruption rate in 1996. This highlights the importance of long-term memory in the coronal magnetic topology. The gradual poleward transport of helicity also explains the relatively infrequent eruptions at higher latitudes. While such eruptions are present, for example around the "detachment" of the polar crowns in 2000, these structures tend to encircle much of the Sun. Once helicity is removed in an eruption, it takes time to build again.

5. Outlook for Space Weather Prediction

Flux-rope eruptions in the NP model can give us insight into the origin of CMEs, although further development is needed before the model can enter the realm of operational space-weather prediction. In the first place, the peak eruption rate of 1.5 per day in the simulation is substantially lower than the rate of observed CMEs, by approximately a factor of three when compared with the manually compiled Coordinated Data Analysis Workshops catalogue (CDAW: Yashiro *et al.*, 2004), which is based on data from *Large Angle and Spectrometric Coronagraph* (LASCO: Brueckner *et al.*, 1995). The CDAW rate, divided by three, is shown by the red curve in Figure 6(c). We have filtered out CMEs with apparent width less than 15° or greater than 270°, and determined error bars by taking into account data gaps (St. Cyr *et al.*, 2000; Yeates, Constable, and Martens, 2010).

The parameter study of Yeates and Mackay (2009) shows that the eruption rate may be increased either by reducing the coronal diffusivity in the simulation (equivalent to η_4 here), or by increasing the hemispheric twist imbalance of the emerging bipolar regions. However, the latter is constrained by observations of filament chirality (Yeates, Mackay, and van Ballegooijen, 2008), and for reasonable values of these parameters the eruption rate remains too low.

Many missed eruptions are likely in active regions. In the present simulations, emerging regions are treated as idealised bipoles, and the effect of this simplification is likely to be most pronounced during the early stages of an active region's lifetime. Our model was designed to follow the long-term evolution of magnetic helicity, and cannot reproduce multiple eruptions in rapid succession from a single active region. Nor are the idealised bipoles a spatially accurate model for more complex δ-spot regions. These limitations prevent us from making either spatial or temporal predictions of individual CMEs (Yeates *et al.*, 2010b).

Nevertheless, more detailed modelling of active-region structure and emergence could in principle be incorporated into the NP model in future. The difficulty is that one needs a description of the three-dimensional structure of emerging regions. The solution may be either to impose time-dependent electric fields on the solar surface (Fan and Gibson, 2007; Fisher *et al.*, 2010), or to locally drive the simulation from higher-cadence magnetograms in place of the flux-transport model (Mackay, Green, and van Ballegooijen, 2011; Cheung and DeRosa, 2012), although these techniques are still under development.

We believe that the time-dependent NP model is worth pursuing because the inherent magnetic memory in the corona implies a degree of predictability. While flares or CMEs from newly emerged active regions are unlikely to be predicted more than a day or two before the active region appears on the surface, CMEs originating from decaying active regions offer scope for longer-term prediction using models such as this. Finally, it should be noted that the NP model includes only the magnetic field and not other plasma properties. Hence it can predict only the initiation of flux-rope ejections, not their subsequent dynamical evolution. For individual events, this evolution can be followed in MHD models (*e.g.* Manchester *et al.*, 2004).

Acknowledgements The author thanks NSO for support to attend the 26th Sac Peak workshop, D.H. Mackay for useful discussions during this research, and the anonymous referee for helpful suggestions. Numerical simulations used the STFC and SRIF funded UKMHD cluster at the University of St Andrews. Magnetogram data from NSO/Kitt Peak were produced cooperatively by NSF/NSO, NASA/GSFC, and NOAA/SEL, and SOLIS data are produced cooperatively by NSF/NSO and NASA/LWS. The observed CME catalogue used is generated and maintained at the CDAW Data Center by NASA and The Catholic University of America in cooperation with the NRL. SOHO is a project of international cooperation between ESA and NASA.

Appendix A: Computational Grid

Our computational grid is divided into latitudinal sub-blocks, each of which has uniform spacing in the stretched variables

$$x = \phi/\Delta, \qquad y = -\log\bigl(\tan(\theta/2)\bigr)/\Delta, \qquad z = \log(r/\mathrm{R}_\odot)/\Delta \tag{16}$$

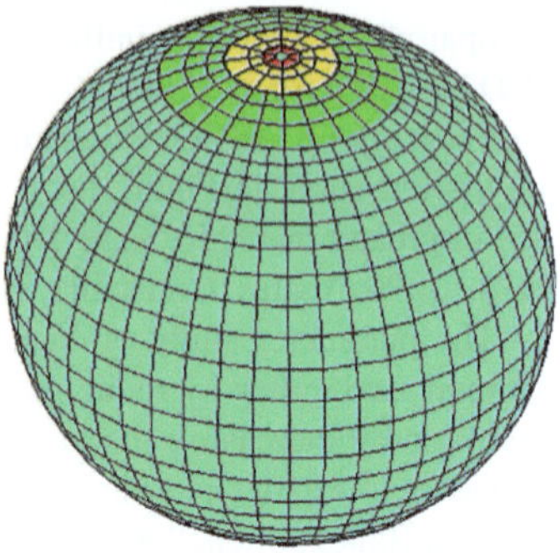

Figure 7 Example of the variable grid with 48 cells at the Equator (compared to 192 in the actual simulations) and seven sub-blocks (compared to nine).

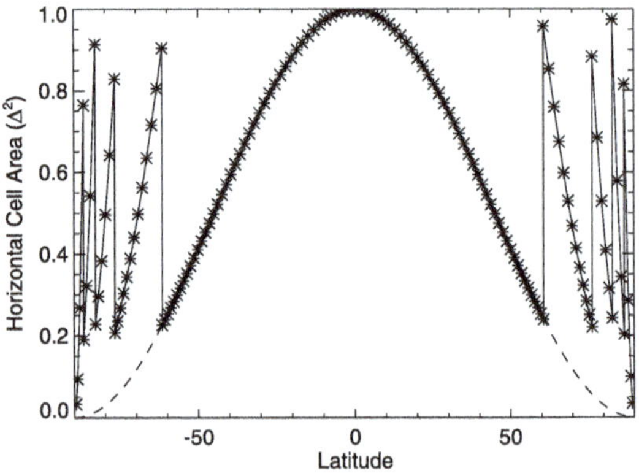

Figure 8 Horizontal cell area (relative to that at the Equator) as a function of latitude, where symbols (joined by solid lines) show the variable grid and dashed lines a uniform grid with $dx = dy = 1$ everywhere.

(van Ballegooijen, Priest, and Mackay, 2000), where Δ is the equatorial grid spacing in longitude [ϕ]. The horizontal cell-sizes are $dx = dy = 1$ for the equatorial sub-block, and double in each sub-block towards the poles (Figure 7). The vertical cell size is $dz = 1$ for all sub-blocks. This introduction of sub-blocks with different spacing counters the problem of grid convergence toward the poles, since the horizontal cell area is $\Delta^2 r^2 \sin^2\theta\, dx dy$. The sub-block boundaries in latitude are determined so that each grid cell is as large in horizontal area as possible, while never exceeding the equatorial cell area $\Delta^2 r^2$ (Figure 8). With a longitudinal resolution of 192 cells at the Equator and 12 at the poles, there are nine sub-blocks. (The number 12 is chosen due to the parallel architecture used.) The total number of grid cells in (x, y) is 18 936, as compared to 63 744 for a uniform, single-block grid with unit spacing.

A complication arising from the variable grid is that the different sub-blocks need to communicate ghost values of B_{0r} and $B_{0\phi}$ with one another at each timestep. This is analogous to the inter-level communications in Adaptive Mesh Refinement (AMR) codes, except that our grid is fixed in time. Restriction (from fine to coarse sub-blocks) is the simpler process, for which we use an area-weighted average over fine grid cells (Balsara, 2001). Prolongation (from coarse to fine sub-blocks) is trickier since the coarse-block solution must be interpolated to get the fine-block ghost values at intermediate locations. We follow the Taylor expansion method of van der Holst and Keppens (2007), using monotonic van Leer slope estimates (Evans and Hawley, 1988). These slopes are also computed throughout the grid and used for slope-limiting in the advection terms, in order to prevent spurious oscillations near sharp gradients. Our numerical tests indicate that numerical diffusion due to this "upwinding" is negligible compared to the physical diffusion D. Finally, after computing the update $\partial \mathbf{A}_0/\partial t$ within each sub-block, we replace boundary values on coarser sub-blocks with those derived from finer sub-blocks. This is analogous to the "flux correction" of Berger and Colella (1989).

Appendix B: Global Boundary Conditions

The staggered grid requires ghost-cell values of two components of $\mathbf{B}_0$ outside each boundary. In longitude the domain is simply periodic. On the photosphere $r = \mathrm{R}_\odot$, we fix ghost cell values of $B_{0\theta}$, $B_{0\phi}$ by requiring that $v_{0r} = 0$ on $r = \mathrm{R}_\odot$ in the magneto-frictional model. On the outer boundary $r = 2.5\mathrm{R}_\odot$, we impose a radial outflow velocity (see Yeates *et al.*, 2010a) that models the effect of the solar wind radially extending field lines, while still allowing horizontal fields to escape during flux-rope ejections. The resulting ghost-cell values of $B_{0\theta}$, $B_{0\phi}$ do not play an important role and are simply set by zero-gradient conditions. On the latitudinal boundaries (located at approximately $\pm 89.33°$ latitude) we need ghost cell values for B_{0r} and $B_{0\phi}$. The latter are simply set to zero, while the ghost values of B_{0r} are chosen to satisfy Stokes' Theorem given the integral of $A_{0\phi}$ around the latitudinal boundary.

References

Altschuler, M.D., Newkirk, G.: 1969, *Solar Phys.* **9**, 131. doi:10.1007/BF00145734.
Arge, C.N., Pizzo, V.J.: 2000, *J. Geophys. Res.* **105**, 10465. doi:10.1029/1999JA900262.
Balsara, D.S.: 2001, *J. Comput. Phys.* **174**, 614.
Basu, S., Antia, H.M.: 2010, *Astrophys. J.* **717**, 488.
Baumann, I., Schmitt, D., Schüssler, M.: 2006, *Astron. Astrophys.* **446**, 307.
Baumann, I., Schmitt, D., Schüssler, M., Solanki, S.K.: 2004, *Astron. Astrophys.* **426**, 1075.
Berger, M.J., Colella, P.: 1989, *J. Comput. Phys.* **82**, 64.
Bhattacharjee, A., Hameiri, E.: 1986, *Phys. Rev. Lett.* **57**, 206.
Boozer, A.H.: 1986, *J. Plasma Phys.* **35**, 133.
Brueckner, G.E., Howard, R.A., Koomen, M.J., Korendyke, C.M., Michels, D.J., Moses, J.D., Socker, D.G., Dere, K.P., Lamy, P.L., Llebaria, A., Bout, M.V., Schwenn, R., Simnett, G.M., Bedford, D.K., Eyles, C.J.: 1995, *Solar Phys.* **162**, 357. doi:10.1007/BF00733434.
Cameron, R.H., Schüssler, M.: 2010, *Astrophys. J.* **720**, 1030.
Cameron, R.H., Jiang, J., Schmitt, D., Schüssler, M.: 2010, *Astrophys. J.* **719**, 264.
Cheung, M.C.M., DeRosa, M.L.: 2012, *Astrophys. J.* **757**, 147.
Cook, G.R., Mackay, D.H., Nandy, D.: 2009, *Astrophys. J.* **704**, 1021.
Craig, I.J.D., Sneyd, A.D.: 1986, *Astrophys. J.* **311**, 451.
Dasi-Espuig, M., Solanki, S.K., Krivova, N.A., Cameron, R., Peñuela, T.: 2010, *Astron. Astrophys.* **518**, A7.
d'Azambuja, L., d'Azambuja, M.: 1948, *Ann. Obs. Meudon* **6** (Fasc. VII).
DeRosa, M.L., Schrijver, C.J.: 2006, In: Fletcher, K. (ed.) *Proceedings of SOHO 18/GONG 2006/HELAS I, Beyond the Spherical Sun* **SP-624**, ESA, Noordwijk.
DeRosa, M.L., Schrijver, C.J., Barnes, G., Leka, K.D., Lites, B.W., Aschwanden, M.J., Amari, T., Canou, A., McTiernan, J.M., Régnier, S., Thalmann, J.K., Valori, G., Wheatland, M.S., Wiegelmann, T., Cheung, M.C.M., Conlon, P.A., Fuhrmann, M., Inhester, B., Tadesse, T.: 2009, *Astrophys. J.* **696**, 1780.
DeVore, C.R., Sheeley, N.R. Jr., Boris, J.P.: 1984, *Solar Phys.* **92**, 1. doi:10.1007/BF00157230.
Downs, C., Roussev, I.I., van der Holst, B., Lugaz, N., Sokolov, I.V., Gombosi, T.I.: 2010, *Astrophys. J.* **712**, 1219.
Evans, C.R., Hawley, J.F.: 1988, *Astrophys. J.* **332**, 659.
Fan, Y., Gibson, S.E.: 2007, *Astrophys. J.* **668**, 1232.
Fisher, G.H., Welsch, B.T., Abbett, W.P., Bercik, D.J.: 2010, *Astrophys. J.* **715**, 242.
Hathaway, D.H., Rightmire, L.: 2010, *Science* **327**, 1350.
Hoeksema, J.T., Wilcox, J.M., Scherrer, P.H.: 1983, *J. Geophys. Res.* **88**, 9910.
Jiang, J., Cameron, R., Schmitt, D., Schüssler, M.: 2009, *Astrophys. J. Lett.* **693**, L96.
Jiang, J., Cameron, R., Schmitt, D., Schüssler, M.: 2010a, *Astrophys. J.* **709**, 301.
Jiang, J., Işik, E., Cameron, R.H., Schmitt, D., Schüssler, M.: 2010b, *Astrophys. J.* **717**, 597.

Jiang, J., Cameron, R.H., Schmitt, D., Schüssler, M.: 2011, *Space Sci. Rev.* **136**.
Leighton, R.B.: 1964, *Astrophys. J.* **140**, 1547.
Lionello, R., Linker, J.A., Mikić, Z.: 2009, *Astrophys. J.* **690**, 902. doi:10.1088/0004-637X/690/1/902.
Luhmann, J.G., Li, Y., Arge, C.N., Gazis, P.R., Ulrich, R.: 2002, *J. Geophys. Res.* **107**, 1154.
Mackay, D.H., van Ballegooijen, A.A.: 2001, *Astrophys. J.* **560**, 445.
Mackay, D.H., van Ballegooijen, A.A.: 2005, *Astrophys. J. Lett.* **621**, L77.
Mackay, D.H., van Ballegooijen, A.A.: 2006, *Astrophys. J.* **641**, 577.
Mackay, D.H., Yeates, A.R.: 2012, *Living Rev. Solar Phys.* **9**(6). http://solarphysics.livingreviews.org/Articles/lrsp-2012-6/. doi:10.12942/lrsp-2012-6.
Mackay, D.H., Gaizauskas, V., van Ballegooijen, A.A.: 2000, *Astrophys. J.* **544**, 1122.
Mackay, D.H., Green, L.M., van Ballegooijen, A.: 2011, *Astrophys. J.* **729**, 97.
Manchester, W.B., Gombosi, T.I., Roussev, I., Ridley, A., De Zeeuw, D.L., Sokolov, I.V., Powell, K.G., Tóth, G.: 2004, *J. Geophys. Res.* **109**, 2107.
Mouradian, Z., Soru-Escaut, I.: 1994, *Astron. Astrophys.* **290**, 279.
Pizzo, V., Millward, G., Parsons, A., Biesecker, D., Hill, S., Odstrcil, D.: 2011, *Space Weather* **9**, 3004.
Riley, P.: 2007, *Astrophys. J. Lett.* **667**, L97.
Rušin, V., Druckmüller, M., Aniol, P., Minarovjech, M., Saniga, M., Mikić, Z., Linker, J.A., Lionello, R., Riley, P., Titov, V.S.: 2010, *Astron. Astrophys.* **513**, A45.
Savcheva, A.S., van Ballegooijen, A.A., DeLuca, E.E.: 2012, *Astrophys. J.* **744**, 78.
Schatten, K.H., Wilcox, J.M., Ness, N.F.: 1969, *Solar Phys.* **6**, 442. doi:10.1007/BF00146478.
Schrijver, C.J.: 2001, *Astrophys. J.* **547**, 475.
Schrijver, C.J., Liu, Y.: 2008, *Solar Phys.* **252**, 19. doi:10.1007/s11207-008-9240-6.
Schrijver, C.J., DeRosa, M.L., Title, A.M.: 2002, *Astrophys. J.* **577**, 1006.
Schüssler, M., Baumann, I.: 2006, *Astron. Astrophys.* **459**, 945.
Sheeley, N.R. Jr.: 2005, *Living Rev. Solar Phys.* **2**(5). http://solarphysics.livingreviews.org/Articles/lrsp-2005-5/. doi:10.12942/lrsp-2005-5.
Snodgrass, H.B.: 1983, *Astrophys. J.* **270**, 288.
St. Cyr, O.C., Plunkett, S.P., Michels, D.J., Paswaters, S.E., Koomen, M.J., Simnett, G.M., Thompson, B.J., Gurman, J.B., Schwenn, R., Webb, D.F., Hildner, E., Lamy, P.L.: 2000, *J. Geophys. Res.* **105**, 18169.
Stenflo, J.O., Kosovichev, A.G.: 2012, *Astrophys. J.* **745**, 129.
Su, Y., van Ballegooijen, A., Schmieder, B., Berlicki, A., Guo, Y., Golub, L., Huang, G.: 2009, *Astrophys. J.* **704**, 341.
van Ballegooijen, A.A.: 2004, *Astrophys. J.* **612**, 519.
van Ballegooijen, A.A., Cranmer, S.R.: 2008, *Astrophys. J.* **682**, 644.
van Ballegooijen, A.A., Martens, P.C.H.: 1989, *Astrophys. J.* **343**, 971.
van Ballegooijen, A.A., Priest, E.R., Mackay, D.H.: 2000, *Astrophys. J.* **539**, 983.
van der Holst, B., Keppens, R.: 2007, *J. Comput. Phys.* **226**, 925.
Wang, Y.-M., Sheeley, N.R.: 1989, *Solar Phys.* **124**, 81. doi:10.1007/BF00146521.
Wang, Y.-M., Sheeley, N.R.: 2002, *J. Geophys. Res.* **107**, 1302.
Wang, Y.-M., Lean, J., Sheeley, N.R.: 2002, *Astrophys. J. Lett.* **577**, L53.
Wang, Y.-M., Sheeley, N.R., Lean, J.: 2000, *Geophys. Res. Lett.* **27**, 621.
Willson, R.C., Hudson, H.S.: 1991, *Nature* **351**, 42.
Worden, J., Harvey, J.: 2000, *Solar Phys.* **195**, 247. doi:10.1023/A:1005272502885.
Yang, W.H., Sturrock, P.A., Antiochos, S.K.: 1986, *Astrophys. J.* **309**, 383.
Yashiro, S., Gopalswamy, N., Michalek, G., St. Cyr, O.C., Plunkett, S.P., Rich, N.B., Howard, R.A.: 2004, *J. Geophys. Res.* **109**, 7105.
Yeates, A.R., Mackay, D.H.: 2009, *Astrophys. J.* **699**, 1024.
Yeates, A.R., Mackay, D.H.: 2012, *Astrophys. J. Lett.* **753**, L34.
Yeates, A.R., Constable, J.A., Martens, P.C.H.: 2010, *Solar Phys.* **263**, 121. doi:10.1007/s11207-010-9546-z.
Yeates, A.R., Mackay, D.H., van Ballegooijen, A.A.: 2007, *Solar Phys.* **245**, 87. doi:10.1007/s11207-007-9013-7.
Yeates, A.R., Mackay, D.H., van Ballegooijen, A.A.: 2008, *Solar Phys.* **247**, 103. doi:10.1007/s11207-007-9097-0.
Yeates, A.R., Mackay, D.H., van Ballegooijen, A.A., Constable, J.A.: 2010a, *J. Geophys. Res.* **115**, 9112.
Yeates, A.R., Attrill, G.D.R., Nandy, D., Mackay, D.H., Martens, P.C.H., van Ballegooijen, A.A.: 2010b, *Astrophys. J.* **709**, 1238.
Zhao, X., Hoeksema, J.T.: 1995, *J. Geophys. Res.* **100**, 19.

DOI 10.1007/978-1-4939-1182-0_15
Reprinted from *Solar Physics* Journal, DOI 10.1007/s11207-013-0392-7

Different Periodicities in the Sunspot Area and the Occurrence of Solar Flares and Coronal Mass Ejections in Solar Cycle 23 – 24

D.P. Choudhary · J.K. Lawrence · M. Norris · A.C. Cadavid

Received: 30 August 2012 / Accepted: 12 August 2013 / Published online: 3 October 2013

Abstract In order to investigate the relationship between magnetic-flux emergence, solar flares, and coronal mass ejections (CMEs), we study the periodicity in the time series of these quantities. It has been known that solar flares, sunspot area, and photospheric magnetic flux have a dominant periodicity of about 155 days, which is confined to a part of the phase of the solar cycle. These periodicities occur at different phases of the solar cycle during successive phases. We present a time-series analysis of sunspot area, flare and CME occurrence during Cycle 23 and the rising phase of Cycle 24 from 1996 to 2011. We find that the flux emergence, represented by sunspot area, has multiple periodicities. Flares and CMEs, however, do not occur with the same period as the flux emergence. Using the results of this study, we discuss the possible activity sources producing emerging flux.

Keywords Sunspot area · Flare · CME · Periodicity

1. Introduction

Coronal mass ejections (CME) and solar flares occur as a result of destabilization of magnetic structures on the Sun. While solar flares are localized phenomena, CMEs occur as larger structures, which may include trans-equatorial magnetic field structures (Khan and Hudson, 2000; Pevtsov, 2004). The number of these events follows the rise and decay of the solar cycle (Gopalswamy, 2008). In the past three decades, there have been extensive studies to find the periodicities of solar flares shorter than 11 years. Solar flares observed in γ-rays, X-rays, and Hα are found to occur with dominant periodicities of about 154 days (Rieger *et al.*, 1984; Bai and Sturrock, 1987; Ichimoto *et al.*, 1985). The long-duration flares, which cause many of the interplanetary proton events, also show periodicities near 155 days

Solar Origins of Space Weather and Space Climate
Guest Editors: I. González Hernández, R. Komm, and A. Pevtsov

D.P. Choudhary (✉) · J.K. Lawrence · M. Norris · A.C. Cadavid
Department of Physics and Astronomy, California State University Northridge, 18111 Nordhoff St., Northridge, CA 91330, USA
e-mail: debiprasad.choudhary@csun.edu

(Antalova, 1994). It has also been found that the periodicity of occurrence of solar flares changes between different solar cycles. For example, the flare periodicity was 154 days in Solar Cycle 19 – 21 and 129 days in Cycle 23. Cycle 23 also showed another additional periodicity of 33.5 days (Bai and Cliver, 1990; Bai, 2003). The influence of periodic occurrences of solar explosive events has been observed in various interplanetary phenomena such as anomalous cosmic-ray fluxes (Hill, Hamilton, and Krimigis, 2001). The periodicity of 153 days in the interplanetary magnetic field and solar-wind speed at 1 AU, and 150 days for various interplanetary phenomena such as solar energetic-particle events and interplanetary coronal mass ejections, are found to be intermittent (Cane, Richardson, and von Rosenvinge, 1998; Richardson and Cane, 2004). Several midterm periodicities in solar phenomena have been found with periods of 100 to 500 days among which the 155-day sunspot area during Solar Cycles 12 – 21 is prominent (Lean and Brueckner, 1989; Lean, 1990). Most of the powerful solar flares that produce interplanetary proton events are associated with emergence of magnetic flux (Priest, 1990; Choudhary, Ambastha, and Ai, 1998; Forbes, 2000). Not surprisingly, a 158-day periodicity in sunspot area and a near-160-day periodicity in photospheric magnetic flux were observed in Cycle 21 (Oliver, Ballester, and Baudin, 1998; Ballester, Oliver, and Carbonell, 2002). However, the periodicity of flux emergence occurs only for a part of the solar cycle and may not repeat during the next cycle (Oliver, Ballester, and Baudin, 1998). Among various possibilities, such periodicities in sunspot area, magnetic-flux emergence, and flare occurrence could be attributed to a Rossby-type-wave induced variation of the magnetic field in the solar surface layer (Lou, 2000).

Even though CMEs occur on large magnetic structures, many of them are triggered by similar phenomena as for solar flares, which include magnetic-flux emergence (Choudhary, Srivastava, and Gosain, 2002). Flux emergence and explosion of magnetic structures are the common features in CME and flares. Recent studies have explored the periodicities in CME occurrence using continuous observations by the *Large Angle and Spectrometric Coronagraph* experiment onboard the *Solar and Heliospheric Observatory* (SOHO/LASCO). For example Lou *et al.* (2003) have found several periodicities in four years of data around the peak of Cycle 23. The CME periodicity of 193 days is found using a longer data set of ten years (Lara *et al.*, 2008). The evidence of periodicity in the CME mass measurement from 2003 to 2009 is observed although not very clearly (Vourlidas *et al.*, 2010, 2011).

Relationships between solar flares and CMEs are not well understood although both events occur due to the destabilization of magnetic field structures (Gosling, 1993). While the flares, sunspot areas, and amount of emerging flux can be observed only on the solar disk, the CMEs from the source regions situated on the farside of the Sun can also be observed. Time-series analysis of the flare, CME, and magnetic field distributions represented by sunspot area would help in understanding the underlying physical processes that produce magnetic field structures above solar photosphere, which may cause the relationship between flares and CMEs. In this study we present a wavelet analysis of the sunspot area and the occurrence of CMEs and solar flares.

2. Data Analysis

We use three time series constructed from daily sunspot area, daily flare counts, and CME counts to study the inter-relationship of their periodicity. The first is the corrected daily total sunspot area obtained from NASA's Marshall Space Flight Center (MSFC). This data set is prepared using the sunspot area from Royal Greenwich Observatory and USAF–NOAA

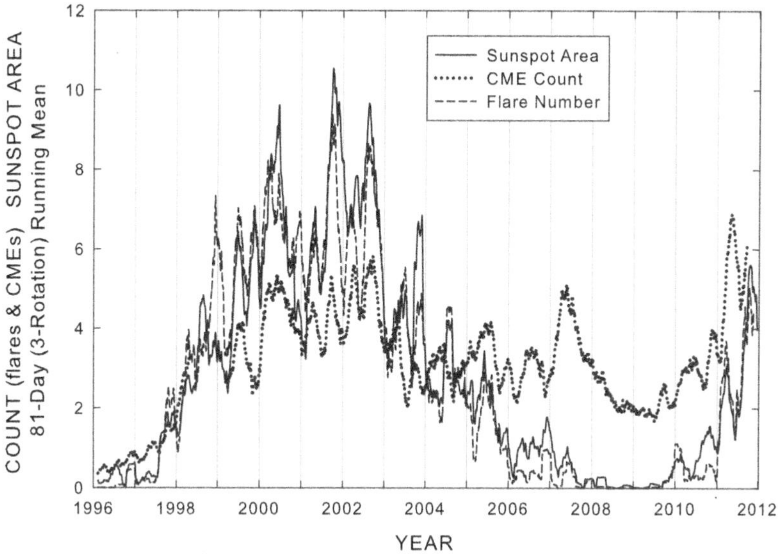

Figure 1 Sunspot area, daily number of flares of X-ray class, and daily CME count for 1996 to 2012.

observations. This uses the whole spot area (corrected for foreshortening) for each sunspot group observed on a given day and then adds them together. The more recent USAF data, from 1977 to the present, have been multiplied by 1.4 to bring them into line with the earlier Greenwich data (Hathaway, 2010). Here, we use the sunspot area as the representative of the solar surface magnetic flux. The second data set is the daily solar-flare count obtained by counting up all flares on a given day appearing in the observations by the *Geostationary Operational Environmental Satellite* (GOES) and stored in the data catalog umbra.nascom.nasa.gov/batse/images/help_batse_catalog.html. As noted earlier, the solar flares represent the explosive events occurred on the "Earth side" of the solar disk. The third data set is a series of daily counts of all CMEs observed by the SOHO/LASCO. The data were cataloged by Gopalswamy (2008). The CMEs are observed by coronagraphs. Therefore, we include the events that have source regions both "far and Earth side" of the solar disk.

The three time series are plotted together in Figure 1. Note the data gap in the CME series in 1998, when the SOHO spacecraft was out of communication. The smoothed time signals for flares, CMEs, and sunspot area in Figure 1 reveal the presence of strong quasi-periodicities. The data have been subjected to an 81-day (three-rotation) smoothing filter to remove fluctuations on the timescale of the solar rotation, but to leave longer ones. Let us consider the interval from 1999.25 to 2005.0. This begins with the end of the influence of the missing 1998 data from the SOHO spacecraft and extends through the main part of Solar Cycle 23. The background noise in these data, even before filtering, is red noise with increasing spectral power at lower frequencies, and this must be taken into consideration in estimates of statistical significance. When the data are pre-whitened by subtraction of a cubic fit, standard Fourier analysis using the Autosignal v1.7 software package reveals a period in the CME signal with a frequency 1.919 ± 0.008 year^{-1}, which is significant to 99.9 %, or about 3.3σ, as shown in Figure 2. This corresponds to a period of 190 ± 1 days. A periodicity of 193 days has been observed in the CME signal by Lara *et al.* (2008). Neither the sunspot area nor the flare frequency show any significant periodicities in this interval.

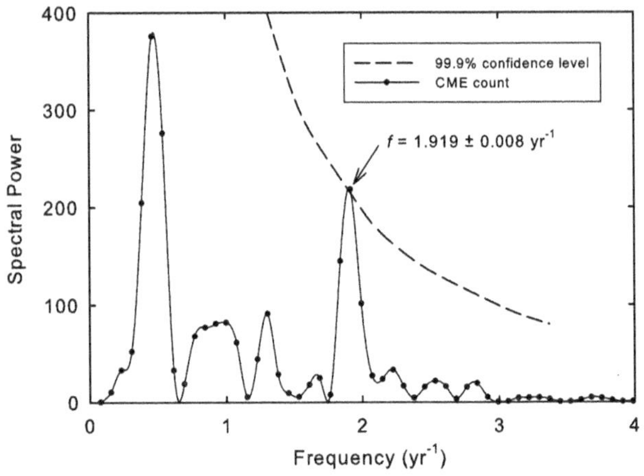

Figure 2 Fourier power spectrum of the CME occurrence data *versus* frequency in year^{-1}. The dashed line shows the 99.9 % confidence level based on a red-noise model.

If we confine our analysis to the interval from 2001.0 to 2004.0, including the peak and some of the decay of Cycle 23, we now find, by the same technique, a periodicity in the CME signal with frequency 2.005 ± 0.009 year^{-1}, which is significant to greater than 99.9 % or about 3.5σ. This corresponds to a period $T_{\mathrm{CME}} = 182 \pm 1$ days. The flare-count signal during this time shows a periodicity with frequency 2.301 ± 0.015 year^{-1} that is significant to 99 % or 2.6σ. This corresponds to a period in the flare occurrence of $T_{\mathrm{flare}} = 159 \pm 1$ days, and it appears to correspond to the 154-day Rieger periodicity associated with hard flares (Rieger *et al.*, 1984). Combining these two periods, which we have observed contemporaneously, gives a difference between the CME and flare periods of $\Delta T = 23 \pm 1$ days. Thus we can definitively reject the hypothesis that the CME and flare periodicities are the same.

To examine and compare the localized periodicities of the data confined to a part of the phase of solar cycle, we calculated wavelet spectra using the Autosignal version 1.7 software package. Wavelet analysis enables the simultaneous study of both the frequency content and the temporal dependence of a signal and is thus ideal for data series that are non-stationary such as ours (see Figure 1). We used a complex Morlet wavelet whose mother function is a plane wave modulated by a Gaussian envelope, and that is similar in some respects to a windowed Fourier transform. By adjusting parameters, it is possible to optimize the resolution either in frequency or in temporal resolution. We have chosen maximum frequency resolution within the limits imposed by the software. We have limited the frequency range of interest to run from 0.3125 year^{-1} (period 114 days) to 0.8333 year^{-1} (304 days). We also have limited the time interval studied to 1999.00 to 2012.00 to avoid the data gap.

While a casual inspection of the time series may indicate that the overall trend of these quantities is in phase, there are significant fine-scale differences that are detected in the wavelet analysis that yield periodograms. The sunspot area represents the spatial coverage and magnitude of magnetic flux on the Sun. Since the magnetic field is the prime source of energy for both solar flares, and CMEs, a strong correlation between these quantities is expected. There is remarkable departure from this expected correlation during the last solar minimum, around 2007, when the CME count was significantly higher than the sunspot area and flare numbers. On several occasions we notice that more flares occur when sunspot number increases but the number of CMEs decreases. An example of this correlation of sunspot number with flare count and anticorrelation with CME count is found in 2000. The occurrence of CMEs without the presence of strong-field regions represented by sunspots show that the physical mechanisms leading to these events may be fundamentally different from those required for solar flares.

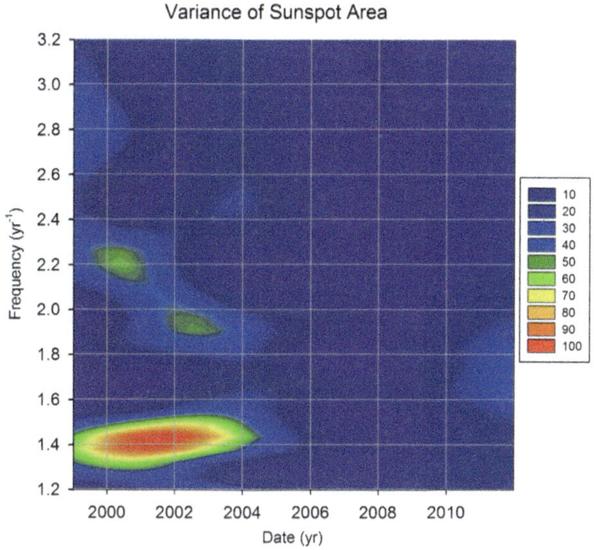

Figure 3 Wavelet power spectra for the daily sunspot area from 1999 to 2012. We have used a Morlet wavelet transform and selected the parameters to emphasize frequency resolution.

3. Results and Discussion

Our goal is to find out if the CMEs occur with any periodicity similar to the occurrence of solar flares and flux emergence, which has been observed earlier. The wavelet power spectra of the sunspot area, flares, and CMEs time series show the dominant periodicities and epoch of occurrence of these events. Figures 3, 4, and 5 show the wavelet power spectra for sunspot area, flare count, and CME numbers, respectively. The periodicity in sunspot area, which represents the appearance of strong magnetic flux at the solar surface, is between 158 – 182 days with a peak at about 178 days during the rising phase of the solar cycle peaking around 2001 – 2002 as seen in Figure 3. This is close to the periodicity of 152 – 160 days obtained earlier, using similar Morlet wavelet analysis, in which varying periodicities and their epoch of occurrence were observed in a different solar cycle (Oliver, Ballester, and Baudin, 1998). The solar flare count periodicity of about 150 days is observed around the solar maximum, which occurred around 2001 as seen in Figure 4. This is also consistent with periodicities in sunspot area, which represent the emergence of localized magnetic flux leading to solar flares (Lean, 1990). Here, we find that neither the epochs nor the periodicities in sunspot area and solar flare occurrence coincide during Solar Cycle 23. The CME spectrum presented in Figure 5 shows that in Cycle 23 there exists a six-month periodicity, which is consistent with the findings of Vourlidas *et al.* (2010, 2011). We also find that there exists a longer periodicity along with the six-month period in Cycle 23 and 150-day period in the rising phase of the current Cycle 24. It is well known that mostly long-duration solar flares are associated with a fraction of CMEs. In the current Cycle 24, however, we notice the periodicity of CME occurrence is about 154 days, which is close to the flare periodicity of several past cycles.

Figures 4 and 5 show that the periodic occurrence of solar flares and coronal mass ejections happen during the same epoch of Cycle 23, which was about the peak of the solar cycle in 2002. At the same time Figure 3 shows that the repetitive flux emergence occurred earlier around late 2000, the year when the rising phase of the Cycle 23 was established. There is a lag between the occurrence of explosive events and flux-emergence periodicity. The occur-

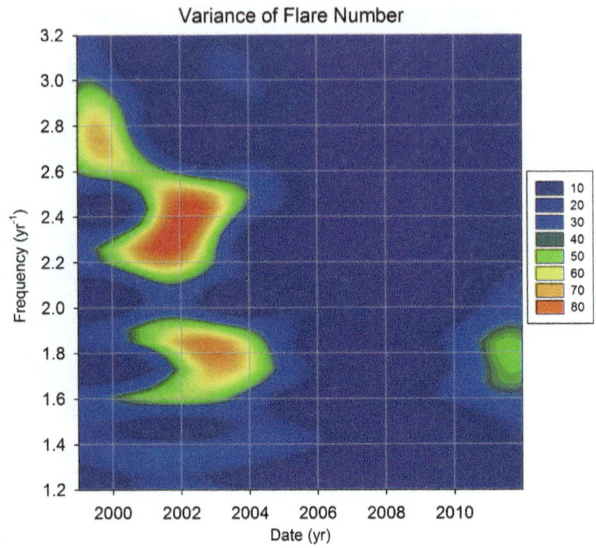

Figure 4 Wavelet power spectra for the daily flare count from 1999 to 2012. We have used a Morlet wavelet transform and selected the parameters to emphasize frequency resolution.

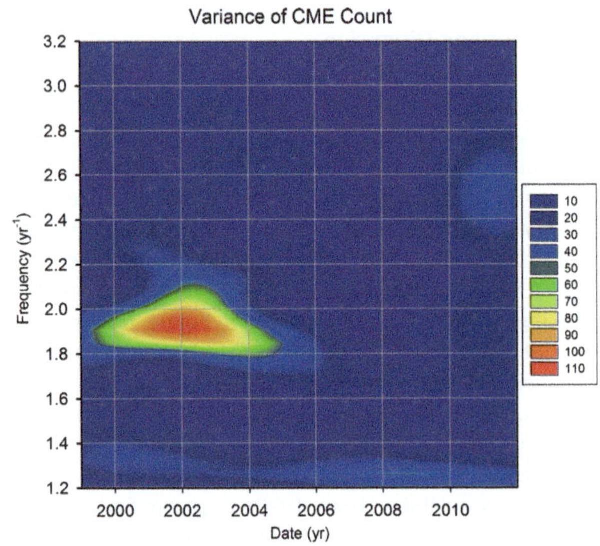

Figure 5 Wavelet power spectra for the daily CME count from 1999 to 2012. We have used a Morlet wavelet transform and selected the parameters to emphasize frequency resolution.

rence of flares and CMEs require that the field configuration be sufficiently complex that it can lead to magnetic reconnection.

Observationally, solar flares are smaller-scale explosive events on the Sun, involving a portion of a single active region, while CMEs occur at large scales often involving multiple active regions. It has been known that more than half of CMEs are associated with the eruption of large quiescent prominences and about 40 % of solar flares do not have CMEs associated with them (Forbes, 2000; Andrews, 2003). It has been often observed that the solar flares that occur after a CME, release a lower amount of energy than their pre-CME counterparts (Green *et al.*, 2001). Although the CMEs associated with major solar flares are on the average faster and broader, there is a continuum of characteristics with non-flare events (Vršnak, Sudar, and Ruždjak, 2005). The independent occurrence of solar flares and

CMEs show that, although both result from the disruption of magnetic field structures, the energy build-up and trigger mechanisms for these events may be different.

Theoretical models describing solar flares and CMEs also present somewhat different scenarios. Solar flares occur due to the loss of equilibrium in the coronal magnetic field as a result of continual flux emergence and the shuffling of the foot-points at the photosphere (Forbes, 2000). The hot coronal plasma produced after the explosion flows downwards and triggers several observed processes such as post-flare loops, flare ribbons, and hard X-ray bright points. CMEs are associated with the previous accumulation of magnetic energy through flux emergence and foot-point motions. Hence apart from short-term variations, the parameters are related to a long-term component evolving from the photosphere to the upper chromosphere (Du, 2012). CMEs can also be triggered as a result of magnetic field reconnection either at coronal heights or in the core of sheared low-lying fields at chromospheric heights. The CMEs initiated as a result of "magnetic breakout" followed by reconnection at coronal heights lead to the removal of unsheared field above the low-lying sheared core flux near the neutral line, thereby allowing this core flux to burst open. The necessary condition for this model is the presence of complex field structures involving multipolar magnetic field configurations (Antiochos, 1998). CMEs can also be initiated in a sheared core of magnetic field structures in the form of runway tether-cutting via implosive/explosive reconnection in the middle of a sigmoid (Sterling, 2000; Moore *et al.*, 2001). In this case a bipolar magnetic field structure is sufficient for the explosion to occur. Both of these models require that large-scale sheared field must be present at the core of a magnetic structure for the occurrence of a coronal mass ejection. Mere flux emergence is not sufficient for initiation of CME. On the other hand, solar flares can happen in a magnetic configuration, which are confined to smaller regions where the flux emergence can serve as the trigger.

The prominence eruption related CMEs can be well understood by the use of MHD simulations that incorporate large flux-rope eruption as a result of the evolution through increasingly energized equilibria until a magnetic-twist threshold is crossed leading to loss of equilibria and eruption (Gibson and Fan, 2006). Eruption of these filaments could be associated with flux emergence or smaller flares (Choudhary, Srivastava, and Gosain, 2002; Zhang and Wang, 2001). Clearly, the CMEs need complex magnetic field configurations on a large scale over a longer period, unlike the flares, which can occur with relatively simpler magnetic field configurations.

We find that in Cycle 23 CMEs and solar flares occurred periodically during the rising phase, while the dominant flux-emergence periodicity was present earlier, around 2001. Besides, the emergence of sunspot area displayed multiple periodicities. The flux emergence at both the 166-day and shorter day periods could be the source of solar flares. As CMEs require organization of emerged field, we notice a delay from the epochs of sunspot area periodicity. The periodicity of CMEs around 2011 and 2012 is similar to the flare periodicity of the last cycle. These CMEs mostly occurred during the solar minimum, which was devoid of flares. It may be possible that these periodicities occur with different frequency and epoch in each solar cycle. The periodicity analysis clearly shows that the solar flares and CMEs are different types of magnetic explosions that require different magnetic configurations. The CMEs occur less often than flares during the maximum phase of the solar cycle when the surface-flux is abundant. At the time of solar minimum, the large-scale surface-flux configuration leading to the formation and disruption of filaments is favorable for CME occurrence.

Acknowledgements This work was partially supported by NSF grant ATM-0548260. MN acknowledges the Department of Physics and Astronomy of California State University Northridge for a grant during this work. AC acknowledges support from the Interdisciplinary Research Institute for the Sciences (IRIS) at CSUN.

References

Andrews, M.D.: 2003, *Solar Phys.* **218**, 261. ADS:2003SoPh..218..261A. doi:10.1023/B:SOLA.0000013039.69550.bf.

Antalova, A.: 1994, *Adv. Space Res.* **14**, 721.

Antiochos, S.K.: 1998, *Astrophys. J. Lett.* **502**, L181.

Bai, T.: 2003, *Astrophys. J.* **591**, 406.

Bai, T., Cliver, E.W.: 1990, *Astrophys. J.* **363**, 299.

Bai, T., Sturrock, P.: 1987, *Nature* **327**, 601.

Ballester, J.L., Oliver, R., Carbonell, J.: 2002, *Astrophys. J.* **566**, 505.

Cane, H.V., Richardson, I.G., von Rosenvinge, T.T.: 1998, *Geophys. Res. Lett.* **25**, 4437.

Choudhary, D.P., Ambastha, A., Ai, G.: 1998, *Solar Phys.* **179**, 133. ADS:1998SoPh..179..133C. doi:10.1023/A:1005063609450.

Choudhary, D.P., Srivastava, N., Gosain, S.: 2002, *Astron. Astrophys.* **359**, 257.

Du, Z.: 2012, *Solar Phys.* **278**, 203. doi:10.1007/s11207-011-9925-0.

Forbes, T.: 2000, *Phil. Trans. Roy. Soc. London A* **358**, 711.

Gibson, S., Fan, Y.: 2006, *J. Geophys. Res.* **111**, 1.

Gopalswamy, N.: 2008, In: Dorotovič, I. (ed.) *Proc. 20th Slovak Nat. Solar Phys. Meeting*, 108, published on DVD.

Gosling, J.T.: 1993, *J. Geophys. Res.* **98**, 18949.

Green, L.M., Hara, L.K., Matthews, S.A., Culhanes, J.L.: 2001, *Solar Phys.* **200**, 189.

Hathaway, D.: 2010, *Living Rev. Solar Phys.* **7**, 1. doi:10.12942/lrsp-2010-1.

Hill, M.E., Hamilton, D.C., Krimigis, S.M.: 2001, *J. Geophys. Res.* **106**, 8315.

Ichimoto, K., Kubata, J., Suzuki, M., Tohmura, I., Kurokawa, H.: 1985, *Nature* **316**, 422.

Khan, J.I., Hudson, H.S.: 2000, *Geophys. Res. Lett.* **27**, 1083.

Lara, A., Borgazzi, A., Mendes, O. Jr., Rosa, R., Domingues, M.O.: 2008, *Solar Phys.* **248**, 155. ADS:2008SoPh..248..155L. doi:10.1007/s11207-008-9153-4.

Lean, J.: 1990, *Astrophys. J.* **363**, 718.

Lean, J., Brueckner, G.E.: 1989, *Astrophys. J.* **337**, 568.

Lou, Y.: 2000, *Astrophys. J.* **530**, 1102.

Lou, Y., Wang, Y., Fan, Z., Wang, S., Wang, J.X.: 2003, *Mon. Not. Roy. Astron. Soc.* **345**, 809.

Moore, R.L., Sterling, A., Hudson, H.S., Lemen, J.R.: 2001, *Astrophys. J.* **552**, 833.

Oliver, R., Ballester, J.L., Baudin, F.: 1998, *Nature* **394**, 552.

Pevtsov, A.A.: 2004, In: Stepnov, A.V., Benevolenskaya, E.E., Kosovichev, A.G. (eds.) *Multi-Wavelength Investigations of Solar Activity*, *Proc. Internat. Astron. Union* **S223**, Cambridge University Press, Cambridge, 521. doi:10.1017/S1743921304005149.

Priest, E.R.: 1990, *Mem. Soc. Astron. Ital.* **61**, 383.

Richardson, I.G., Cane, H.V.: 2004, *J. Geophys. Res.* **1029**, 1.

Rieger, E., Share, G.H., Forrest, D.J., Kanbach, G., Reppin, C., Chupp, E.L.: 1984, *Nature* **312**, 623.

Sterling, A.C.: 2000, *J. Atmos. Solar-Terr. Phys.* **62**, 1427.

Vourlidas, A., Howard, R.A., Esfandiari, E., Patsourakos, S., Yoshiro, S., Michalek, G.: 2010, *Astrophys. J.* **722**, 1522.

Vourlidas, A., Howard, R.A., Esfandiari, E., Patsourakos, S., Yoshiro, S., Michalek, G.: 2011, *Astrophys. J.* **730**, 59.

Vršnak, B., Sudar, D., Ruždjak, D.: 2005, *Astrophys. J.* **435**, 1149.

Zhang, J., Wang, J.: 2001, *Astrophys. J.* **554**, 474.

DOI 10.1007/978-1-4939-1182-0_16
Reprinted from *Solar Physics* Journal, DOI 10.1007/s11207-013-0427-0

Do Solar Coronal Holes Affect the Properties of Solar Energetic Particle Events?

S.W. Kahler · C.N. Arge · S. Akiyama · N. Gopalswamy

Received: 2 November 2012 / Accepted: 4 October 2013 / Published online: 8 November 2013

Abstract The intensities and timescales of gradual solar energetic particle (SEP) events at 1 AU may depend not only on the characteristics of shocks driven by coronal mass ejections (CMEs), but also on large-scale coronal and interplanetary structures. It has long been suspected that the presence of coronal holes (CHs) near the CMEs or near the 1-AU magnetic footpoints may be an important factor in SEP events. We used a group of 41 $E \approx 20$ MeV SEP events with origins near the solar central meridian to search for such effects. First we investigated whether the presence of a CH directly between the sources of the CME and of the magnetic connection at 1 AU is an important factor. Then we searched for variations of the SEP events among different solar wind (SW) stream types: slow, fast, and transient. Finally, we considered the separations between CME sources and CH footpoint connections from 1 AU determined from four-day forecast maps based on Mount Wilson Observatory and the National Solar Observatory synoptic magnetic-field maps and the Wang–Sheeley–Arge model of SW propagation. The observed *in-situ* magnetic-field polarities and SW speeds at SEP event onsets tested the forecast accuracies employed to select the best SEP/CH connection events for that analysis. Within our limited sample and the three analytical treatments, we found no statistical evidence for an effect of CHs on SEP event peak intensities, onset

Solar Origins of Space Weather and Space Climate
Guest Editors: I. González Hernández, R. Komm, and A. Pevtsov

S.W. Kahler (✉) · C.N. Arge
Air Force Research Laboratory, Space Vehicles Directorate, 3550 Aberdeen Ave., Kirtland AFB, NM 87117, USA
e-mail: stephen.kahler@kirtland.af.mil

C.N. Arge
e-mail: nick.arge@kirtland.af.mil

S. Akiyama
The Catholic University of America, Washington, DC 20064, USA
e-mail: sachiko.akiyama@nasa.gov

N. Gopalswamy
NASA Goddard Space Flight Center, Greenbelt, MD 20771, USA
e-mail: nat.gopalswamy@nasa.gov

times, or rise times. The only exception is a possible enhancement of SEP peak intensities in magnetic clouds.

Keywords Energetic particles – acceleration · Magnetic fields – models · Coronal mass ejections – low coronal signatures

1. Introduction

1.1. Coronal Hole Locations and Retarded Solar Energetic Particle Event Onsets

One of the largest $E > 10$ MeV solar energetic particle (SEP) events of Solar Cycle 21 began on 6 June 1979 at about 1850 UT. The associated X2 flare peaked at 0516 UT on 5 June, so the time from flare peak to SEP onset at 1 AU was more than 37 hours: a surprisingly long time considering its source location near the central meridian at N20E16 in NOAA Active Region (AR) 1781. The absence of a prompt onset in this SEP event was first noted by von Rosenvinge and Reames (1983), who pointed out the presence of a coronal hole (CH) to the West of AR 1781 and conjectured that SEPs diffusing westward from the AR were intercepted by the open fields of the CH. An associated large, but poorly observed, east-limb CME was reported for this event in the P78-1 *Solwind* transient list (Howard *et al.*, 1985), but Bravo (1993, 1995) cited this event as an example of the view that CMEs with interplanetary shocks were only by-products of fast solar wind (SW) eruptions in adjacent CHs and were not drivers of the shocks. Although this idea was not accepted by the community, the possibility of some CH connection to SEP events and CMEs remained. The SEP onset on 6 June also occurred as the Earth moved from a negative to a positive polarity SW sector with its source in the nearby CH. This was the basis of an alternative interpretation by Kahler, Kunches, and Smith (1995), who proposed that the open fields of the adjacent CH were filled with SEPs, but only up to the interplanetary current sheet, which acted as a barrier to SEP propagation into the negative polarity region. These authors retracted this interpretation when their statistical study (Kahler, Kunches, and Smith, 1996) showed that SEP event properties are independent of whether the SEP source is in the same or a different sector as the observer at 1 AU.

A more convincing argument for CH effects on SEP events was made by Kunches and Zwickl (1999) using the NOAA Space Environment Solar Catalog (SESC) of large $E > 10$ MeV events and their associated flares. For suitable events, they examined He 10 830 Å disk images to determine whether a CH lay on a line between the flare AR and the footpoint of Earth's magnetic-field line (hereafter 1 AU footpoint) calculated kinematically from the local SW speed. In nearly all cases with interposed CHs, including the 5 June 1979 flare, the flare AR lay in the eastern hemisphere. Plots of the times from X-ray flare peak to 1 AU SEP onset of the ≈ 30 MeV proton events *versus* either solar longitude or azimuthal separations of flare AR from 1 AU footpoint yielded a population of generally longer onset times for events with interposed CHs. For the physical explanation of this effect, the authors suggested only that coronal shocks were somehow retarded in their passages through the CH high-speed streams before reaching the field lines connecting to Earth.

In a more recent test of CH effects on SEP events, Shen *et al.* (2006) selected as candidates for SEP production a sample of 56 CMEs with projected speeds and widths exceeding 1000 km s^{-1} and 130°, respectively, and originating from the western hemisphere. These authors found no dependence of $E > 10$ MeV or $E > 50$ MeV SEP production on the proximity of CHs, determined from 284 Å solar images, to the CME sources. In the spirit of the Kunches and Zwickl (1999) result, they separated cases with and without CHs extending into the longitudes between the CME sources and the 1 AU footpoints, again finding no significant difference in SEP production between the two groups. A subsequent study (Shen *et al.*, 2010) with an updated list of 76 fast and wide western-hemisphere CMEs and CHs, now based on photospheric-field extrapolations, confirmed their earlier result of no effect of CHs on SEP production.

A problem arises, however, in comparing the apparently conflicting Shen *et al.* (2010) results with those of Kunches and Zwickl (1999). While Shen *et al.* (2010) determined that 61 of 76 western-hemisphere CMEs were separated by CHs from 1 AU footpoints, Kunches and Zwickl (1999) in their sample of 87 SEP events from all solar longitudes found interposed CHs for only 21 events, all in the eastern hemisphere. Considering this hemispherical difference, the two studies are not strictly incompatible, but it is puzzling that in one study (Shen *et al.*, 2010) most western-hemisphere CMEs had interposed CHs and in the other (Kunches and Zwickl, 1999) none did. This might be the result of using solar He 10 830 Å images *versus* photospheric-field extrapolations for the CH determinations, but it leaves unresolved the basic question of CH influence on SEP events.

1.2. CH Deflections of Fast CMEs and Possible Effects on SEP Events

The CH–SEP relationship has thus far been considered in the context of a given injection of SEPs at or near the CME source AR. Another view is that the nonradial field of an adjacent CH may deflect the CME in a direction away from the CH location. This has been shown to be the case for interplanetary shocks from CMEs near the central meridian, but without observed accompanying CME drivers (driverless) (Gopalswamy *et al.*, 2009). Studies using a CH influence vector parameter (CHIP) based on measured properties of observed disk CHs to compare with driverless ICMEs (Gopalswamy *et al.*, 2010), position angles of fastest CME propagation (Mohamed *et al.*, 2012), and presence of magnetic clouds (MCs) (Mäkelä *et al.*, 2013) strongly support the concept of CH deflections of CMEs. This and other evidence of nonradial CME propagation raises the possibility that CH deflections of CMEs may lead to modulations of SEP events.

In our previous work (Kahler, Akiyama, and Gopalswamy, 2012, hereafter KAG) we compared the parameters of 41 SEP events with the expected deflections of CMEs originating within 20° of the central meridian. We not only found no effects of the CME deflections on the onset and rise times and on the peak intensities of SEP events, but also no relationship between these SEP parameters and the initial directions of CME propagations. This suggested that the SEPs may be produced in shock sources much larger than the CMEs. In accord with this possibility, Wood *et al.* (2012) recently reported an observation by the *Solar Terrestrial Relations Observatory/Sun Earth Connection Coronal and Heliospheric Investigation* (STEREO/SECCHI) instruments of an eastward deflection of a fast CME by a CH located on its western flank. The CME-driven shock, however, readily expanded into the adjacent fast stream region of the CH and was observed *in situ* more than a day earlier at STEREO-A than at *Wind* at 1 AU. Although not discussed by Wood *et al.* (2012), this CME was associated with an $E > 10$ MeV 50 proton flux unit [pfu = 1 proton $\mathrm{cm^{-2}\,s^{-1}\,sr^{-1}}$] event observed by the GOES spacecraft. The AR 11164 source of the CME was well connected to Earth at N24W59, so while that event might show that CHs are not impacting SEP events by CME deflections, CHs might still be important for SEP propagation to 1 AU, as found by Kunches and Zwickl (1999).

1.3. SEP Events and SW Streams

The related question of SEP production by shocks in high-speed streams from CHs was taken up by Kahler (2004), who argued that two factors mitigate against the production of SEPs in high-speed streams. The first is that both the Alfvén and flow speeds of high-speed streams exceed those of the low-speed streams, making it less likely for fast CMEs to drive shocks in the high-speed streams. The second is that if suprathermal ions with speeds extending to > ten times the SW speed are the seed populations of SEP events, those populations are much less intense in the high-speed streams (Gloeckler, 2003). Despite these arguments, Kahler (2004) found not only the presence of SEP events in high-speed streams, but also no requirement for the associated CMEs to be any faster than those with SEP events in low-speed streams. Expanded studies using either SW O^{+7}/O^{+6} values (Kahler, 2005) or the SW stream types (Kahler, 2008) classified by Richardson, Cane, and Cliver (2002) again showed no dependence of SEP event timescales or intensities on SW-stream type.

1.4. Magnetic Connectivity to CHs and Effects on SEP Events

The angular separation between the 1 AU footpoints and the source CMEs is assumed to be an important determinant of SEP events. The locations of these footpoints based on simple kinematic extrapolation of 1 AU SW speeds can be misleading because interplanetary field lines invariably converge to CHs (*e.g.* Luhmann *et al.*, 2009; Wang *et al.*, 2010) that may be substantially displaced in latitude and/or longitude from an assumed W60° (Shen *et al.*, 2006, 2010) or kinematic SW-speed (Kunches and Zwickl, 1999) source. Here we perform another SEP–CH comparison using the Wang–Sheeley–Arge (WSA) model (Arge and Pizzo, 2000; Arge *et al.*, 2004). WSA is a combined empirical and physics-based representation of the corona and quasi-steady global SW flow. The coronal portion of the WSA model is a coupling of the Potential Field Source Surface model (PFSS) and the Schatten Current Sheet (SCS) model with the source surface set to 2.5 $R_\odot$ and the SCS model solution used only out to 5 $R_\odot$ (hereafter referred to as the outer coronal boundary). The SW portion of the model is a simple 1D modified kinematic model that takes into account stream interactions in an *ad-hoc* manner. The model propagates SW parcels out to Earth (or any other desired point in space), keeping track of their source regions back at the Sun (*i.e.* the latitude and longitudes of their photospheric-field footpoints) along with other parameters such as the polarity and field strengths of the footpoints. Hence, each WSA SW prediction at 1 AU comes with a direct mapping of its magnetic-field line back to its 1 AU footpoint at the Sun. Since stream interactions are taken into account in the WSA approach, it is more reliable than the traditional method of mapping the SW back to the Sun assuming a constant speed. We approach the relationship of SEP events and CHs from a more general viewpoint by asking whether the CME location relative to the WSA-based 1 AU footpoint has any bearing on SEP event characteristics.

2. Data Analysis

2.1. Selection of SEP Events and Parameters

For this work we used the 41 SEP events selected by KAG from Kahler (2013) for their analysis of CME deflections by CH magnetic fields. The solar sources of the 20 MeV SEP events lay within 20° of the central meridian, with the basic parameters taken as the peak SEP intensity I_p, the SEP onset time T_O, defined as the time from CME onset to SEP onset at the *Wind* spacecraft, and T_R: the time from SEP onset to the half maximum of I_p. The SEP onset date and event parameters are given in the first four columns of Table 1. The CME source region, determined by the associated flare location, is given in the fifth column.

Table 1 SEP Event sources and properties and matching ch footpoint sources.

Date	SEP[a]			CME site	WSA[b] Lat	WSA[b] Long	Δ[c] Lat	Δ[c] Long	Δ[c] Ang.	SW[d] Vel	Notes[e]	SW[f] Type	CH Type
	T_O	T_R	I_p										
01-April-97	6.7	2.0	−2.70	S25E16	−62	365	37	75	61	400	N, both pol	2	no
07-April-97	1.9	1.5	−2.05	S30E18	−63	312	33	104	69	400	Y	3	no
12-May-97	1.7	1.5	−1.70	N21W08	−61	194	82	55	93	310	N, both pol	3	no
21-May-97	1.0	1.0	−2.15	N05W12	56	61	51	45	62	310	Y, NS pol	3	no
03-November-97	2.5	1.5	−2.22	S20W13	−58	392	38	45	50	318	N, NS pol	3	no
02-May-98	1.2	2.0	0.30	S15W15	..	..	..	..	..	600	N, TRAN	MC	no
05-November-98	1.8	11.0	−1.52	N22W18	18	213	4	30	28	430	Y	3	P
29-June-99	5.0	27.0	−2.70	N18E04	18	352	0	60	57	520	Y	1	no
28-August-99	3.3	1.0	−3.00	S26W14	−22	265	4	32	29	520	Y	2	no
18-January-00	2.8	2.0	−1.70	S19E11	−12	192	7	68	66	300	Y, MW DG	1	no?
17-February-00	1.9	2.0	−1.70	S29E07	−13	165	16	76	72	400	Y	3	no
14-July-00	0.6	4.5	2.48	N22W07	..	..	..	..	..	590	N, TRAN	MC	no
12-September-00	1.7	5.5	0.30	S17W09	−19	282	28	39	48	360	N, both pol	1	N
09-October-00	8.0	7.0	−2.30	N01W14	21	266	20	25	32	330	Y	3	P
24-November-00	1.3	1.5	−0.70	N20W05	−15	435	35	76	82	330	N, both pol	2	P
24-November-00	0.8	1.0	0.18	N22W07	3	420	19	64	65	450	N, both pol	2	no
29-March-01	2.1	4.5	−0.30	N14W12	..	..	..	..	..	530	N, TRAN	MC	no?
09-April-01	2.5	1.0	−1.00	S21W04	11	64	32	63	70	560	Y	1	no?
10-April-01	3.1	4.5	0.30	S23W09	11	424	34	68	75	570	Y, NS jump	1	no?
09-August-01	7.2	6.5	−1.52	N15W18	45	240	30	39	44	350	N, NS pol	3	no
17-September-01	9.9	3.0	−2.52	S14E04	−27	98	13	74	70	410	N, both pol	2	N
09-October-01	8.3	0.5	−1.52	S28E08	16	162	44	74	84	430	Y	2	no?

Table 1 (*Continued.*)

Date	SEP[a]			CME	WSA[b]	WSA[b]	Δ^c	Δ^c	Δ^c	SW[d]	Notes[e]	SW[f]	CH
	T_O	T_R	I_p	site	Lat	Long	Lat	Long	Ang.	Vel		Type	Type
19-October-01	2.6	2.0	−1.05	N16W18	−24	430	40	82	89	285	N, MW jump	3	no
22-October-01	1.2	3.5	−0.52	S21E18	..	..	..	..	..	540	N, TRAN	MC	no?
04-November-01	1.3	6.5	1.70	N06W18	−23	124	29	21	36	300	Y	3	no?
15-March-02	3.6	6.0	−1.82	S08W03	−45	288	37	102	93	335	Y, MW pol	3	no
15-July-02	13.0	9.0	0.00	N19W01	4	38	15	33	36	330	Y	2	no?
18-August-02	1.7	0.5	−1.22	S12W19	..	..	..	..	..	570	N, TRAN	MC	no
19-December-02	1.1	1.0	−1.05	N15W09	−25	157	40	45	59	520	Y	1	P
21-April-03	3.0	2.0	−2.22	N18E02	−3	327	21	46	50	530	Y	2	no
28-October-03	0.9	1.0	1.30	S16E08	6	355	22	67	70	770	Y	1	no?
29-October-03	0.8	1.0	1.48	S15W02	..	..	..	..	..	1090	N, TRAN	MC	no?
18-November-03	2.3	1.5	−2.00	N00E18	−5	443	5	84	84	700	Y	2	no
07-November-04	1.7	2.0	0.70	N09W17	..	..	..	..	..	490	N, TRAN	MC	no?
03-December-04	9.2	6.0	−1.30	N09E03	−7	106	16	77	78	415	Y	3	P
15-January-05	0.9	2.5	−0.82	N11E06	22	249	11	68	66	640	Y	2	no
15-January-05	1.3	11.0	0.78	N15W05	23	248	8	65	62	540	Y	2	N
13-May-05	0.8	22.0	0.70	N12E12	−3	145	15	94	95	550	N, pol unc	2	P

Table 1 (*Continued.*)

Date	SEP[a]			CME site	WSA[b] Lat	WSA[b] Long	Δ^c Lat	Δ^c Long	Δ^c Ang.	SW[d] Vel	Notes[e]	SW[f] Type	CH Type
	T_O	T_R	I_p										
13-September-05	3.7	4.0	0.48	S09E10	..	..	..	..	..	630	N, TRAN	MC	N
16-August-06	4.2	4.0	−3.00	S14W13	57	177	71	55	84	300	Y	3	no
19-May-07	3.1	1.0	−2.82	N01W05	−22	119	23	43	48	510	Y, dif. lats	2 1	

[a] T_O and T_R: SEP onset and rise times in hours; I_p: log of peak 20 MeV SEP intensity in $p[cm^2 \text{ sr sec MeV}]^{-1}$.

[b] WSA: Wang–Sheeley–Arge footpoint mapping to Carrington latitudes and longitudes in degrees.

[c] Latitudinal, longitudinal, and angular separations between the CME and WSA source regions in degrees.

[d] Estimated solar wind speed at *Wind* in $\text{km}\,\text{s}^{-1}$.

[e] Y, used in the SEP comparisons with latitude, longitude, and angular separations. N, not used in SEP comparison for reasons: TRAN – MC present; MW or NS pol – *Wind* SW polarity did not match WSA forecasts with MW or NS maps: DG – data gap; jump – jump in predicted solar source region near SEP onset; dif.lats – MWO and NSO predictions with very different source latitudes.

[f] SW stream type from Richardson, Cane, and Cliver (2002). 1, transient; 2, fast wind; 3, slow wind; MC, magnetic cloud.

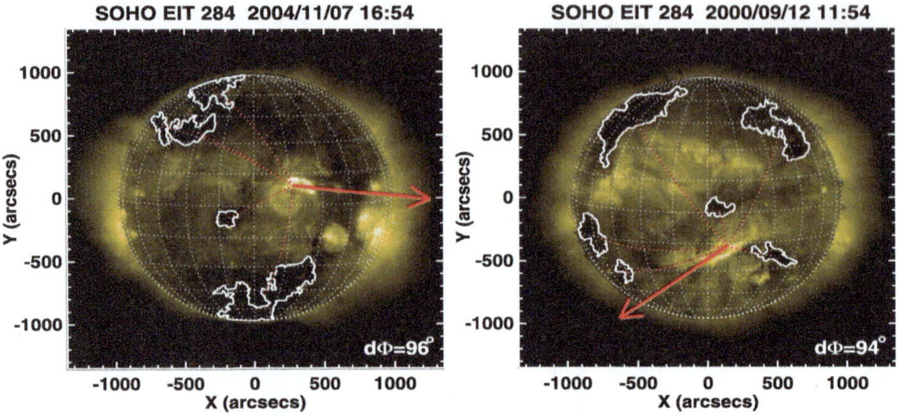

Figure 1 SOHO/EIT 284 Å images used by KAG to determine the CH properties and boundaries, outlined in white. The red arrows indicate the expected CME deflections from the CH source regions at the tails of the arrow. Left: The EIT image of 7 November 2004 without an intervening CH between CME source and the 1 AU footpoint. Right: The EIT image of 12 September 2000 with an intervening negative polarity CH West of the CME source region.

2.2. CH Locations and SEP Events

Our first search for CH effects on SEP events was to determine whether SEP event parameters are dependent on the presence of CHs located between the central-meridian SEP CME source ARs and the 1 AU footpoints determined from the SW speeds of Table 1. We used the locations of CHs identified by KAG from SOHO/EIT full-disk 284 Å images (Figure 1). This was done without regard to either the CH size or field intensity, or to the interplanetary magnetic-field configuration during the SEP onset. We ignored high-latitude ($> 35°$) and eastern-hemisphere CHs to select CHs only in the same latitude ranges as the CME-source ARs. The CH configurations are given in the last column of Table 1. A P or N indicates an interposed CH of positive or negative polarity, such as that shown in the southwest quadrant in the right panel of Figure 1, or no CH when there is no CH, but with an additional (?) for 11 uncertain cases. In the latter category were CHs at about the same longitude range as the CME source or those extending only slightly below a latitude of 35°, as in the example in the left panel of Figure 1.

We matched the three SEP parameters of the 11 cases of interposed CHs first with all 30 of the no-CH cases and then with only the 19 certain cases of no CH. The median values for each case are given in the first part of Table 2: Solar CH Configuration. Differences among these median values are dwarfed by the large standard deviations of the parameters of the full set of 41 events given in the last line of Table 2. Both groups of SEP events, those with and without intervening CHs, display a broad range of overlapping SEP parameters. We show the onset times T_O for the two groups as a function of solar source longitude in Figure 2, which can be compared with Figure 1 of Kunches and Zwickl (1999). The plotted longitude range of these authors extends from 0° to E90°, but even in the range of 0° to E20°, common to both plots, their SEP events with CHs have clearly longer T_O than those without, while our Figure 2 shows no significant difference between the two groups.

Table 2 Median values of SEP event parameters.

SEP/CME or SW	$\log I_{\mathrm{p}}$	T_{O}[a]	T_R[a]
Solar CH Configuration			
With CH (11 events)	−1.05	1.8	5.5
All no CH (30 events)	−1.37	2.2	2.0
Certain no CH (19 events)	−1.82	2.3	2.0
SW Stream Type			
Class 1 (ICME, 15 events)	0.30	1.7	2.0
MCs (8 events subset)	0.39	1.5	2.7
Class 2 (fast SW, 13 events)	−1.52	3.0	2.0
Class 3 (slow SW, 13 events)	−1.70	2.5	4.0
All 41 SEP Events	−1.22	2.1	2.0
Standard Dev (41 events)	1.42	2.9	5.4

[a]Onset and rise times in hours.

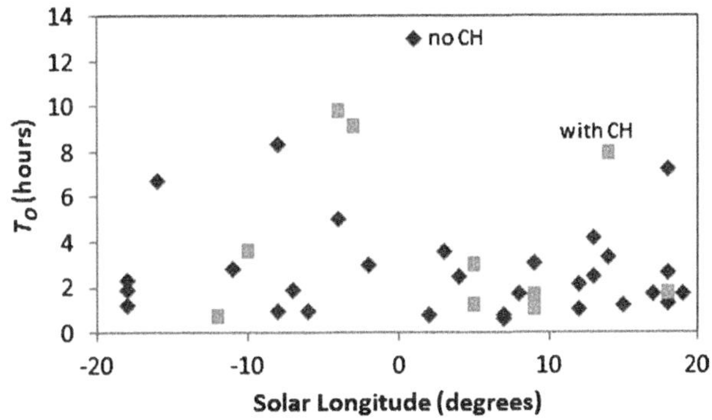

Figure 2 The SEP onset times [T_{O}] as a function of solar source longitude for the 11 SEP events with (squares) and 30 SEP events without (diamonds) CHs lying between the CME source ARs and the assumed 1 AU magnetic footpoints at ≈ W60°. Although the number of SEP events with interposed CHs is limited, the distributions of the two groups appear to overlap, which is inconsistent with the idea of a role for CHs in SEP events.

2.3. Solar Wind Stream Types and SEP Events

We sorted the SEP event onset times into the three SW stream classes first discussed by Richardson, Cane, and Cliver (2002). These are i) transient structures, including interplanetary CMEs (ICMEs), shocks, and postshock flows; ii) fast SW streams; and iii) slow SW streams. The criteria for ICME selections were updated (Richardson and Cane, 2005) and the revised list of ICME periods posted at www.ssg.sr.unh.edu/mag/ace/ACElists/ICMEtable.html. The three-stream classification periods updated throughout 2008 (I. Richardson, private communication, 2013) were used here. From the SW stream class i) we selected the times of magnetic clouds (MCs) for separate consideration. The distribution of the 41 SEP events was 15 ICMEs, of which 8 were MCs, and 13 events in each of classes ii) and iii). Under SW Stream Type of Table 2 we show the median values of the SEP parameters for each group.

We found slightly lower values of T_{O} for the ICME group than for the other SW stream groups, consistent with what Kahler (2008) found for his larger group of SEP events, but in view of the large standard deviation of 2.9 hours for all 41 events, the result is not significant. On the other hand, the median $\log I_{\mathrm{p}}$ of 0.30 for the ICME streams is more than a standard deviation above those of the other two SW groups. To visualize that difference, we plot the

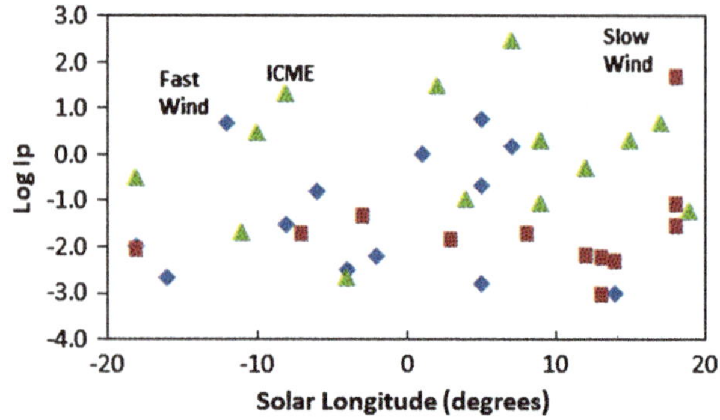

Figure 3 Logs of the peak 20 MeV SEP event intensities [I_P] *versus* the parent solar CME longitudes. No significant difference in I_P is found between events in fast SW regions (blue diamonds) and in slow SW regions (red squares), but SEP events in transient SW regions (green triangles) are somewhat enhanced.

logs of I_p *versus* solar longitude in Figure 3. Note that three of the eight SEP events in MCs (2 May 1998, 14 July 2000, and 4 November 2001) were ground-level events (GLEs: Reames, 2009). We conclude that there is an indication, still not statistically significant, that SEP events may be more intense in the ICME streams. Returning to our goal of searching for CH effects via the fast SW streams of group ii), we do not find from Table 2 that these events are distinguished from the SEP events of the other SW groups.

2.4. WSA Solar Footpoint Connections and SEP Events

We began with maps of photospheric CHs derived from the WSA model using as inputs to the model Carrington photospheric magnetic-field maps available from Mount Wilson Observatory (MWO) and the National Solar Observatory (NSO). Figure 4 shows two Carrington maps of CH fields with the connections from a 5 $R_\odot$ outer coronal boundary to the photosphere. We generated four-day advance forecasts of 1 AU SW speeds and magnetic-field polarities based on the WSA model (Section 1.4) using photospheric magnetic-field observations of both MWO and NSO. The forecast pairs were generated by letting SW parcels leave the Sun at uniform cadence (one parcel for every 2.5° of solar rotation), but due to their varying departing speeds, they arrive at Earth at nonuniform times, with approximately five predictions per day using 2.5° resolution maps. Each forecast included not only the 1 AU SW speed and magnetic-field polarity, but also the Carrington longitude and latitude of the 1 AU footpoint. The calculated MWO and NSO footpoints typically agreed to within 5° to 10°; the averaged source latitudes and longitudes for the SEP events are given in Columns 6 and 7 of Table 1, except for the eight cases when MCs at 1 AU rendered the forecasts moot. In Columns 8 and 9 we give the latitudinal and longitudinal separations between the 33 CME source regions and the WSA 1 AU footpoints, and in Column 10 their angular separations computed from the law of cosines for spherical surfaces (Smart, 1977).

In typical analyses of SEP events (*e.g.* Kunches and Zwickl, 1999), the 1 AU footpoint is calculated from a simple kinematic extrapolation based on the 1 AU SW speed. We give in Column 11 the SW speeds observed on the *Wind* spacecraft and obtained from the NASA Coordinated Data Analysis Workshop (CDAW) website. We computed the longitudinal and latitudinal differences between the WSA and kinematic 1 AU footpoints and plot these differences in Figure 5. The large $> 50°$ latitudinal differences are due to the polar CH connections during 1997, as shown in the top of Figure 4. Taking the WSA model as definitive, from Figure 5 we see that the characteristic error in determining footpoints by tracking back from SW speeds alone is $\approx 20°$ in both latitude and longitude.

The MWO and NSO four-day advance forecasts of SW speeds and magnetic-field polarities were validated in a comparison with values observed at the *Wind* spacecraft. The

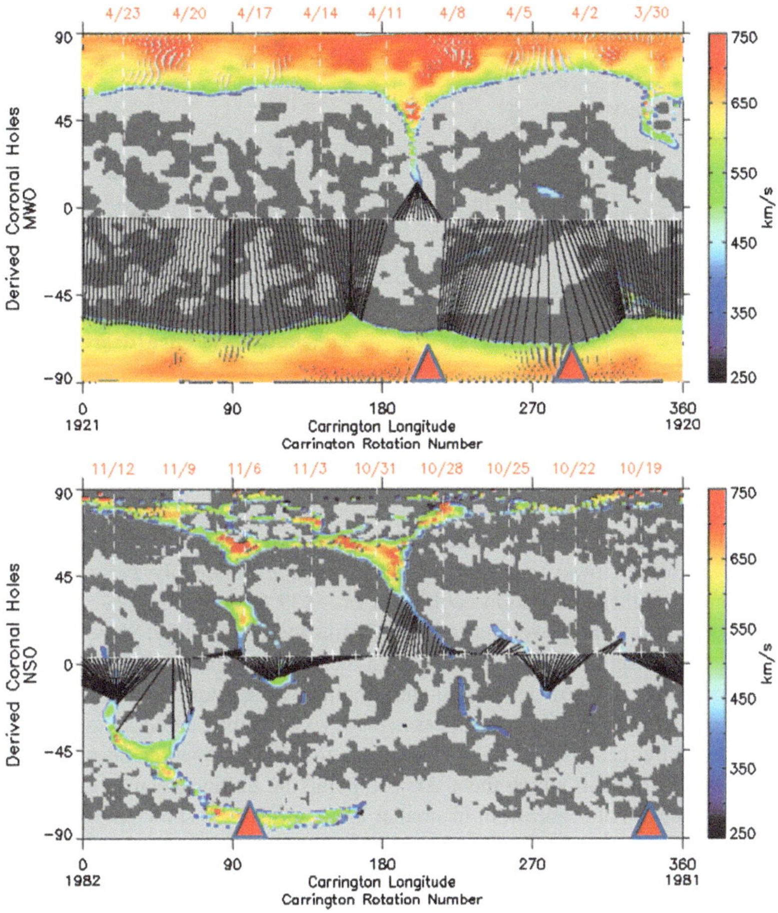

Figure 4 Carrington synoptic maps of derived CHs color-coded for SW speeds. Diagonal lines show the magnetic connections from the sub-Earth points on the 5 $R_{\odot}$ outer coronal boundary to the photospheric CHs at the dates at the top of each map. Gray areas are closed fields. The footpoint connections at the times of the SEP events correspond to the dates several days preceding the SEP events. Top: the MWO map of CR 1921 showing the 1 AU high-latitude footpoint connections. The CME sources of the SEP events of 1 and 7 April 1997 lie at CR longitudes of 290° and 208° (red triangles), respectively, while the modeled 1 AU connections lie at longitudes 346° and 314°. Bottom: the NSO map of CR 1982 showing the 1 AU low- and intermediate-latitude footpoint connections. The CME sources of the SEP events of 19 October and 4 November 2001 lie at CR longitudes of 348° and 103° (red triangles), respectively, and the modeled 1 AU connections lie at longitudes 70° of the preceding CR 1981 and 124°.

polarity agreements based on both the observed magnetic-field azimuthal directions and the electron heat-flux directions observed with the UC Berkeley 3DP instrument were the primary test of the forecasts, and the SW speed was a secondary consideration. When both the MWO and NSO forecasted polarities disagreed with the 1 AU data, we eliminated the SEP events from the subsequent consideration. These and the MC events were eliminated and listed as N, with notations, in Column 12 of Table 1. Forecasts and 1 AU footpoints of the remaining 23 SEP events were considered validated, listed as Y, and used in the analysis here, although five Y events are qualified with some uncertainty, as noted in Column 12.

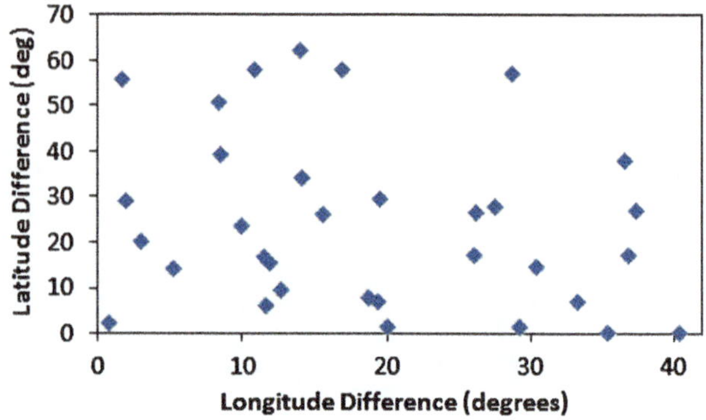

Figure 5 Absolute values of longitudinal and latitudinal differences between the WSA and the SW speed 1 AU footpoints for the 33 non-MC events of Table 1.

The question that we address here is whether the SEP event parameters depend on the separations of the CME source regions from the 1 AU footpoints of the WSA model. We calculated the correlation coefficients between the three SEP parameters and the latitudinal, longitudinal, and angular separations between the CME sources and 1 AU footpoints and found that in eight of the nine matches the correlation coefficients are < 0.25, with significance probabilities lower than 80 % for the 23 SEP events. Only T_R *versus* latitudinal separation had a higher correlation coefficient of –0.39 (a decreasing T_R with increasing latitudinal separation), still lower than a 95 % significance probability. Since we might have expected the SEP onset time T_O to increase with increasing separation between CME source and 1 AU footpoint, we show latitudinal and longitudinal separations of that parameter for the 23 SEP events in Figure 6. In both cases these separations ranged up to $\approx 80°$, but T_O was not ordered by these parameters. This means that SEP properties do not significantly vary between situations in which the 1 AU footpoints lie in high polar-latitude CHs, as in the example at the top of Figure 4, or in smaller low-latitude CHs shown in the bottom panel of Figure 4.

3. Summary and Discussion

3.1. Null Effects of CH Fields on SEP Events

SEP events are observed over wide ranges of both latitude (Dalla *et al.*, 2003; Malandraki *et al.*, 2009) and longitude (Cliver *et al.*, 2005) and exhibit broad ranges of T_O and T_R as well as peak intensities I_p (Kahler, 2005, 2013), which are currently unexplained. It has long been known that the intensity–time profiles of gradual SEP events are ordered by at least two important factors. The first is the proximity of the source region, which we assumed to be the AR source of a fast and wide CME, to the 1 AU footpoint, typically in the longitudinal range of $\approx$W40°, –W70° (*e.g.* van Hollebeke, Ma Sung, and McDonald, 1975; Cane and Lario, 2006; Reames, 2009; Gardini, Laurenza, and Storini, 2011). The second factor is the propagation of the CME-driven shock (Cane, Reames, and von Rosenvinge, 1988; Reames, Barbier, and Ng, 1996). SEP events may be additionally modulated by closed interplanetary magnetic topologies (Richardson, Cane, and von Rosenvinge, 1991) or through SEP reflections at magnetic-field enhancements that enhance for downstream observer SEP intensities (Kocharov *et al.*, 2009; Tan *et al.*, 2009) and retard or diminish intensities for upstream observers (Lario *et al.*, 2008). Whether a fast and wide CME will even produce an observable SEP event at 1 AU also appears to depend on the CME interaction

Figure 6 The 23 SEP event onset times [T_O] *versus* longitudinal (top) and latitudinal (bottom) separations of the CME source region and the 1 AU footpoint connection. Correlation coefficients were −0.13 and −0.03 for the top and bottom plots.

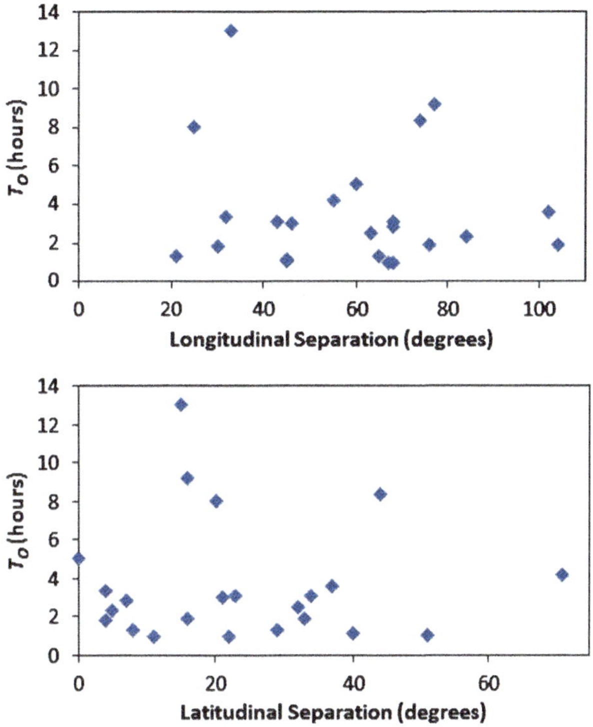

with a streamer or previous CME (Gopalswamy *et al.*, 2003; Kahler and Vourlidas, 2005; Ding *et al.*, 2013). CME widths and speeds correlate weakly with T_R, although not with T_O (Kahler, 2005, 2013).

We might expect additional organizing factors based on the large-scale coronal and solar structures encountered by the fast and wide CMEs and by their preceding fast shocks that are responsible for producing SEPs. As we discussed in Section 1, CHs have long been suspected to be a factor, although the proposed mechanism of simply blocking or retarding the SEPs released in the solar corona seems simplistic in light of the current paradigm of widespread shock acceleration of SEPs.

The goal of this and the companion work (KAG) has been to search for a statistical effect in the basic time–intensity profiles of SEP events that could be related to CHs. We selected a sample of SEP events with source regions around the central meridian to search for such effects. The relevant associated CHs were required to be well observed on the solar disk and in locations such that they could affect the SEP propagation to 1 AU footpoint field-lines. We used three different techniques to search for these possible CH effects. First, we followed the approach of Kunches and Zwickl (1999) to divide SEP events into two groups depending on whether a CH appeared to lie on a line connecting the SEP source and the 1 AU footpoint. This simple concept belies the complexity of determining i) the SEP source region, which involves the large-scale CME, ii) the extent of the CH field lines, which extend nonradially from their apparent wavelength-dependent source boundaries, and iii) the 1 AU footpoints, which generally lie tens of degrees in latitude and longitude away from their kinematically computed sources (Figure 5). Contrary to Kunches and Zwickl (1999), we did not find T_O delayed for SEP events with interposed CHs (Figure 2) in the E20° – 0° longitude regions common to the two studies. We do not understand (Section 1.2) why they found no intervening CHs for their western-hemisphere SEP events, but their stricter requirement that a CH must lie on a line connecting the SEP and 1 AU footpoint regions, as opposed to lying in an intermediate longitude, may be a factor. A previous comparison of He 10 830 Å used by Kunches and Zwickl (1999) and soft X-ray CH boundaries (Kahler, Davis, and Harvey, 1983) showed poor agreement for low-latitude CH boundaries. Our results, that peak intensities [I_p] are not dependent on CH locations, agree with those of Shen *et al.* (2006).

Comparisons of SEP event properties in different types of SW streams were our second approach in searching for evidence of CH effects, this time through their associated high-speed streams. While we did find higher SEP I_p values among the ICME MC group of events, including three GLEs, SEP events of the fast SW group ii) streams were not distinguished from the others in terms of their SEP events (Table 2). This result is consistent with the lack of any significant variation of SEP event characteristics with SW stream type in the broader survey of Kahler (2008). The additional lack of a significant SW stream variation in SEP elemental composition (Kahler, Tylka, and Reames, 2009) supports the conclusion that the SW stream structure is simply not a determining factor for SEP propagation.

Since the work of Kunches and Zwickl (1999), we recognize that the 1 AU footpoints invariably trace the edges or interiors of CHs, which may be far removed from the nominal 1 AU ecliptic projections. Although the PFSS model accuracy may be limited (Nitta and De Rosa, 2008), a footpoint connection to the CH vicinity of a flaring AR is crucial for observing impulsive SEP events (Rust *et al.*, 2008) and for determining the seed-particle population for shock-accelerated events (Ko *et al.*, 2012). We aimed at a more accurate comparison between the CME source regions and the 1 AU footpoints (Figure 4). We searched for effects on SEP events that could be attributed to variations of these footpoint locations by sorting the SEP events on the basis of the angular separations between their source regions and the calculated 1 AU footpoints. We interpreted the null result of Figure 6 as an indication that the low coronal magnetic-field connection does not order SEP events, perhaps because the shock propagation and SEP injection are occurring above the outer coronal boundary 5 $R_\odot$ source height where the PFSS fields are presumed to be nearly radial. Our study has been confined to SEP events originating within a 40° band of the central meridian, so the possibility of CH effects on SEPs from other regions, particularly eastern-hemisphere sources (Kunches and Zwickl, 1999), cannot be ruled out. We did not attempt to include elemental composition as an SEP event property, but other studies give no indication that stream structures play a role in SEP composition (Kahler, Tylka, and Reames, 2009, 2011).

3.2. SEP Events and Latitudinal Separations of 1 AU Footpoints

The latitudinal separations between associated AR flares and the 1 AU footpoints rarely exceed about 30° for SEP events and are generally ignored in comparison with the much wider range of longitudinal separations. However, the observation of *Ulysses* high-latitude SEP events prompted a new search for possible latitude effects on SEP event characteristics. Dalla *et al.* (2003) found that $E \approx 30$ MeV SEP event times to maximum, roughly equivalent to our $T_O + T_R$, increased with increasing latitudinal separations for nine *Ulysses* high-latitude events. For the corresponding nine SEP events observed at 1 AU, however, their SEP times to maxima showed no latitudinal dependence, similar to our Figure 6. Dalla and Agueda (2010) followed this *Ulysses* 1 AU comparison with a larger 1 AU study based on the nominal latitudinal separations $\Delta\theta$ of 496 well-connected $\geq$C8 solar flares. The probability of detecting any associated SEP event at 1 AU peaked in the range of $\Delta\Theta = 4° - 12°$, but as in the Dalla *et al.* (2003) study, the SEP times to maximum showed no dependence on $\Delta\Theta$, suggesting that it is the larger-scale interplanetary and not the coronal latitudinal and longitudinal field-line separations that are important for SEP propagation. This conclusion is supported by modeling of SEP profiles for different latitudinal separations between the observer and the progenitor CME (Rodríguez-Gasén *et al.*, 2011). This result may not apply at the highest energies of ground-level enhancement (GLE) events, however. Gopalswamy *et al.* (2013) found that while fast CMEs from well-connected solar longitude regions produced strong SEP events in Solar Cycle 24, it was necessary for the CME nose to be close ($\leq 5^\circ$) to the ecliptic plane to produce a GLE. Otherwise, only lower-energy SEPs, presumably from the shock flanks, reached Earth.

Since large-scale coronal and interplanetary structures seem to give only rough guidance to SEP event timescales, the strong variability of SEP event intensities and timescales observed at 1 AU may be due to spatial and temporal variations inherent in the shocks themselves (Kóta, 2010) and to their time-dependent connections to the field lines at 1 AU. We may expect better understanding of SEP profiles from more detailed modeling efforts based on multiple-imaging observations of interplanetary shock fronts and SEPs, such as those of Rouillard *et al.* (2011).

Acknowledgements SWK was funded by AFOSR Task 2301RDZ4. NG and SA were supported by NASA's LWS TR&T program. CME data were taken from the CDAW LASCO catalog. This CME catalog is generated and maintained at the CDAW Data Center by NASA and The Catholic University of America in cooperation with the Naval Research Laboratory. SOHO is a project of international cooperation between ESA and NASA. EIT images of Figure 1 were obtained from the EIT instrument webpage. We thank Ian Richardson for providing the SW stream listings and Don Reames for the use of the EPACT proton data. We used *Wind* data provided by J.H. King, N. Papatashvilli, and R. Lepping at the NASA/GSFC CDAW website.

References

Arge, C.N., Pizzo, V.J.: 2000, Improvement in the prediction of solar wind conditions using near-real time solar magnetic field updates. *J. Geophys. Res.* **105**, 10465 – 10480. doi:10.1029/1999JA900262.

Arge, C.N., Luhmann, J.G., Odstrcil, D., Schrijver, C.J., Li, Y.: 2004, Stream structure and coronal sources of the solar wind during the May 12th, 1997 CME. *J. Atmos. Solar-Terr. Phys.* **66**, 1295 – 1309. doi:10.1016/j.jastp.2004.03.018.

Bravo, S.: 1993, The SC event of 6 June 1979 and related solar and interplanetary observations. *Adv. Space Res.* **13**, 371 – 374. doi:10.1016/0273-1177(93)90508-9.

Bravo, S.: 1995, A solar scenario for the associated occurrence of flares, eruptive prominences, coronal mass ejections, coronal holes, and interplanetary shocks. *Solar Phys.* **161**, 57 – 65. doi:10.1007/BF00732084.

Cane, H.V., Lario, D.: 2006, An introduction to CMEs and energetic particles. *Space Sci. Rev.* **123**, 45 – 56. doi:10.1007/s11214-006-9011-3.

Cane, H.V., Reames, D.V., von Rosenvinge, T.T.: 1988, The role of interplanetary shocks in the longitude distribution of solar energetic particles. *J. Geophys. Res.* **93**, 9555 – 9567. doi:10.1029/JA093iA09p09555.

Cliver, E.W., Thompson, B.J., Lawrence, G.R., Zhukov, A.N., Tylka, A.J., Dietrich, W.F., Reames, D.V., Reiner, M.J., MacDowall, R.J., Kosovichev, A.G., Ling, A.G.: 2005, The solar energetic particle event of 16 August 2001: 400 MeV protons following an eruption at $\approx$ W180. In: Acharya, B.S., Gupta, S., Jagadeesan, P., Jain, A., Karthikeyan, S., Morris, S., Tonwar, S. (eds.) *Proc. 29th Int. Cosmic Ray Conf. 1*, Tata Inst. Fund. Res., 121 – 124.

Dalla, S., Agueda, N.: 2010, Role of latitude of source region in solar energetic particle events. In: Maksimovic, M., Issautier, K., Meyer-Vernet, N., Moncuquet, M., Pantellini, F. (eds.) *Proc. 12th Int. Solar Wind Conf.* **CP-1216**, AIP, New York, 613 – 616. doi:10.1063/1.3395941.

Dalla, S., Balogh, A., Krucker, S., Posner, A., Müller-Mellin, R., Anglin, J.D., Hofer, M.Y., Marsden, R.G., Sanderson, T.R., Tranquille, C., Heber, B., Zhang, M., McKibben, R.B.: 2003, Properties of high heliolatitude solar energetic particle events and constraints on models of acceleration and propagation. *Geophys. Res. Lett.* **30**, 8035. doi:10.1029/2003GL017139. ULY 9-1.

Ding, L., Jiang, Y., Zhao, L., Li, G.: 2013, The "twin-CME" scenario and large solar energetic particle events in solar cycle 23. *Astrophys. J.* **763**, 30. doi:10.1088/0004-637X/763/1/30.

Gardini, A., Laurenza, M., Storini, M.: 2011, SEP events and multi-spacecraft observations: constraints on theory. *Adv. Space Res.* **47**, 2127–2139. doi:10.1016/j.asr.2011.01.025.

Gloeckler, G.: 2003, Ubiquitous suprathermal tails on the solar wind and pickup ion distributions. In: Velli, M., Bruno, R., Malara, F. (eds.) *Proc. 10th Int. Solar Wind Conf.* **CP-679**, AIP, New York, 583–588. doi:10.1063/1.1618663.

Gopalswamy, N., Yashiro, S., Michalek, G., Kaiser, M.L., Howard, R.A., Leske, R., von Rosenvinge, T., Reames, D.V.: 2003, Effect of CME interactions on the production of solar energetic particles. In: Velli, M., Bruno, R., Malara, F. (eds.) *Proc. 10th Int. Solar Wind Conf.* **CP-679**, AIP, New York, 608–611. doi:10.1063/1.1618668.

Gopalswamy, N., Mäkelä, P., Xie, H., Akiyama, S., Yashiro, S.: 2009, CME interactions with coronal holes and their interplanetary consequences. *J. Geophys. Res.* **114**, A00A22. doi:10.1029/2008JA013686.

Gopalswamy, N., Mäkelä, P., Xie, H., Akiyama, S., Yashiro, S.: 2010, Solar sources of "driverless" interplanetary shocks. In: Maksimovic, M., *et al.* (eds.) *Proc. Twelfth Int. Solar Wind Conf.* **CP-1216**, 452–455. doi:10.1063/1.3395902.

Gopalswamy, N., Xie, H., Akiyama, S., Yashiro, S., Usoskin, I.G., Davilla, J.M.: 2013, The first ground level enhancement of solar cycle 24: direct observation of shock formation and particle release heights. *Astrophys. J.* **765**, L30. doi:10.1088/2041-8205/765/2/L30.

Howard, R.A., Sheeley, N.R. Jr., Michels, D.J., Koomen, M.J.: 1985, Coronal mass ejections – 1979–1981. *J. Geophys. Res.* **90**, 8173–8191. doi:10.1029/JA090iA09p08173.

Kahler, S.W.: 2004, Solar fast-wind regions as sources of shock energetic particle production. *Astrophys. J.* **603**, 330–334. doi:10.1086/381358.

Kahler, S.W.: 2005, Characteristic times of gradual solar energetic particle events and their dependence on associated coronal mass ejection properties. *Astrophys. J.* **628**, 1014–1022. doi:10.1086/431194.

Kahler, S.W.: 2008, Time scales of solar energetic particle events and solar wind stream types. In: Caballero, R., D'Olivo, J.C., Medina-Tanco, G., Nellen, L., Sánchez, F.A., Valdés-Galicia, J.F. (eds.) *Proc. 30th Int. Cosmic Ray Conf.* **1**, Univ. Nac. Auto. de Mexico, Mexico, 143–146.

Kahler, S.W.: 2013, A comparison of solar energetic particle event timescales with properties of associated coronal mass ejections. *Astrophys. J.* **769**, 110. doi:10.1088/0004-637X/769/2/110.

Kahler, S.W., Akiyama, S., Gopalswamy, N.: 2012, Deflections of fast coronal mass ejections and the properties of associated solar energetic particle events. *Astrophys. J.* **754**, 100. doi:10.1088/0004-637X/754/2/100 (KAG).

Kahler, S.W., Davis, J.M., Harvey, J.W.: 1983, Comparison of coronal holes observed in soft X-ray and He I 10 830 Å spectroheliograms. *Solar Phys.* **87**, 47–56. doi:10.1007/BF00151159.

Kahler, S.W., Kunches, J., Smith, D.F.: 1995, Coronal and interplanetary magnetic sector structure and the modulation of solar energetic particle events. In: *24th Int. Cosmic Ray Conf.* **4**, 325–328.

Kahler, S.W., Kunches, J.M., Smith, D.F.: 1996, Role of current sheets in the modulation of solar energetic particle events. *J. Geophys. Res.* **101**, 24383–24392. doi:10.1029/96JA02446.

Kahler, S.W., Tylka, A.J., Reames, D.V.: 2009, A comparison of elemental abundance ratios in SEP events in fast and slow solar wind regions. *Astrophys. J.* **701**, 561. doi:10.1088/0004-637X/701/1/561.

Kahler, S.W., Vourlidas, A.: 2005, Fast coronal mass ejection environments and the production of solar energetic particle events. *J. Geophys. Res.* **110**, A12S01. doi:10.1029/2005JA011073.

Kahler, S.W., Cliver, E.W., Tylka, A.J., Dietrich, W.F.: 2011, A comparison of ground level event e/p and Fe/O ratios with associated solar flare and CME characteristics. *Space Sci. Rev.* **171**, 121–139. doi:10.1007/s11214-011-9768-x.

Ko, Y.-K., Tylka, A.J., Ng, C.K., Wang, Y.-M.: 2012, On the relationship between heavy-ion composition variability in gradual SEP events and the associated IMF source regions. In: Hu, Q., Li, G., Zank, G.P., Ao, X., Verkhoglyadova, O., Adams, J.H. (eds.) *Space Weather: The Space Radiation Environment* **1500**, AIP, New York, 26–31. doi:10.1063/1.4768740.

Kocharov, L., Laitinen, T., Al-Sawad, A., Saloniemi, O., Valtonen, E., Reiner, M.J.: 2009, Gradual solar energetic particle event associated with a decelerating shock wave. *Astrophys. J. Lett.* **700**, L51–L55. doi:10.1088/0004-637X/700/1/L51.

Kóta, J.: 2010, Particle acceleration at near-perpendicular shocks: the role of field-line topology. *Astrophys. J.* **723**, 393–397. doi:10.1088/0004-637X/723/1/393.

Kunches, J.M., Zwickl, R.D.: 1999, The effects of coronal holes on the propagation of solar energetic protons. *Radiat. Meas.* **30**, 281–286.

Lario, D., Decker, R.B., Malandraki, O.E., Lanzerotti, L.J.: 2008, Influence of large-scale interplanetary structures on energetic particle propagation: September 2004 event at Ulysses and ACE. *J. Geophys. Res.* **113**, A03105. doi:10.1029/2007JA012721.

Luhmann, J.G., Lee, C.O., Li, Y., Arge, C.N., Galvin, A.B., Simunac, K., Russell, C.T., Howard, R.A., Petrie, G.: 2009, Solar wind sources in the late declining phase of cycle 23: effects of the weak solar polar field on high speed streams. *Solar Phys.* **256**, 285–305. doi:10.1007/s11207-009-9354-5.

Mäkelä, P., Gopalswamy, N., Xie, H., Mohamed, A.A., Akiyama, S., Yashiro, S.: 2013, Coronal hole influence on the observed structure of interplanetary CMEs. *Solar Phys.* **284**, 59 – 75. doi:10.1007/s11207-012-0211-6.
Malandraki, O.E., Marsden, R.G., Lario, D., Tranquille, C., Heber, B., Mewaldt, R.A., Cohen, C.M.S., Lanzerotti, L.J., Forsyth, R.J., Elliott, H.A., Vogiatzis, I.I., Geranios, A.: 2009, Energetic particle observations and propagation in the three-dimensional heliosphere during the 2006 December events. *Astrophys. J.* **704**, 469 – 476. doi:10.1088/0004-637X/704/1/469.
Mohamed, A.A., Gopalswamy, N., Yashiro, N., Akiyama, S., Mäkelä, P., Xie, H., Jung, H.: 2012, The relation between coronal holes and coronal mass ejections during the rise, maximum, and declining phases of solar cycle 23. *J. Geophys. Res.* **117**, A01103. doi:10.1029/2011JA016589.
Nitta, N.V., De Rosa, M.L.: 2008, A comparison of solar open field regions found by type III radio bursts and the potential field source surface model. *Astrophys. J. Lett.* **673**, L207 – L210. doi:10.1086/527548.
Reames, D.V.: 2009, Solar energetic-particle release times in historic ground-level events. *Astrophys. J.* **706**, 844 – 850. doi:10.1088/0004-637X/706/1/844.
Reames, D.V., Barbier, L.M., Ng, C.K.: 1996, The spatial distribution of particles accelerated by coronal mass ejection-driven shocks. *Astrophys. J.* **466**, 473 – 486. doi:10.1086/177525.
Richardson, I.G., Cane, H.V.: 2005, A survey of interplanetary coronal mass ejections in the near-Earth solar wind during 1996 – 2005. In: Fleck, B., Zurbuchen, T.H., Lacoste, H. (eds.) *Proc. Solar Wind 11/SOHO 16 Conf.* **SP-592**, ESA, Noordwijk, 755 – 758.
Richardson, I.G., Cane, H.V., Cliver, E.W.: 2002, Sources of geomagnetic activity during nearly three solar cycles (1972 – 2000). *J. Geophys. Res.* **107**, SSH 8-1. doi:10.1029/2001JA000504.
Richardson, I.G., Cane, H.V., von Rosenvinge, T.T.: 1991, Prompt arrival of solar energetic particles from far eastern events – the role of large-scale interplanetary magnetic field structure. *J. Geophys. Res.* **96**, 7853 – 7860. doi:10.1029/91JA00379.
Rodríguez-Gasén, R., Aran, A., Sanahuja, B., Jacobs, C., Poedts, S.: 2011, Why should the latitude of the observer be considered when modeling gradual proton events? An insight using the concept of cobpoint. *Adv. Space Res.* **47**, 2140 – 2151. 10.1016/j.asr.2010.03.021.
Rouillard, A.P., Odstrcil, D., Sheeley, N.R., Tylka, A., Vourlidas, A., Mason, G., Wu, C.-C., Savani, N.P., Wood, B.E., Ng, C.K., Stenborg, G., Szabo, A., St. Cyr, O.C.: 2011, Interpreting the properties of solar energetic particle events by using combined imaging and modeling of interplanetary shocks. *Astrophys. J.* **735**, 7. doi:10.1088/0004-637X/735/1/7.
Rust, D.M., Haggerty, D.K., Georgoulis, M.K., Sheeley, N.R., Wang, Y.-M., De Rosa, M.L., Schrijver, C.J.: 2008, On the solar origins of open magnetic fields in the heliosphere. *Astrophys. J.* **687**, 635 – 645. doi:10.1086/592017.
Shen, C., Wang, Y., Ye, P., Wang, S.: 2006, Is there any evident effect of coronal holes on gradual solar energetic particle events? *Astrophys. J.* **639**, 510 – 515. doi:10.1086/499199.
Shen, C.-L., Yao, J., Wang, Y.-M., Ye, P.-Z., Zhao, X.-P., Wang, S.: 2010, Influence of coronal holes on CMEs in causing SEP events. *Res. Astron. Astrophys.* **10**, 1049 – 1060. doi:10.1088/1674-4527/10/10/008.
Smart, W.M.: 1977, *Textbook on Spherical Astronomy*, Cambridge Univ. Press, Cambridge.
Tan, L.C., Reames, D.V., Ng, C.K., Saloniemi, O., Wang, L.: 2009, Observational evidence on the presence of an outer reflecting boundary in solar energetic particle events. *Astrophys. J.* **701**, 1753 – 1764. doi:10.1088/0004-637X/701/2/1753.
van Hollebeke, M.A.I., Ma Sung, L.S., McDonald, F.B.: 1975, The variation of solar proton energy spectra and size distribution with heliolongitude. *Solar Phys.* **41**, 189 – 223. doi:10.1007/BF00152967.
von Rosenvinge, T.T., Reames, D.V.: 1983, The delayed energetic particle event of June 6 – 10, 1979. In: Durgaprasad, N., Ramadurai, S., Ramana Murthy, P.V., Rao, M.V.S., Sivaprasad, K. (eds.) *18th Int. Cosmic Ray Conf.* **10**, Tata Inst. Fund. Res., 373 – 376.
Wang, Y.-M., Robbrecht, E., Rouillard, A.P., Sheeley, N.R. Jr., Thernisien, A.F.R.: 2010, Formation and evolution of coronal holes following the emergence of active regions. *Astrophys. J.* **715**, 39 – 50. doi:10.1088/0004-637X/715/1/39.
Wood, B.E., Wu, C.-C., Rouillard, A.P., Howard, R.A., Socker, D.G.: 2012, A coronal hole's effects on coronal mass ejection shock morphology in the inner heliosphere. *Astrophys. J.* **755**, 43. doi:10.1088/0004-637X/755/1/43.

DOI 10.1007/978-1-4939-1182-0_17
Reprinted from *Solar Physics* Journal, DOI 10.1007/s11207-013-0308-6

From Predicting Solar Activity to Forecasting Space Weather: Practical Examples of Research-to-Operations and Operations-to-Research

R.A. Steenburgh · D.A. Biesecker · G.H. Millward

Received: 7 November 2012 / Accepted: 8 April 2013 / Published online: 7 May 2013

Abstract The successful transition of research to operations (R2O) and operations to research (O2R) requires, above all, interaction between the two communities. We explore the role that close interaction and ongoing communication played in the successful fielding of three separate developments: an observation platform, a numerical model, and a visualization and specification tool. Additionally, we will examine how these three pieces came together to revolutionize interplanetary coronal mass ejection (ICME) arrival forecasts. A discussion of the importance of education and training in ensuring a positive outcome from R2O activity follows. We describe efforts by the meteorological community to make research results more accessible to forecasters and the applicability of these efforts to the transfer of space-weather research. We end with a forecaster "wish list" for R2O transitions. Ongoing, two-way communication between the research and operations communities is the thread connecting it all.

Keywords Coronal mass ejections · Magnetohydrodynamics

Invited Review.

Solar Origins of Space Weather and Space Climate
Guest Editors: I. González Hernández, R. Komm, and A. Pevtsov

R.A. Steenburgh (✉) · D.A. Biesecker · G.H. Millward
NOAA/Space Weather Prediction Center, Boulder, CO, USA
e-mail: robert.steenburgh@noaa.gov

D.A. Biesecker
e-mail: doug.biesecker@noaa.gov

G.H. Millward
e-mail: george.millward@noaa.gov

G.H. Millward
Cooperative Institute for Research in Environmental Sciences (CIRES), University of Colorado, Boulder, CO, USA

1. Introduction

Over the past decade, forecast operations at the US National Oceanic and Atmospheric Administration's (NOAA) Space Weather Prediction Center (SWPC) were revolutionized. Among other factors, the revolution was propelled by the availability of new data, such as those from NASA's *Solar TErrestrial RElations Observatory* (STEREO: Kaiser *et al.*, 2008) and *Solar Dynamics Observatory* (SDO: Pesnell, Thompson, and Chamberlin, 2012) missions, and from the introduction of physics-based numerical models such as the Wang–Sheely–Arge (WSA)–Enlil model (Odstrcil *et al.*, 2004).

This revolution was facilitated by successful research-to-operations (R2O) transitions of researcher knowledge and modeling to the forecasting office. A key component of successful transition is for the space-weather research community to become familiar with operational space-weather forecasting. Consequently, we will begin by describing the current environment and challenges faced by the practitioners. Then we will examine three examples of productive space-weather R2O transitions and the interactions between the operations and research personnel (*i.e.*, operations to research: O2R) that contributed to the success. When these interactions are a two-way street, success becomes much more likely. A discussion of steps to achieve a successful R2O transition, an example of how the SWPC implements R2O/O2R, and the current needs within the operations community follows.

2. The Practice of Space Weather Forecasting

Like their meteorologist counterparts, space-weather forecasters must cope with the rigors of shift work. They must assimilate several disparate, and often interrupted, data streams as well as a growing number of numerical models. From these, they are expected to create an accurate description of the current and expected state of the geospace environment. They are confronted with situations in which they must make significant decisions with incomplete data and inadequate time. They are simultaneously confronted with the pressure not to miss a warning and the pressure not to trigger a false alarm. They must be cognizant of a variety of conceptual models that describe space-weather phenomena such as coronal mass ejections (CMEs) and flares. They must know how and when to apply these models. If the observations do not fit the models, forecasters must determine why and what impact that will have on the forecast. Then, they must clearly communicate the forecast to a large and varied customer base, keeping in mind the impacts of greatest importance to each customer. Forecasters must strive to stay abreast of the latest developments in research. They must continually gain and maintain proficiency in new concepts, tools, and observations introduced to the forecast process. The professionals who confront these challenges are drawn from a variety of educational and occupational backgrounds.

While the skills and education required to become a scientific researcher are well known (*e.g.* a Ph.D. degree in an associated discipline), less well known is the typical skill set and education of a National Weather Service (NWS) Space Weather Forecaster. For the two communities to work together successfully, it is paramount that each understand the capabilities and language of the other. The SWPC forecasters are classified as physical scientists. This classification requires a four-year undergraduate degree in physical science, engineering, or mathematics including 24 semester hours (six to eight university courses) of physical science and/or related engineering science coursework; or an equivalent combination of education and experience. Currently, the educational background of the forecasters ranges from the undergraduate through the doctoral level. Experience levels range from over three decades to less than a year. Such a diverse group presents significant challenges to researchers attempting to design and deliver education and training. For instance, the familiarity and level of comfort with mathematics varies among forecasters. Effective training translates the mathematics into words and concepts without the "gory details"; however, these details should be included for those forecasters who are comfortable with them and wish to know more.

The educational background of forecasters can vary significantly, but the demands on space-weather forecasters are nearly identical to those of meteorologists; thus, it is not surprising that many space-weather forecasters have backgrounds in meteorology. The skills needed to forecast terrestrial weather and provide timely watches, warnings, and alerts are easily transferred to space weather. Physicists, environmental scientists, and others round out the current forecast team. Additionally, the United States Air Force maintains a small contingent of solar observers and space-weather forecasters, and some have become forecasters at the SWPC following their military service.

3. Recent R2O, O2R Examples

Creating operational products and services from the fruits of research has been a renewed focus of the SWPC since it was renamed from the Space Environment Center in 2007. Recent examples include two ionospheric products, D-RAP (Sauer and Wilkinson, 2008), which provides HF propagation forecasts, and US-TEC (Fuller-Rowell *et al.*, 2006), which provides total electron content information over the United States, and one interplanetary propagation product, WSA–Enlil, which will be covered in detail in later sections. However, R2O and O2R activities had been ongoing since the SWPC's genesis as the Interservice Radio Propagation Laboratory in the 1940s.

Physics-based numerical models are not the only research to be transitioned to operations. New models (*both* conceptual and numerical), new observation platforms, and new tools are all candidates for R2O activities. In the next section, we will examine the convergence of three of these elements: a new observation platform, a new visualization tool, and a new physics-based numerical model.

3.1. The STEREO Mission

NASA's *International Sun–Earth Explorer 3* (ISEE-3: Tsurutani and Baker, 1979) and NASA's *Advanced Composition Explorer* (ACE: Zwickl *et al.*, 1998) are among the first successful examples of enlisting research spacecraft to provide real-time space-weather monitoring to the operational community. This concept became an integral part of subsequent endeavors such as the ESA/NASA *Solar and Heliospheric Observatory* (SOHO), and NASA's STEREO and SDO missions. The idea of using STEREO data operationally was present from its inception (St. Cyr and Davila, 2001).

Imagery, solar-wind, and particle-beacon data from STEREO became available in the forecast office in March 2007 (Biesecker, Webb, and St. Cyr, 2008). An overview of the various instruments and products produced by STEREO is provided by Kaiser *et al.* (2008). Data from the *In-situ Measurements of Particles and CME Transients* (IMPACT) and *PLAsma and SupraThermal Ion Composition* (PLASTIC) instruments were displayed in a format identical to similar data from the ACE spacecraft, making it easy for forecasters to begin using the data. Forecasting the arrival time and potential impact of coronal-hole high-speed streams is one example of the application of the data. Forecasters also used IMPACT data to identify regions with potential to produce energetic-particle events, then monitored these suspect regions as they entered threatening longitudes. The *Sun Earth Connection Coronal and Heliospheric Investigation* (SECCHI) imagery was used to monitor active regions about to rotate onto the visible solar disk, and to locate the origin of CMEs. These predictions and observations made their way into several SWPC products including the *Report of Geophysical and Solar Activity*, a 24-hour summary and three-day forecast of space-weather conditions, and the *Preliminary Report and Forecast of Solar Geophysical Data*, also known as *The Weekly*.

STEREO imagery and data were a welcome addition to the forecast office, but they were not without challenges, including one of the earliest: the changing perspective of the spacecraft as the mission unfolded. To overcome this, one of the forecasters created a wooden model which included the Sun, Earth, and approximate positions of the STEREO and SOHO spacecraft. The STEREO spacecraft could be repositioned on the model as required. This model helped both forecasters and SWPC visitors interpret the *Extreme UltraViolet Imager* EUVI and *Outer Coronagraph* COR2 imagery. Recently, one of the SWPC's new forecasters has developed code to overlay a Stonyhurst grid and Earth-relative limb position on the STEREO/EUVI imagery, making interpretation even easier.

While the changing perspective was constant, the receipt of imagery and data from STEREO was not; occasional gaps in both were common. Forecasters, particularly those with military experience, were used to operating with such limitations and were thankful for any data they could get. However, plans had already been made for some of the data long before forecasters got their first look at it.

Approximately two years before the STEREO mission was launched, Pizzo and Biesecker (2004) described a geometric localization technique to locate and characterize CMEs using STEREO imagery, and de Koning, Pizzo, and Biesecker (2009) described the first application of this technique using STEREO beacon-quality coronagraph imagery. SWPC forecasters who had previously relied on plane-of-sky speed estimates derived from the SOHO/*Large Angle and Spectrometric Coronagraph Experiment* (LASCO) imagery soon began augmenting their estimates using analysis requested from, and provided by, de Koning. These refined estimates provided forecasters another data point with which to compare forecasts from Shock Time of Arrival (STOA: Moon *et al.*, 2002) and Hakamada–Akasofu–Fry (Fry *et al.*, 2001) models as well as empirical prognostic methods. The advent of the STEREO era coincided with the appearance of a time-dependent, three-dimensional, magnetohydrodynamic model of the heliosphere called WSA–Enlil. It would be the convergence of these two technologies that would pave the way for improved geomagnetic forecasts at SWPC.

3.2. The WSA–Enlil Model

Pizzo *et al.* (2011) provide a concise description of the model and its initial integration into operations. The WSA–Enlil model first appeared in the forecast office in 2008. The WSA portion of the model (Arge and Pizzo, 2000) provides a semiempirical characterization of the base solar-wind field from the National Solar Observatory's (NSO) *Global Oscillation Network Group* (GONG) magnetogram measurements accumulated over a solar rotation. WSA output, in turn, drives the Enlil (Odstrcil *et al.*, 2004, 2005) code, which provides the ambient solar-wind outflow. CMEs can then be injected into this ambient flow and the subsequent evolution observed. Contributions from the entire community, including support for the development of WSA from the Naval Research Laboratory (NRL) and Air Force Research Laboratory (AFRL), and support for WSA–Enlil model development and testing from the National Science Foundation (NSF), the Center for Integrated Space Weather Modeling

(CISM), and the NASA/NSF Community Coordinated Modeling Center (CCMC), built the foundation for the WSA–Enlil model transition.

Initially, the model was run on a development platform at NOAA's Environmental Modeling Center, and the output was displayed on a "test bed" machine in the corner of the forecast office. The test-bed environment was utilized to first address general model robustness and to assess key run statistics (such as the wallclock run time) that would have an impact on future operational use. This system also provided an environment to test and develop suitable graphical products, a process that involved scientists and developers with feedback from forecasters. It also provided an ideal environment for forecasters to become acquainted with the abilities of the model and for initial attempts at validating the output.

Transitioning the model into operations began by constructing a concept of operations (CONOPS). For WSA–Enlil, the CONOPS consisted of running the model on a repeating two-hourly cycle. In its simplest "ambient" mode, the model run would proceed completely automatically on the National Weather Service operational supercomputers using the latest GONG magnetogram as sole input (the GONG input having been pulled automatically by ftp from NSO servers). On completion, the model output was automatically pulled back to the SWPC and placed into a database.

This system would be augmented in the event of a potentially geoeffective CME. In "CME-based" mode, analysis of the CME by forecasters (described later) would yield key parameters, which would then be stored in the database as a "CME analysis" via a web-based interface. The parameters from this analysis would then be parsed by automatic scripts running at the SWPC into an input file which was then pushed over to the supercomputers in time for the next two-hourly run cycle. Run scripts operating on the NWS supercomputers would detect the CME input file and WSA–Enlil would run in "CME based" mode, with inputs from both GONG and the CME (or CMEs; the system would have the ability to handle multiple CME events). As before, on completion, model output would be pulled back to the SWPC for analysis by forecasters.

Having arrived at a viable CONOPS, the transition process then essentially involves constructing and testing robust scripts to control all aspects of the run process: the networking of inputs and outputs, the running of the model, the creation of graphical products, and the ingesting of final data into the database.

One technical element of implementing a research model on operational supercomputers that is not obvious requires adapting the model to use a predefined directory structure and set of scripts. The operational supercomputer hosts dozens of models, and it would be impractical from a support and maintenance perspective to allow them to each do things their own way. Therefore, these are all standardized, enabling more robust real-time support in case any problems arise. By early 2011, WSA–Enlil processing had moved to operational supercomputers at the National Centers for Environmental Prediction facilities at Gaithersburg, Maryland and Fairmont, West Virginia, for a year-long trial (Pizzo *et al.*, 2011). It ran four times each day in "ambient mode"; no CMEs were injected. Instead, the model was used to forecast the arrival and departure of high-speed wind streams associated with recurrent coronal holes. On-demand CME runs also were provided at this time, with the analysis of the candidate CME and model initiation handled by researchers at the SWPC.

Once WSA–Enlil became routinely available, both modelers and forecasters began evaluating its performance. A critique of model performance was added to a daily briefing conducted each morning by the on-duty forecaster (this is described in more detail in Section 4.2). Because Enlil was dependent on WSA input, forecasters learned to link errors in WSA characterization of the base of the solar wind with subsequent shortcomings in the overall model output. (Recognizing when and why a model run has gone "off the rails" is an essential part of integrating the output into the forecast routine.) Conversations between researchers and forecasters figured greatly in the development of this ability in the forecast team.

As well as evaluating model performance, forecasters also provided feedback on the depiction of model output. Changes to the depiction based on human-factors considerations often happened in less than a week. This was rewarding to both the forecasters and the modelers. The forecasters were better able to interpret model output, and the modelers knew their work was appreciated and, more importantly, becoming part of the forecast process. Refinements to the output depiction are ongoing, but the pace has slowed since the initial fielding.

3.3. The CME Analysis Tool (CAT)

Since the ultimate aim was to use WSA–Enlil to improve the accuracy of CME arrival forecasts, the focus then became determining how best to inject CMEs into the model output. Xie, Ofman, and Lawrence (2004) described a cone model for halo CMEs. A graphical user interface (GUI), called the Cone Tool, was built to allow CMEs to be characterized using this cone model. Joint validation and verification (V&V) efforts were subsequently carried out by researchers and forecasters. Among other things, it became apparent that the relationship between cone angle and radial distance was not constrained in any way, leading to significant variations in estimates of the velocity and angular width for a given CME.

These problems led to the development of a new version of the GUI called the CME Analysis Tool (CAT). This tool modeled the CME as a three-dimensional (3D) lemniscate rather than a cone, and used simultaneous comparison of STEREO-A and -B and SOHO coronagraph imagery to fit the ejecta. The inclusion of the STEREO data and the choice of the lemniscate had their origins in the work of Pizzo and Biesecker (2004) and de Koning, Pizzo, and Biesecker (2009). V&V using the CAT followed, with results more promising than those achieved with the Cone Tool. The success of CAT suggested that the 3D lemniscate model of the CME was a significant improvement over the simple cone model. However, since STEREO data will eventually be unavailable for WSA–Enlil input, our eventual goal is to collect enough data using the CAT to effectively constrain the cone angle and radial distance obtained with a successor to the Cone Tool. A detailed description of the development of the CAT and the results of V&V efforts will be presented elsewhere (Millward *et al.*, 2013). Here we will concentrate on the evolution of the CAT through forecaster and researcher interaction.

The CAT in its present state is shown in Figure 1. The CAT GUI consists of the (1) Start/End Time widget; (2) Image Timeline; (3 – 5) STEREO-B, LASCO, and STEREO-A displays; (6) Animation Controls; (7) Image Adjust widget; (8) CME Controls; (9) CME Leading Edge *vs.* Time Plot; (10) Enlil Parameters and Export widget. CME analysis and fitting begins with loading the imagery using the Start/End Time widget. The Image Timeline is populated with symbols indicating the available imagery from STEREO and LASCO while the images themselves populate the displays. The analysis step, described below, is often complicated by the delayed receipt of sufficient SOHO or STEREO imagery. In these cases, a preliminary analysis is conducted with the available imagery and refined as more imagery becomes available. We emphasize that, with the exception of NOAA's *Polar* and *Geostationary Operational Environmental Satellite* (POES, GOES) vehicles, all of the space-based platforms upon which space-weather forecasters depend are primarily research platforms. Consequently, uninterrupted operation and data availability are not guaranteed.

The forecaster can then animate or enhance the imagery as needed to best reveal the CME structure. Most of the forecaster's time is spent manipulating the CME controls to

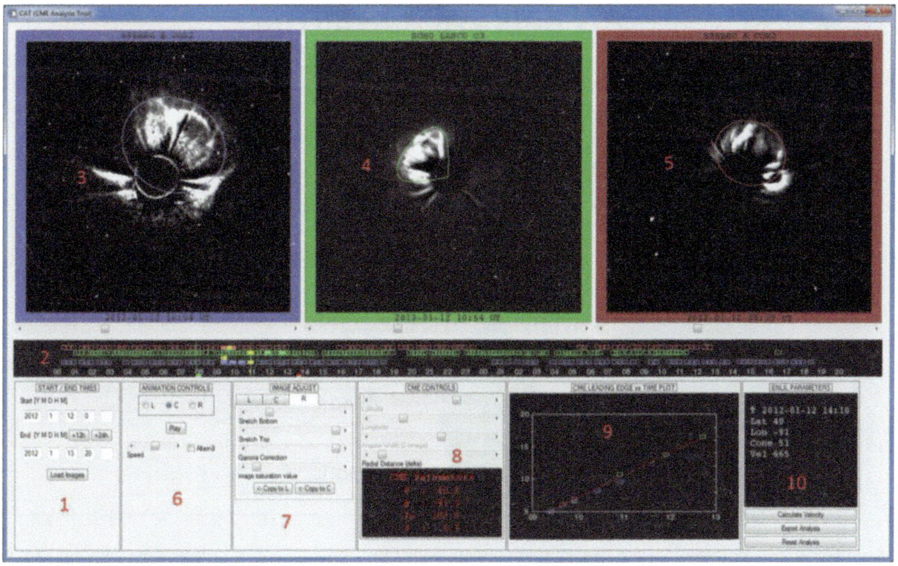

Figure 1 The CAT GUI consists of the (1) Start/End Time widget; (2) Image Timeline; (3–5) STEREO B, LASCO, and STEREO A displays; (6) Animation Controls; (7) Image Adjust widget; (8) CME Controls; (9) CME Leading Edge *vs.* Time Plot; (10) Enlil Parameters and Export widget

encompass the ejecta in several frames, that is, adjusting the size and orientation of the 3D lemniscate to match it to the CME images from SOHO and STEREO, simultaneously. Once a fit has been accomplished in at least two images at different time steps, the forecaster can produce a leading-edge *versus* time plot and obtain an estimated velocity. Typically, fits are made to several images. Once the forecaster is satisfied, the model parameters can be exported. These parameters are entered into a web-based Solar Predictions Interface (SPI), which prepares the next CME-based model run at the National Centers for Environmental Prediction (NCEP). In the current concept of operations, the model runs in ambient mode every two hours beginning at 00:00 UTC. The parameters must be entered before the next even hour, or the forecaster will have to wait until the current "ambient run" cycle finishes before the model run can be completed: a maximum of four hours.

Initially, two forecasters were selected to evaluate the CAT and attempt V&V fits for 75 different CMEs. During the course of the V&V, the forecasters provided feedback to the researchers regarding the GUI, providing an example of O2R in action. For example, forecasters soon determined that they needed a way to easily return to default values as they experimented with enhancements to draw out the CMEs in the imagery. Similarly, they needed a functionality that permitted them to quickly discard an unsatisfactory analysis and begin another attempt. Their suggestions led to the addition of a "reset" button on the Image Adjust widget (7) as well as a "reset analysis" button on the Enlil Parameters and Export widget (10). While not essential to CAT functionality, these enhancements made the tool easier to employ, particularly when the office was busy. There were also the typical glitches that accompany any R2O effort, and these were captured and reported by the forecasters. The forecasters were given short-term workarounds until the code could be remedied. The CAT was eventually installed in the forecast office, promoting experimentation with the tool and practice on CMEs that were not part of the V&V set. The parameters that they obtained

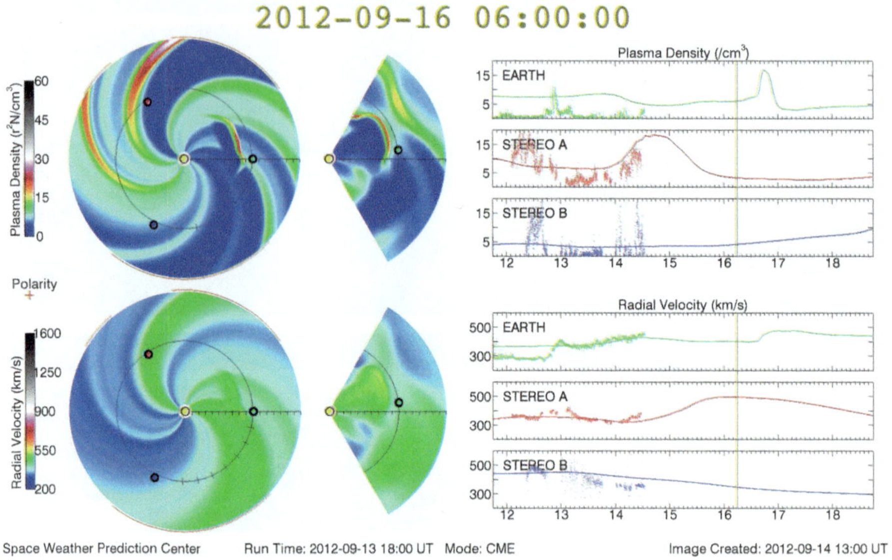

Figure 2 An example of ACE data superposed on WSA–Enlil model output. The date is depicted on the x-axis. Earth is depicted on the right side of the plan-view plots, and a circle is drawn at the 1 AU range. STEREO-A and -B spacecraft are also depicted. In this example, WSA–Enlil solar-wind speed predictions were performing well at STEREO-A and the ACE spacecraft based on the observed data.

could then be compared to the researchers' results and any systemic problems identified and corrected.

With the advent of routine ambient WSA–Enlil runs on the NCEP computers and the capability to extract model parameters using the CAT, real-time CME events were injected and output retrieved and displayed using the SPI. At first, researchers conducted the analysis and prepared CME-based WSA–Enlil runs. They were on call around the clock and would be notified by forecasters when candidate CMEs were observed. Forecasters eventually gained more experience and began initiating model runs independently. Researchers would also initiate independent runs using the characteristics gleaned from their own CAT results. The forecasters and researchers discussed the confidence in their respective analyses. Once the WSA–Enlil run was complete, the output was scrutinized, and the solution deemed most probable became the official forecast. The multiple runs effectively functioned as a mini-ensemble – a valuable contribution to the overall forecast. Postmortems were subsequently conducted during which forecasters and researchers discussed the results. Topics might include strategies for dealing with missing or poor-quality imagery; determining what features are, or are not, part of the CME; and coping with overlapping events. Exchanges helped forecasters gain confidence in the tool and their ability to use it. Similarly, researchers were able to learn how forecasters were using the tool and correct any misunderstandings or errors.

The model run(s) were typically available one to four days before the CME was predicted to arrive. Superposing observed ACE solar-wind data and the model prediction during the transit allowed forecasters to determine how well WSA–Enlil was performing and adjust the forecast accordingly. Figure 2 shows an example in which WSA–Enlil solar-wind speed predictions were performing well at STEREO-A and the ACE spacecraft. The ability to easily and rapidly compare model predictions with observed values is a key element in crafting an accurate forecast. Figure 3 shows a similar overlay over three days (24, 25, and 27 February 2012). The predicted speed was higher than the observed speed until late on the 25th, when the observed speed began to match the model forecast. The ICME arrived at Earth as predicted by the model, but arrived at STEREO-A earlier than expected.

By developing the CAT to incorporate STEREO imagery in the specification of CME parameters for WSA–Enlil, we effectively constrained the relationship between the cone angle and radial distance. Anecdotal evidence suggests that model output has improved as a result, and data are being collected and analyzed for a more formal analysis to be presented elsewhere (Millward *et al.*, 2013).

4. Keys to Successful R2O

From the field of meteorology, Doswell (1986) suggested several actions for the successful transition of science and technology into operations. Among them, he recommended having "researchers work together with forecasters in developing new science and technology to *suit the needs of the forecasters* [emphasis added]." This recommendation captures the essence of the entire transition of STEREO, WSA–Enlil, and the CAT to operations at the SWPC.

4.1. Lessons Learned

The importance of communication was echoed almost a quarter century later by Araujo-Pradere (2009) and comes at the end of his list of "building blocks" for successful R2O:

i) Identifying the customer's wish list [*i.e.*, avoid a model in search of a customer]
ii) Conducting verification and validation
iii) Specifying failures and events
iv) Documenting errors and uncertainties
v) Flagging – identifying compromised quality
vi) Providing thorough and understandable documentation
vii) Communicating – interaction between researchers and forecasters

While this article has largely focused on the interactions between researchers and forecasters, it is important to note that the foundation for success was built on the preceding blocks. The SWPC works closely with the operational user community to anticipate their needs and leverage the work of the research community to meet those needs. The development and implementation of WSA–Enlil grew out of the SWPC's desire to move from specification to prediction, and to provide better forecast products for our customers (Item i).

Items ii – iv led to the abandonment of the Cone Tool and the development of the CAT as described in Section 3.3. WSA–Enlil model V&V has been complicated by the variability of forecasters; given the same CME, each forecaster will produce a slightly (or sometimes wildly) different analysis and resulting model parameters. One goal of training and V&V was to move the analysis solutions towards a reasonable ensemble and eliminate the outliers. Since resource limitations meant model runs could not be completed for every set of parameters generated, it was impossible to know which analysis was "right." When multiple runs were accomplished for an operational event, as described in Section 3.3, a less than satisfactory analysis could still result in a better forecast, because the ambient solar wind was not being handled well by the model; *i.e.* the forecast was right for the wrong reasons. While WSA–Enlil model output is not flagged (Item v), the overlay of observed data from

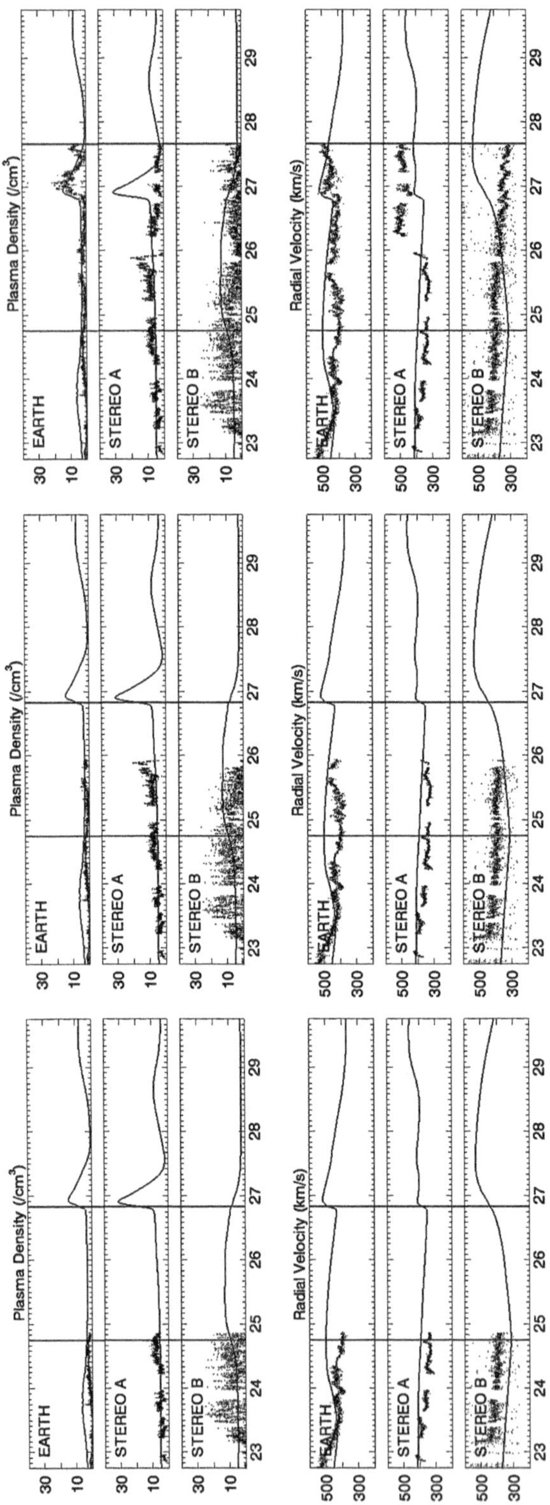

Figure 3 An example of ACE data superposed on WSA–Enlil model output on 24 (left), 25 (middle), and 27 (right) February 2012. The date is indicated on the x-axis. The forecast for arrival at Earth verified well; however the ICME arrived at STEREO-A sooner than anticipated.

the ACE spacecraft described in Section 3.3 above alerts the forecaster when model output is in disagreement with reality. Collaboration between the forecast and research staff resulted in an accurate and usable guide to the CAT (Item vi).

There is a temptation to call model output "the forecast," but it is important to remember that model output is only one ingredient in the forecast. The forecast is built from observations and analysis, empirical tools, and expertise. When a forecaster succumbs to the temptation to call model output the "forecast," it leads to an abdication of responsibility and an erosion of expertise. This phenomenon was described by Snellman (1977) as "meteorological cancer" after numerical weather prediction (NWP) models became operational. This topic will be revisited in Section 4.4.

A final lesson learned – worthy of its own section – is the importance of a robust training program to R2O success. Such a program will enable forecasters to take full advantage of any new observational platforms, tools, or numerical models.

4.2. The Importance of Education and Training

Training forecasters with a diverse educational background who work a rotating shift schedule is a nontrivial, but absolutely essential, undertaking. Throughout the development of WSA–Enlil, the Cone Tool, the CAT, and their integration into forecast-office operations, formal training was conducted with the SWPC forecasters. Knowledge-based education and training included a review of the quiet corona, structures in the solar wind, near-Sun magnetic fields, interplanetary propagation, and the topic of "how derived cone parameters relate to WSA–Enlil physical inputs and the subsequent interplanetary evolution." Once the relevant background had been established, the focus shifted to learning to use the CAT.

A combination of task-based and knowledge-based training was required for the CAT. Two training sessions were conducted by the researchers. During the initial session, they provided background information about the shortcomings of the Cone Tool and the application of the geometric localization technique. Forecasters were then given the opportunity to fit some example CMEs. As they practiced, the researchers would guide the forecasters through the process and correct any errors or misunderstandings along the way. The two forecasters chosen for the CAT V&V described in Section 3.3 were also present to assist. One of them wrote a concise CAT users' guide that was available during and after the training. Because the guide was written by a forecaster for forecasters, it was focused on the operational application of the tool rather than the theoretical underpinnings.

After this initial session, the trainees were provided a list of ten candidate CMEs to fit. Their solutions were stored and evaluated by the researchers. A second training session was conducted during which the various solutions were discussed, the outliers identified and any problems corrected, and questions answered.

In addition to the in-house training, forecasters have attended the Center for Integrated Space Weather Modeling (CISM) Summer School (Lopez and Gross, 2008) since its inception, typically within their first year at the SWPC. This course provides an up-to-date overview of both the theoretical and applied aspects of space weather. Forecasters are introduced to space-weather modeling, and their classmates are primarily future researchers. Thus, construction of the bridge between research and operations begins early in the careers of both populations.

Since forecasters and researchers are located in the same facility, daily interactions provide additional, mostly informal, training opportunities. One formal interaction is the daily space-weather briefing mentioned in Section 3.2. Every weekday morning, the forecaster provides a review of activity over the past 24 – 72 hours, and a rationale for the day's forecast. This allows researchers to see what tools, models, and data a forecaster is focused on as well as to judge the rationale being offered. They also offer insight into what the latest research might say about the situation. The forecaster provides an evaluation of tools, model performance, and data quality while receiving immediate feedback on the intricacies of the same.

Finally, Stuart, Schultz, and Klein (2007) note that forecasters and researchers approach learning from two distinct perspectives. They described researchers as "knowledge seekers," suggesting that they "prefer to understand theoretical concepts." Forecasters were described as "goal seekers," suggesting that they "prefer concrete examples from the real world." While it is dangerous to paint entire communities with such a broad brush, including real-world examples in education and training efforts facilitates understanding and application among forecasters. For instance, practicing on real CMEs during the CAT training described above enabled forecasters to immediately grasp the value brought by the new tool.

Twenty-six years ago, Doswell (1986) wrote of the importance of training to an improved meteorological forecast system:

> It is inconceivable that we spend billions of dollars on hardware and virtually nothing on meaningful training. Technology is supposed to enable us to do as well or better than our present performance without *any* substantial investment in our people. I think that any real *improvements* in the performance of our forecasting system must come from advances in the concepts and tools provided to the *people* operating the system. Giving a word processor to someone whose knowledge of English is deficient does not make the writing any better than when it was done with pencil and paper. And the word processor cannot do the writing by itself.

4.3. Making Research Accessible

As noted in Section 3, numerical models are not the only candidates for R2O activities. Siscoe (2007) compared the evolution of meteorology and space-weather forecasting, noting that the most significant gains in forecast accuracy came with the introduction of numerical weather prediction into operations. However, these advancements in model sophistication would not have been possible without an ever-expanding network of observations and robust data-assimilation schemes with which to initialize the models. Nor would the advancements have been possible without sustained research directed towards understanding the evolution of synoptic and mesoscale phenomena.

The responsibility for making research accessible rests with both the research and the operations communities. Doswell, Lemon, and Maddox (1981) described the challenges faced by weather forecasters in this regard:

> Forecasters are not generally trained or encouraged to read the current meteorological literature, so typically they can't even try to ferret out those articles that actually have a bearing on their job. Rotating shift work makes it difficult for forecasters to participate actively in any but the most brief of training programs, much less maintain currency in their profession. (Having experienced the rigors of shift work, we can only admire those rare individuals who can conduct applied research and/or stay up-to-date under the handicap of rotating shifts.)

In the three decades since that article was written, the NWS has taken steps to improve the situation. One step was the creation of the Science and Operations Officer (SOO) position at forecast offices and national centers. Among their many duties, "The [SOO] is expected to:

i) Initiate and oversee the transfer of new technologies from the research community to the operational environment...
ii) Lead and/or participate in significant joint research projects and developmental efforts conducted in a collaborative manner with ... science experts in the collocated or nearby university, other Federal agencies, and/or related professional societies and organizations.
iii) Assess continuing and future training needs required to successfully incorporate new technology and science into the ... operations.
iv) Coordinate and consult with scientists in the NWS, NOAA, other agencies, academia and the private sector to identify development opportunities for enhanced forecast procedures and techniques to be used at the [forecast office]. Integrate new scientific/technological advances and techniques into... operational procedures and operations. (Department of Commerce, 2012)"

The SOO is only required to work rotating shifts 25 % of the time, opening up the possibility that they will be able to focus on the primary duties described above. Space-weather forecast centers would do well to emulate this model.

In addition to the creation of an SOO position, another effort within the meteorological community to bring research to operators is the establishment of the Cooperation in Meteorological Education and Training (COMET) program (Spangler *et al.*, 1994; Johnson and Spayd, 1996) under the auspices of the University Corporation for Atmospheric Research and NWS. COMET provides a variety of Internet-based and in-residence courses designed to bring the fruits of research to the operational community. The on-demand Internet-based courses allow forecasters to complete training that they might otherwise miss because of rotating shift work. The COMET program originally began with a focus on mesoscale meteorology. However, in the past few years, the SWPC has collaborated with COMET developers to create and offer space-weather tutorials on the COMET website. These tutorials were designed to provide meteorologists with enough background in space weather to field routine questions from the public.

Finally, traditional methods of making research available such as conferences and publications should not be overlooked. Assuming at least one forecaster is able to attend a given conference, the possibility exists for mutually beneficial interactions between representatives from the research and operational communities.

4.4. An Operator's Wish List

A discussion of R2O and O2R would not be complete without the inclusion of a wish list from the operations side of the house. This short wish list incorporates suggestions from operations personnel both inside and outside the forecast office.

- Siscoe (2007) called the stage of meteorology that followed the advent of numerical weather prediction the storm-tracking stage. The capability to track ICMEs across the heliosphere and, in particular, interrogate the magnetic structure throughout its transit, is a key to evolving to this stage. Although we are improving the arrival-time predictions through numerical models, we are no closer to being able to forecast the magnitude of the ensuing geomagnetic storm after the ICME reaches Earth. This is a huge gap.
- As numerical weather prediction advanced, grid spacing became smaller, and mesoscale models emerged. Similarly, the creation of geospace models, and subsequently, *regional* geospace models would be a welcome development. Several large-scale models have emerged over the past decade and are being evaluated (Pulkkinen *et al.*, 2011).

- Current flare probabilities are based on empirical studies. Being able to accurately forecast regions capable of producing M- and X-class solar flares, CMEs, and energetic particle events well in advance would be a great step forward (see, *e.g.*, Falconer, Moore, and Gary, 2003).
- Closely related is the ability to forecast regions of emerging flux (see, *e.g.*, Ilonidis *et al.*, 2012).
- The ability to nowcast/forecast the radiation environment in interplanetary space and at GEO, MEO, and HEO would be helpful.
- The creation of tools to help *quantitatively* analyze and track active-region creation, evolution, and decay using space-based imagery would be valuable. This includes areal extent, McIntosh classification, Mt. Wilson classification, gradients, and any other characteristics that may lead to forecast techniques.
- Finally, the ability to run displaced-real-time simulations of historic space-weather events in an environment that mirrors the forecast office would be beneficial. This will enable forecasters to maintain proficiency and prepare for the next solar maximum during solar minimum. Magsig, Page, and Page (2003) describe such a system being used to train terrestrial weather forecasters.

Any technology or automation introduced to the forecast office should support the continued development of expertise, not hinder it. It should make the forecast process and the resulting forecast better, not worse. This seems obvious but, unfortunately, has not always been the case in the field of meteorology (Pliske, 1997 and Klein, 2011). Klein (2011) describes the situation:

> The NWP [numerical weather prediction] system was so awkward to use that forecasters just accepted what the system said unless it really made a blunder. The experts forced to use it found themselves accepting forecasts that were sort of good enough. They know they could improve on the system, but they didn't have the time to enter the adjustments ... This problem will probably disappear once they retire. The newest forecasters won't know any better, and so expertise will be lost to the community because systems like NWP get in the way of building expertise.

Wrapped up in each of these "wishes" is an increased understanding of the phenomena we are trying to forecast.

5. Conclusion

New observational capabilities (STEREO), methods to exploit these capabilities (Geometric Localization and CAT), and a robust magnetohydrodynamic model (WSA–Enlil) were integrated to form a powerful new forecast tool. Throughout the creation of this new tool, dialogue between the researchers developing it and the forecasters who would wield it contributed greatly to its smooth integration into operations. Some lessons for successful R2O/O2R activities can be drawn from this endeavor.

We have presented three examples of the successful transfer of research to operations. STEREO, WSA–Enlil, and CAT can each stand alone as a case study in R2O/O2R interaction. However, by integrating these technologies into a new operational concept, we hope to be able to continuously improve the forecast process and our resulting products. Doswell, Lemon, and Maddox (1981) point out that the great contributions in meteorology came "at those times and in those places [where] the interactions and mutual respect between theoreticians and forecasters was substantial." This climate is being actively cultivated at the SWPC, leading to positive outcomes within the organization and extending to the larger community we serve.

In another encouraging development, Lanzerotti (2011) noted that the establishment of a "Research to Operations/Operations to Research" working group for the recent Decadal Survey (National Research Council, 2012) signaled the recognition of the importance of including the O2R requirements from "designers, systems operators, modelers, [and] forecasters" in the "thinking and discussions by the decadal survey steering committee for new solar-terrestrial research programs, initiatives, and activities." This acknowledgement contributes to community cohesion and sets the stage for continued contributions to science and society.

Acknowledgements The authors thank Vic Pizzo, Eduardo Araujo-Pradere, Howard Singer, and Chris Balch for their helpful contributions.

References

Araujo-Pradere, E.A.: 2009, Transitioning space weather models into operations: the basic building blocks. *Space Weather* **7**, 10006. doi:10.1029/2009SW000524.

Arge, C.N., Pizzo, V.J.: 2000, Improvement in the prediction of solar wind conditions using near-real time solar magnetic field updates. *J. Geophys. Res.* **105**, 10465 – 10480. doi:10.1029/1999JA900262.

Biesecker, D.A., Webb, D.F., St. Cyr, O.C.: 2008, Stereo space weather and the space weather beacon. *Space Sci. Rev.* **136**(1 – 4), 45 – 65.

de Koning, C.A., Pizzo, V.J., Biesecker, D.A.: 2009, Geometric localization of CMEs in 3D space using STEREO beacon data: first results. *Solar Phys.* **256**, 167 – 181. doi:10.1007/s11207-009-9344-7.

Department of Commerce: 2012, Science and operations officer position description. hr.commerce.gov/Practitioners/ClassificationAndPositionManagement/PDLibrary/prod01_000645. Office of Human Resource Management, Position Description Library. Accessed: 15/10/2012.

Doswell, C.A.: 1986, The human element in weather forecasting. *Natl. Weather Dig.* **11**(2), 6 – 17.

Doswell, C.A., Lemon, L.R., Maddox, R.A.: 1981, Forecaster training – a review and analysis. *Bull. Am. Meteorol. Soc.* **62**(7), 983 – 988.

Falconer, D.A., Moore, R.L., Gary, G.A.: 2003, A measure from line-of-sight magnetograms for prediction of coronal mass ejections. *J. Geophys. Res.* **108**(A10). doi:10.1029/2003JA010030.

Fry, C.D., Sun, W., Deehr, C.S., Dryer, M., Smith, Z., Akasofu, S.I., Tokumaru, M., Kojima, M.: 2001, Improvements to the HAF solar wind model for space weather predictions. *J. Geophys. Res.* **106**(A10), 20985 – 21001. doi:10.1029/2000JA000220.

Fuller-Rowell, T., Araujo-Pradere, E., Minter, C., Codrescu, M., Spencer, P., Robertson, D., Jacobson, A.R.: 2006, US-TEC: a new data assimilation product from the space environment center characterizing the ionospheric total electron content using real-time GPS data. *Radio Sci.* **41**(6), RS6003. doi:10.1029/2005RS003393.

Ilonidis, S., Zhao, J., Kosovichev, A., Hartlep, T.: 2012, Helioseismic measurements of emerging magnetic flux in the solar convection zone. In: *Am. Astron. Soc. Meeting Abs.* **220**, 109.02. ADS: 2012AAS...22010902I.

Johnson, V.C., Spayd, L.E.: 1996, The COMET(r) outreach program: cooperative research to improve operational forecasts. *Bull. Amer. Meteor. Soc.* **77**(10).

Kaiser, M.L., Kucera, T.A., Davila, J.M., St. Cyr, O.C., Guhathakurta, M., Christian, E.: 2008, The STEREO mission: an introduction. *Space Sci. Rev.* **136**(1 – 4), 5 – 16. doi:10.1007/s11214-007-9277-0.

Klein, G.: 2011, *Streetlights and Shadows: Searching for the Keys to Adaptive Decision Making*, 122 – 125. Bradford, Cambridge. ISBN 0262516721. www.worldcat.org/isbn/0262516721.

Lanzerotti, L.J.: 2011, Research to operations paired with operations to research. *Space Weather* **9**(2), S02003. doi:10.1029/2010SW000653.

Lopez, R., Gross, N.: 2008, Active learning for advanced students: the center for integrated space weather modeling graduate summer school. *Adv. Space Res.* **42**(11), 1864 – 1868. doi:10.1016/j.asr.2007.06.056.

Magsig, M., Page, E.: 2003, Weather event simulator implementation and future development. In: *Preprints, 19th Conf. on Interactive Information Processing Systems (IIPS) for Meteorology, Oceanography, and Hydrology, Long Beach, CA*, Am. Meteorol. Soc., Boston, **12**. CD-ROM

Millward, G., Biesecker, D., Pizzo, V., de Koning, C.A.: 2013, An operational software tool for the analysis of coronagraph images: determining CME parameters for input into the WSA–Enlil heliospheric model. *Space Weather*, doi:10.1002/swe.20024.

Moon, Y.J., Dryer, M., Smith, Z., Park, Y.D., Cho, K.S.: 2002, A revised shock time of arrival (STOA) model for interplanetary shock propagation: STOA-2. *Geophys. Res. Lett.* **29**(10), 1390. doi:10.1029/2002GL014865.

National Research Council: 2012, *Solar and Space Physics: A Science for a Technological Society*. National Academies Press, Washington, ISBN 9780309263986. sites.nationalacademies.org/SSB/CurrentProjects/SSB_056864. Accessed: 15 October 2012.

Odstrcil, D., Pizzo, V.J., Arge, C.N.: 2005, Propagation of the 12 May 1997 interplanetary coronal mass ejection in evolving solar wind structures. *J. Geophys. Res.* **110**(A2), A02106. doi:10.1029/2004JA010745.

Odstrcil, D., Pizzo, V.J., Linker, J.A., Riley, P., Lionello, R., Mikic, Z.: 2004, Initial coupling of coronal and heliospheric numerical magnetohydrodynamic codes. *J. Atmos. Solar-Terr. Phys.* **66**(15 – 16), 1311 – 1320. doi:10.1016/j.jastp.2004.04.007.

Pesnell, W.D., Thompson, B.J., Chamberlin, P.C.: 2012, The Solar Dynamics Observatory (SDO). *Solar Phys.* **275**(1 – 2), 3 – 15. doi:10.1007/s11207-011-9841-3.

Pizzo, V.J., Biesecker, D.A.: 2004, Geometric localization of STEREO CMEs. *Geophys. Res. Lett.* **31**, 21802. doi:10.1029/2004GL021141.

Pizzo, V., Millward, G., Parsons, A., Biesecker, D., Hill, S., Odstrcil, D.: 2011, Wang–Sheeley–Arge–Enlil cone model transitions to operations. *Space Weather* **9**, 3004. doi:10.1029/2011SW000663.

Pliske, R.: 1997, Understanding skilled weather forecasting: implications for training and the design of forecasting tools. Human Resources Directorate, Armstrong Laboratory, Air Force Materiel Command, San Antonio, Texas.

Pulkkinen, A., Kuznetsova, M., Ridley, A., Raeder, J., Vapirev, A., Weimer, D., Weigel, R., Wiltberger, M., Millward, G., Rastätter, L., *et al.*: 2011, Geospace environment modeling 2008 – 2009 challenge: ground magnetic field perturbations. *Space Weather* **9**(2), S02004.

Sauer, H.H., Wilkinson, D.C.: 2008, Global mapping of ionospheric HF/VHF radio wave absorption due to solar energetic protons. *Space Weather* **6**, 12002. doi:10.1029/2008SW000399.

Siscoe, G.: 2007, Space weather forecasting historically viewed through the lens of meteorology. In: Bothmer, V., Daglis, I.A. (eds.) *Space Weather-Physics and Effects*, Springer, Berlin, 5 – 30.

Snellman, L.W.: 1977, Man–machine mix. In: *Proc. 7th Tech. Exch. Conf.*, Atmospheric Sciences Laboratory, White Sands Missile Range, 37.

Spangler, T.C., Johnson, V.C., Alberty, R.L., Heckman, B.E., Spayd, L., Jacks, E.: 1994, COMET(r) – an education and training program in mesoscale meteorology. *Bull. Am. Meteorol. Soc.* **75**(7).

St. Cyr, O.C., Davila, J.M.: 2001, The STEREO space weather broadcast. In: Song, P., Singer, H.J., Siscoe, G.L. (eds.) *Geophys. Monogr. Ser.* **125**, AGU, Washington, 205 – 209. doi:10.1029/GM125p0205.

Stuart, N.A., Schultz, D.M., Klein, G.: 2007, Maintaining the role of humans in the forecast process: analyzing the psyche of expert forecasters. *Bull. Am. Meteorol. Soc.* **88**(12), 1893 – 1898. doi:10.1175/BAMS-88-12-1893.

Tsurutani, B., Baker, D.: 1979, Substorm warnings: an ISEE-3 real time data system. *Eos Trans. AGU* **60**(41), 71. doi:10.1029/EO060i041p00701.

Xie, H., Ofman, L., Lawrence, G.: 2004, Cone model for halo CMEs: application to space weather forecasting. *J. Geophys. Res.* **109**, 3109. doi:10.1029/2003JA010226.

Zwickl, R., Doggett, K., Sahm, S., Barrett, W., Grubb, R., Detman, T., Raben, V., Smith, C., Riley, P., Gold, R., Mewaldt, R., Maruyama, T.: 1998, The NOAA Real-Time Solar-Wind (RTSW) system using ACE data. *Space Sci. Rev.* **86**, 633 – 648. doi:10.1023/A:1005044300738.

GPSR Compliance

The European Union's (EU) General Product Safety Regulation (GPSR) is a set of rules that requires consumer products to be safe and our obligations to ensure this.

If you have any concerns about our products, you can contact us on ProductSafety@springernature.com

In case Publisher is established outside the EU, the EU authorized representative is:

Springer Nature Customer Service Center GmbH
Europaplatz 3
69115 Heidelberg, Germany

Batch number: 10621434

Printed by Printforce, the Netherlands